SILICON NANOELECTRONICS

Edited by

Shunri Oda • David Ferry

CRC Press is an imprint of the
Taylor & Francis Group, an **informa** business
A TAYLOR & FRANCIS BOOK

CRC Press
Taylor & Francis Group
6000 Broken Sound Parkway NW, Suite 300
Boca Raton, FL 33487-2742

First issued in paperback 2019

ISBN-13: 978-0-8247-2633-1 (hbk)
ISBN-13: 978-0-367-39253-6 (pbk)
Library of Congress Card Number 2005005007

Library of Congress Cataloging-in-Publication Data

Silicon nanoelectronics / edited by Shunri Oda and David Ferry.
p. cm.
ISBN 0-8247-2633-2
1. Molecular electronics. [DNLM: 1. Nanotechnology. 2. Silicon Compounds.] I. Oda, Shunri. II. Ferry, David K.

TK7874.8.S55 2005
621.381--dc22 2005005007

Visit the Taylor & Francis Web site at
http://www.taylorandfrancis.com

and the CRC Press Web site at
http://www.crcpress.com

Preface

The advances in ultra-large-scale integration (ULSI) technology mainly have been based on downscaling of the minimum feature size of complementary metal-oxide semiconductor (CMOS) transistors. The limit of scaling is approaching and there are unsolved problems such as the number of electrons in the device's active region. If this number is reduced to less than 10 electrons (or holes), quantum fluctuation errors will occur and the gate insulator thickness will become too small to block quantum mechanical tunneling, which may result in unacceptably large leakage currents. On the other hand, the recent evolution of nanotechnology may provide opportunities for novel devices, such as single-electron devices, carbon nanotubes, Si nanowires, and new materials, which may solve these problems. Utilization of quantum effects and ballistic transport characteristics also may provide novel functions for silicon-based devices. Among various candidate materials for nanometer scale devices, silicon nanodevices are particularly promising because of the existing silicon process infrastructure in semiconductor industries, the compatibility to CMOS circuits, and a nearly perfect interface between the natural oxide and silicon.

The goal of this book is to give an update of the current state of the art in the field of silicon nanoelectronics. This book is a compact reference source for students, scientists, engineers and specialists in various fields including electron devices, solid-state physics and nanotechnology.

Shunri Oda and David Ferry

About the Editors

Shunri Oda is a professor at the Quantum Nanoelectronics Research Center and the chair of the Department of Physical Electronics at the Tokyo Institute of Technology in Tokyo, Japan, where he obtained his doctorate in physical information processing. He is the director of the CREST and SORST NeoSilicon projects, which are sponsored by the Japan Science and Technology Agency. His recent research interests include formation of well-controlled silicon quantum structures and nanoscale silicon devices. He has authored more than 200 papers published in journals and conference proceedings.

David K. Ferry is the Regents' Professor of Electrical Engineering at the Arizona State University in Tempe, Arizona, where he is actively involved in thesis and postdoctoral mentoring. He received his doctorate in elecrical engineering from The University of Texas at Austin. He has coauthored many recent articles relevant to nanotechnology. In 2000, he received Arizona State University's Outstanding Graduate Mentor Award, and in 1999 he received the Institute of Electrical and Electronics Engineers's Cledo Brunetti Award, for advances in nanoelectronics theory and experiment.

Contributors

Richard Akis
Department of Electrical Engineering
Arizona State University
Tempe, Arizona

Haroon Ahmed
Microelectronics Research Centre
Cambridge, United Kingdom

David K. Ferry
Department of Electrical Engineering
Arizona State University
Tempe, Arizona

David J. Frank
IBM Watson Research Center
Yorktown Heights, New York

Akira Fujiwara
NTT Basic Research Laboratories
NTT Corporation
Kanagawa, Japan

Matthew J. Gilbert
Department of Electrical Engineering
Arizona State University
Tempe, Arizona

L. Jay Guo
Department of Electrical Engineering and Computer Science
University of Michigan
Ann Arbor, Michigan

Toshiro Hiramoto
Institute of Industrial Science
University of Tokyo
Tokyo, Japan

Hiroya Ikeda
Research Institute of Electronics
Shizuoka University
Hamamatsu, Japan

Hiroshi Inokawa
NTT Basic Research Laboratories
NTT Corporation
Kanagawa, Japan

Yasuhiko Ishikawa
Research Institute of Electronics
Shizuoka University
Hamamatsu, Japan

Hisao Kawaura
Fundamental Research Laboratories
NEC Corporation
Ibaraki, Japan

Hiroshi Mizuta
Department of Physical Electronics
Tokyo Institute of Technology
Tokyo, Japan

Kazuo Nakazato
Department of Electrical Engineering and Computer Science
Nagoya University
Nagoya, Japan

Katsuhiko Nishiguchi
Quantum Nanoelectronics Research Center
Tokyo Institute of Technology
Tokyo, Japan

Shunri Oda
Tokyo Institute of Technology
Quantum Nanoelectronics Research Center
Tokyo, Japan

Yukinori Ono
NTT Basic Research Laboratories
NTT Corporation
Kanagawa, Japan

Stephen M. Ramey
Department of Electrical Engineering
Arizona State University
Tempe, Arizona

Michiharu Tabe
Research Institute of Electronics
Shizuoka University
Hamamatsu, Japan

Yasuo Takahashi
Graduate School of Information Science and Technology
Hokkaido University
Sapporo, Japan

Sandip Tiwari
School of Electrical and Computer Engineering
Cornell University
Ithaca, New York

Kazuo Yano
Hitachi Central Research Laboratory
Tokyo, Japan

Contents

1 Physics of Silicon Nanodevices

David K. Ferry, Richard Akis, Matthew J. Gilbert, and Stephen M. Ramey

1.1 INTRODUCTION

For the past several decades, miniaturization in silicon integrated circuits has progressed steadily with an exponential scale described by Moore's Law.[1] This incredible progress has generally meant that critical dimensions are reduced by a factor of two every three years, while chip density increases by a factor of four over this period. However, modern chip manufacturers have been accelerating this pace recently, and currently chips are being made with gate lengths in the 45 to 65 nm range. More scaling is expected, however, and 15-nm gate lengths are scheduled for production before the end of this decade. Such devices have been demonstrated by Intel[2] and AMD,[3] and IBM has recently shown a 6-nm gate length *p*-channel FET.[4] While the creation of these very small transistors is remarkable enough, the fact that they seem to operate in a quite normal fashion is perhaps even more remarkable.

Almost 25 years ago, the prospects of making such small transistors was discussed, and a suggested technique for a 25-nm gate length, Schottky source-drain device, was proposed.[5] At that time, it was suggested that the central feature of transport in such small devices would be that the microdynamics could not be treated in isolation from the overall device environment (of a great many similar devices). Rather, it was thought that the transport would by necessity be described by quantum transport and that the array of such small devices on the chip would lead to considerable coherent many-device interactions. Although this early suggestion does not seem to have been fulfilled, as witnessed by the quite normal behavior of these devices, there have been many subsequent suggestions for treatment via quantum transport.[6–10] Moreover, there is ample suggestion that the transport will not be normal, but will have significant ballistic transport effects[11] and this, in turn, will lead to quantum transport effects.

In this first chapter, the concept of ballistic transport will be reviewed, starting in the next section. We then turn to the most important aspect of small devices, and that is the breakdown of ensemble averaging, so that the role of discrete, localized impurities and fluctuations in sizes becomes important. Following this, we begin to discuss the role of quantization. First, we will review how it is found in large metal-oxide-semiconductor field-effect transistors (MOSFETs) and then turn to the much more important role in small transistors. We follow this with a discussion of the ultimately small device—the quantum dot and single-electron tunneling. Finally, a discussion is given of many-body effects in such small devices. Each of these topics will be discussed in far greater detail in subsequent chapters, but here we hope to give an overall unifying view to these topics.

1.2. SMALL MOSFETS

The MOSFET is created when the electric field between the gate and the semiconductor is such that an inverted carrier population is created and forms a conducting channel. This channel extends between the source and drain regions, and the transport through this channel is modulated by the gate potential. This much has been known since the first descriptive patent on the topic.[12] Indeed, the operation of the MOSFET is almost exactly as described in a simple one-dimensional semiclassical treatment, and this approach has been modified and adapted continuously over the past few decades. However, it has become understood that there is quantization in the basic MOSFET, even for quite large gate lengths. This is because the gate field pulls the inversion channel carriers quite close to the oxide-semiconductor interface, and these carriers are confined between this interface and the potential in the bulk. This confinement is sufficient to cause quantization to occur in the direction normal to the oxide-semiconductor interface.[13] This quantization leads to a quasi-two-dimensional carrier gas in the plane of the channel.[14] While this effect is quite important, it is equally important to understand that the transport is in the plane of this quantized layer, and so is not directly affected by this quantization. We will discuss this in more detail in a subsequent section.

As the channel length has gotten smaller, there has been considerable effort to incorporate a variety of new effects into the simple (as well as the more complex) models. These include short-channel effects, narrow width effects, degradation of the mobility due to surface scattering, hot carrier effects, and velocity overshoot.[13] However, as gate lengths have become less than ca. 100 nm, the issue is becoming one of *ballistic* transport rather than these other problems. By ballistic transport, we refer to the situation in which the channel length is less than the mean-free path of the carriers, so that very little scattering occurs within the channel itself. If we take the thermal velocity of a carrier in Si as 2.5×10^7 cm/s at room temperature, a channel mobility of 300 cm^2/Vs leads to a relaxation time of 5×10^{-14} sec and a mean-free path of the order of 12×10^{-7} cm, or 12 nm. Thus, we might expect only a few scattering events in a channel length of 20 to -30 nm. While this is a very crude approximation, it points out that the properties of the carriers in these very small devices will be quite different than those in larger devices. In this case, the "theory" of the device is actually much closer to that of the simple approach discussed in the Simple One-Dimensional Theory section, at least in conceptual detail. For this reason, we will review some simple interpretations of the one-dimensional current equation, and then develop the ballistic device theory. This becomes important, because the same intuitive ideas carry over to the Landauer formula,[15] which is often invoked in pure quantum transport situations.

1.2.1 The Simple One-Dimensional Theory

In general, the current through a semiconductor device is found by writing an equation for the differential voltage drop along a point in the channel in terms of the current and local conductance (this may be found in most elementary textbooks; see, e.g.,[16]). This expression is then integrated over the length of the channel, with the result (for the MOSFET)

$$I_D = \frac{WeC\mu}{L}\left(V_G - V_T - \frac{V_D}{2}\right)V_D \tag{2.1}$$

where I_D is the drain current, W is the width of the channel, C is the gate capacitance per unit area, μ is the mobility of the carriers, L is the electrical channel length, V_G is the gate-source voltage, V_T is the threshold voltage (at which the channel begins to form), and V_D is the drain-source voltage. From this expression, the current rises almost linearly for small drain voltage, and then saturates at a value of drain voltage given by

$$V_{D,sat} = V_G - V_T \tag{2.2}$$

which may be found by taking the derivative of Equation (2.1) and setting it to zero.

A more intuitive view of the current may be obtained by rewriting Equation (2.1) to separate the source originating current and the drain originating current as

$$I_D = \frac{WeC\mu}{2L}\left[\left(V_G - V_T\right)^2 - \left(V_G - V_T - V_D\right)^2\right] = I_{SD} - I_{DS} \tag{2.3}$$

Now, it is clear that saturation sets in when the second term in the square brackets, the drain originating current (or reverse current), vanishes for the condition of Equation (2.2). In this equation, we can connect parts of the formula with particular physical effects. Here, we may connect

$$C\left(V_G - \ldots\right) \tag{2.4}$$

with the local carrier density (in carriers per unit area) in the channel, and

$$\mu\frac{V}{2L} \tag{2.5}$$

is the (average) velocity in the channel. Hence, we may rewrite Equation (2.3) once again as

$$I_D = We\left[n_S v_S - n_D v_D\right] \tag{2.6}$$

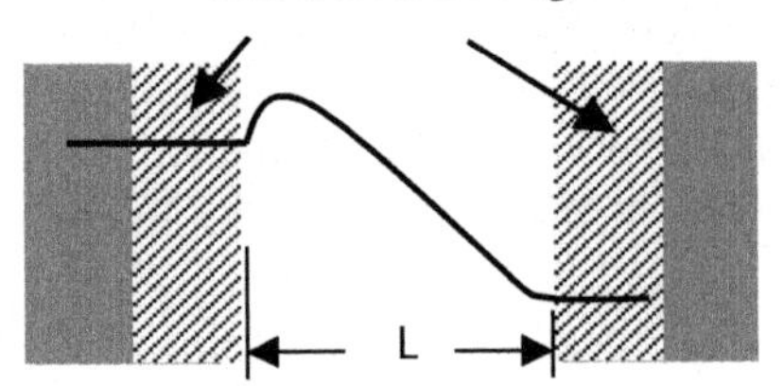

FIGURE 1.1 A conceptual device under bias. The source is at the left and the drain at the right, as indicated by the two gray areas, which may be considered to be the "contacts." The areas to the left and right of the traditional active length L, indicated here as the decoherence regions, must now be considered part of the active device.

Here, n_S and n_D are the two-dimensional densities at the source and drain, respectively, and v_S and v_D are the velocities at these two points. W is the width of the channel. This particular form will be the basis for developing the ballistic treatment in the next section.

1.2.2 Ballistic Transport in the MOSFET

In general, the potential profile through a MOSFET looks somewhat like that shown in Figure 1.1. From the source end, there is a small potential barrier between the source and the channel, and then the potential falls to the level of the drain potential (the energy is shown, this has a negative sign from the voltage). Lundstrom[11] then identifies two major scattering regions: (a) the barrier between the channel and the source, which gives a reflection r_s, and (b) within the channel, which gives a reflection r_c. In both cases, the reflection coefficients are related to transmission coefficients t by

$$r_s = 1 - t_s, \quad r_c = 1 - t_c \tag{2.7}$$

The steady-state flux which reaches the drain can now be written in terms of the entering flux a_s (which is a function of the depth y) as

$$a_D = a_s t_s t_c \tag{2.8}$$

At the entrance to the channel (which is taken to be $x = 0$, with x the axis aligned from source to drain), the density of carriers can be written as[11]

$$n(0,y) = \left[\frac{a_s t_s + r_c t_s a_s}{v_T} \right] = \frac{a_s t_s (1 + r_c)}{v_T} \tag{2.9}$$

The numerator accounts for particles which come from the source, as well as those that are reflected in the channel and return to $x = 0$. Here, v_T is the velocity of the positively and negatively directed fluxes, and y is the direction of the channel depth (normal to the oxide-semiconductor interface). Solving for t_s in Equation (2.9) and using this in Equation (2.8) yields

$$a_D = v_T n(0,y)\frac{t_c}{1+r_c} = v_T n(0,y)\frac{1-r_c}{1+r_c} \tag{2.10}$$

The sheet carrier density is given by integrating over the y coordinate, as

$$n_s = \int_0^{y_{max}} n(0,y)dy = \frac{C}{e}(V_G - V_T) \tag{2.11}$$

With this result, the drain current can be written as

$$I_{D,sat} = CWv_T\left(\frac{1-r_c}{1+r_c}\right)(V_G - V_T) \tag{2.12}$$

which may be compared with Equation (2.3) or Equation (2.6). Here, the reverse current is represented by the r_c term in the equation, but the form is quite similar to that of the simple theory. However, here we do not define a mobility, but instead discuss the transport in terms of the velocity and the transmission and reflection coefficients within the device. The task is to estimate just what these parameters should be. Price[17] has suggested that carriers cannot be back-scattered to the $x = 0$ point once they have traveled down a potential drop equal to the thermal energy, from which one may estimate the reflection coefficient as

$$r_c = \frac{\xi}{\xi+\lambda}, \quad \xi = \frac{k_B T}{eE(0)} \tag{2.13}$$

where $E(0)$ is the electric field on the channel side of the origin and λ is the mean free path. This has become the most quoted version of Lundstrom's theory, in which any carriers that make it past this first energy drop will ultimately appear at the drain. In this simple approach, nothing that happens beyond this point is important in the drain current, which is simplistic.

In fact, the nature of the barrier in Figure 1.1 is that of a self-consistent potential subject to a constraint of the applied gate and drain voltages. The exact distribution of charge in the channel and in the drain will affect this potential barrier due to the nonlinear feedback of solving Poisson's equation. This has been shown already in some detail.[18] Nevertheless, the Lundstrom theory represents a good zero-order approximation that is useful in estimating the amount of ballistic transport present in the transistor.

Natori[9] has given another version of a ballistic transport treatment for the MOS-FET, and has used this to some success in fitting to experimental data[19] Although

Natori developed his expression with a full quantum mechanical basis, the approach is an outgrowth of the Duke tunneling formula,[20] and we can follow a variation of the semiclassical approach.[21] We will assume that the direction normal to the oxide-semiconductor interface (the y-direction) is quantized,[14] and concern ourselves with integrations over the other two directions in reciprocal space. Then, the forward current may be written as

$$J_{SD} = 2e \sum_{valleys} \sum_{n_y} \int \int \frac{dk_z dk_x}{4\pi^2} v_x(k_x) T(k_x) f(\varphi_{FS}, E) \left[1 - f(\varphi_{FD}, E)\right] \quad (2.14)$$

The integer n_y runs over the occupied subbands in the inversion layer, the first summation runs over the six equivalent valleys of the conduction band, and the total energy is

$$E = E_x + E_z = \frac{\hbar^2}{2} \left(\frac{k_x^2}{m_x} + \frac{k_z^2}{m_z} \right) \quad (2.15)$$

The valley summation is necessary, since the mass that is appropriate for the two coordinate axes is different in each of the three pairs of valleys (this will be discussed further in a later section).

In a similar manner to Equation (2.14), we may also write the reverse current (that flowing from the drain to the source) as

$$J_{DS} = 2e \sum_{valleys} \sum_{n_y} \int \frac{dk_z dk_x}{4\pi^2} v_x(k_x) T(k_x) f(\varphi_{FD}, E) \left[1 - f(\varphi_{FS}, E)\right] \quad (2.16)$$

We may then write the total current as

$$I_{SD} = 2eW \sum_{valleys} \sum_{n_y} \int \frac{dk_z dk_x}{4\pi^2} v_x(k_x) T(k_x) \left[f(\varphi_{FS}, E) - f(\varphi_{FD}, E)\right] \quad (2.17)$$

In general, the treatment of ballistic transport is that for which the carriers move over the barrier, so that we may take $T = 1$. We now rescale the energy through the introduction of the scaled k vectors as

$$k_x'^2 = \frac{\sqrt{m_x m_z}}{m_x} k_x^2, \quad k_z'^2 = \frac{\sqrt{m_x m_z}}{m_z} k_z^2 \quad (2.18)$$

so that

$$E = \frac{\hbar^2}{2m^*}\left(k_x'^2 + k_z'^2\right), \quad m^* = \sqrt{m_x m_z} \tag{2.19}$$

With this transformation, we may change the variables in Equation (2.17) as

$$dk_x dk_z = \sqrt{\frac{m_x}{m^*}} dk_x' \sqrt{\frac{m_y}{m^*}} dk_z' = dk_x' dk_z' = k' dk' d\vartheta = \frac{m^*}{\hbar^2} dE d\vartheta \tag{2.20}$$

The angular integration can be carried out immediately, and Equation (2.17) becomes

$$I_{SD} = \frac{eWm^*}{\pi\hbar^2} \sum_{valleys} \sum_{n_y} \int v_x(k_x)\left[f(\varphi_{FS}, E) - f(\varphi_{FD}, E)\right] dE \tag{2.21}$$

The velocity can be assumed to be a thermal velocity, which is isotropic, so that

$$v_x \sim \sqrt{\frac{v^2}{2}} \sim \frac{\sqrt{2m_z}}{m^*}\sqrt{E} \tag{2.22}$$

where the scaled coordinates have been incorporated. If we now introduce the reduced coordinates

$$\eta = \frac{\varphi_{FS}}{k_B T}, \quad \chi_n = \frac{E_{n_y}}{k_B T}, \quad \varphi = \frac{eV_D}{k_B T} \tag{2.23}$$

the current can be written as

$$I = \frac{\sqrt{2} eW (k_B T)^{3/2}}{\pi\hbar^2} \sum_{valleys} \sum_{n_y} \sqrt{m_z} \left[F_{1/2}(\eta - \chi_n) - F_{1/2}(\eta - \chi_n - \phi)\right] \tag{2.24}$$

The functions $F_{1/2}$ are the Fermi-Dirac integrals of half-integer order.[22]

However, there is a problem with Equation (2.24) and the development leading up to it. This problem lies in the fact that MOSFETs dissipate a significant amount of heat. If we use two thermal distribution functions at the lattice temperature, then these must be evaluated well into the reservoirs.[23,24] That is, we must use the distribution function in the metallic interconnects rather than in the drain region near the channel. If we want to use this latter region, which is the obvious point of discussion in the above derivations, then we must account for the higher electron temperature in this region. Each carrier that exits the channel into the drain brings with it an excess, directed energy of eV_D. This extra energy is rapidly thermalized by carrier-carrier scattering,[25] which provides an elevated electron temperature $T_e > T$ in the drain. It is no simple task to determine this electron temperature, and clearly gives a rationale for the use of detailed Monte Carlo simulations (classical)[26] or nonequilibrium Green's functions[8] in order to find the detailed distribution function that should be utilized in Equation (2.24). Moreover, the number of occupied subbands (in the y-direction) will be different in the drain end than in the source end. Hence, we should rewrite Equation (2.24), using primes to denote the expressions of Equation (2.23) evaluated with the electron temperature, as

$$I = \frac{\sqrt{2}eW(k_BT)^{3/2}}{\pi\hbar^2} \sum_{valleys} \left\{ \sum_{n_{y,S}} \sqrt{m_z} F_{1/2}(\eta - \chi_n) - \left(\frac{T_e}{T}\right)^{3/2} \sum_{n_{y,D}} \sqrt{m_z} F_{1/2}(\eta' - \chi'_n - \phi') \right\} \tag{2.25}$$

It is clear that a good model for the electron temperature in the drain, near the channel, is necessary to really apply these ballistic formulas.

When the width of the device begins to get small as well, then quantization also occurs in this direction. While Natori[9] has mentioned this, it is relatively easy to incorporate this into Equation (2.24), leading to the Landauer formula, as is shown in Ferry.[21] We will not deal with this here, as the full quantum treatment is discussed in a later section.

1.3 GRANULARITY

By granularity, we refer to the failure of thermodynamic averaging in small devices. If we consider a silicon-on-insulator (SOI) MOSFET, with the silicon channel 10 nm thick, 20 nm wide and 10 nm long, and doped to 10^{19} cm^{-3}, then there are only 20 dopant atoms in the channel. If the carrier density is 10^{13} cm^{-2}, then there are only 20 carriers in the channel at any one time. With such a small number of dopants and carriers, it is impossible to use average densities and statistics. Instead, the position of each impurity is quite important and device performance depends not only upon this number, but also upon the exact position of each of the impurities. Keyes[27] was the first to warn about threshold voltage fluctuations arising from variations in the number of dopant atoms in the channel, but did no simulations to evaluate the problem.

Perhaps the first to study the role of discrete dopants on transport were Boudville and McGill,[28] who studied ohmic contacts to GaAs. Then, Joshi and Ferry[29] showed that, in heavily doped GaAs, an electron was typically interacting with three or more impurities at the same time. Wong and Taur[30] subsequently studied the role of discrete dopants in a Si MOSFET, and Zhou and Ferry[31–33] discussed the problem in MESFETs and HEMTs. Later, Vasileska et al.[34] and Asenov[35] reviewed MOSFET behavior, and the field has blossomed since then.

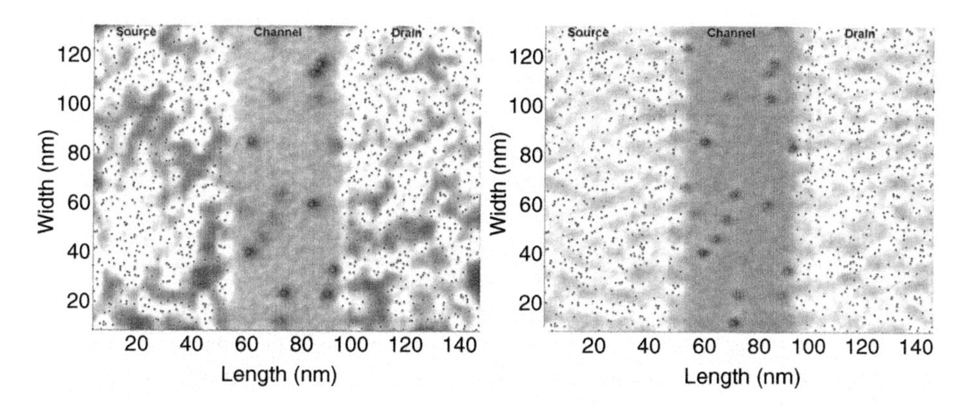

FIGURE 1.2 Electron density from a Monte Carlo simulation using molecular dynamics for the carrier-carrier interaction. (a) Without the effective potential included to simulate quantum confinement, and with $V_G = 0.4$ V, $V_D = 0.1$ V. (b) With the effective potential included in the simulation, and with $V_G = 0.6$ V and $V_D = 0.1$ V. The higher gate voltage was used to get more electrons into the channel for image clarity. The lighter shades represent higher carrier densities, and the dots indicate the position of the impurities (donors in the source and drain, and acceptors in the channel). It is clear that the density tends to cluster around the impurities due to the lower potentials in this region.

We can illustrate the problems inherent with the granularity, by looking, for example, at a simulation of a thin SOI MOSFET. In Figure 1.2, we plot the carrier density in an *n*-channel SOI MOSFET. The density is indicated by the grey scale of the plot, and we are looking down into the plane of the device. Panel (a) shows the case in which a purely classical simulation is incorporated, and it is quite clear that the variations in the carrier density are large. On the other hand, this device is small, and quantization should begin to occur. Panel (b) shows how the density fluctuations are reduced by introducing an effective potential (discussed in the next section) to account for quantum effects. While the density fluctuation has been reduced, it is still significant. Simulations such as these point out that each device, which will have a different number of actual donor and acceptor atoms with different configurations of these atoms, will have its own characteristic performance. While having millions of such devices on a chip can be viewed as an ensemble averaging process, it is important to note that the performance depends upon each individual device and not upon their average behavior. The variations in individual device behavior arise from the failure of thermodynamic averaging within the device, and we cannot invoke ensemble averaging when each device is important.

Dopant atoms are not the only problem that arises from the granularity of the device. Linton et al.[36,37] have pointed out that device variations can occur due to the line edge roughness of the gate polysilicon line. Variations in performance with top surface roughness (variations in thickness) for MOS structures[38] and for MOSFETs[39] have also been considered. Roughness at the oxide-semiconductor interface has usually been treated as a scattering process,[40] but Brown et al.[41] have recently directly incorporated a model of the surface height variation to study thickness variations in SOI MOSFETs. It is quite clear that a truly small semiconductor device can no longer be considered as a generic entity. It will have its own characteristic performance that will depend upon the configuration of the dopants, the variations of the oxide thickness and gate lines, and the variations in the "thickness" induced by roughness at the top and bottom (in SOI device) oxides. Limitation on the ultimate scalability may in the end depend upon the ability to control these fluctuations to a degree that allows the fabrication of billions of reasonably reliable devices.

1.4 QUANTUM BEHAVIOR IN THE DEVICE

As noted previously, channel quantization in the direction normal to the oxide-semiconductor interface has been a fact of life for many years. This leads to important modifications which are readily seen in smaller devices. Two such effects are a shift in the threshold voltage, due to the rise of the lowest occupied subband above the conduction minimum, and a reduction in the gate capacitance, due to the setback of the maximum in the inversion density away from the interface. This latter produces a so-called quantum capacitance which is effectively in series with the normal gate capacitance.[42] If these are the major effects produced by the quantization, then they can be readily handled in a normal semiclassical theory by the introduction of an effective potential.[43] On the other hand, if the individual quantum levels in the inversion layer become resolved, or if the lateral quantization (in either width or thickness of an SOI layer) becomes important, then a full quantum mechanical model is required to handle the device. In the following, we first discuss the effective potential approach, and then turn to the description of a full quantum mechanical simulation for ultrasmall SOI MOSFETs.

1.4.1 The Effective Potential

In recent years, it has become of interest to include a *quantum potential* as a correction to the solutions of the Poisson equation in self-consistent simulations.[44] The quantum potential has a rich history (which will be discussed later), but recently has come to be called the "density-gradient" approach, since the quantum potential is often defined in terms of the second derivative of the square root of local density. Such an approach is highly sensitive to noise in the local carrier density, and the methodology is highly suspect in cases of strong quantization.[45]

We have developed a different approach, which introduces an *effective* potential. Here, the natural non-zero size of an electron wave packet in the quantized system, is used to introduce a smoothing of the local potential (found from Poisson's equation).[46] This approach naturally incorporates the quantum potentials, which are *approximations* to the effective potential. The introduction of an effective potential follows two trends that have been prominent in statistical physics during most of the twentieth century and into the current century. These are the non-zero size of an electron wave packet and the use of a modified potential to describe quantum effects within classical statistical mechanics. Here, we review these two approaches and show how they combine to give a form for the effective potential. We then show how the quantum potential derives from the effective potential as an approximation, and finally provide results from simulations to compare these approaches. We also estimate the problems in incorporating tunneling via this approach.

1.4.1.1 Effective Carrier Wave Packet

In order to describe the packet of a carrier in real space, one must account for the contributions to the wave packet from all occupied plane wave states.[47] That is, the states that exist in momentum space are the Fourier components of the real-space wave packet. If we want to estimate the size of this wave packet, we must utilize all Fourier components, not just a select few. (This approach is familiar from the definition of Wannier functions and their use to evaluate the size of a bound electron orbit near an impurity.) This is not the first attempt to define the nature of the quantum wave packet corresponding to a (semi)classical electron. Indeed, the study of the classical-quantum correspondence has really intensified over the past few decades, due in no small part to the rich nature of chaos in classical systems and the search for the quantum analog of this chaos. This has led to a number of studies of the manifestation of classical phase-space structure.[48] These have shown that meaningful sharp structure can exist in quantum phase-space representations, and these can profitably be used to explain (or to interpret) quantum dynamics; for example, to study the quantum effects that arise in otherwise classical simulations for semiconductor devices. The use of a Gaussian wave packet as a representation of the classical particle is the basis of the well-known coherent-state representation. However, if we have two such wave packets, there is a problem. When we take the two real-space wave packets and create a phase-space Wigner representation, then there is a superposition wave between the two phase-space packets. This represents coherence between the two packets. We can approach the classical regime only by first destroying this decoherence.[49] Then, one can pass to the classical limit and the packets become discrete points in phase space. We shall return to this point shortly.

In the coherent state (Gaussian packet) approach, the phase-space representation of the quantum density localized at point $\mathbf{x}$ is given by[50,51]

$$\langle \mathbf{x} \mid \mathbf{p}, \mathbf{q} \rangle = \frac{1}{(\pi\sigma^2)^{N/4}} \exp\left[-\frac{(\mathbf{x}-\mathbf{q})^2}{2\sigma^2} + i\frac{\mathbf{p}\cdot(\mathbf{x}-\mathbf{q})}{2\hbar} \right] \tag{4.1}$$

In Equation (4.1), $\mathbf{p}$ is the momentum of the wave packet, $\mathbf{q}$ is the centroid position and $\mathbf{x}$ is the general coordinate. As in most cases, the problem is to find the value of the spatial spread of the wave packet, which is defined by the parameter σ, which is related to the width of the wave packet. In this representation, the quantum particle has a phase-space extent determined by the parameter σ, and this goes to zero as we pass to the classical limit. Hence, σ must be related to $\hbar$ in some manner. It was found earlier[47] that σ is given approximately by the thermal de Broglie wavelength.

For this approach to be valid, we must have wave packets that do not have coherence among the packets. This really means that the eigenvalue spectrum of the

Schrödinger equation must be washed out by the thermal smearing. If this spectrum is distinguishable, then a single wave packet for each particle is not a valid approach, and our effective potential method will fail. When the approach is valid, we can then examine how the Gaussian wave packet leads to a smoothing of the classical potential. The scalar potential is related to the charge density through the static Lienard-Wiechert potential[52]

$$V(\mathbf{r}) = \frac{1}{4\pi\varepsilon}\int \frac{\rho(\mathbf{r}')}{|\mathbf{r}-\mathbf{r}'|} d^3\mathbf{r}' \tag{4.2}$$

If we now introduce the discrete charge, this latter equation can be written as

$$V(\mathbf{r}) = \frac{1}{4\pi\varepsilon}\int \frac{e}{|\mathbf{r}-\mathbf{r}'|}\left\{\sum_i \delta(\mathbf{r}'-\mathbf{r}_i^D) - \sum_j \langle \mathbf{r}' \,|\, \mathbf{p}_j, \mathbf{r}_j\rangle\right\} d^3\mathbf{r}' \tag{4.3}$$

The first summation (index i) runs over the ionized donors, and the second summation (index j) runs over the free electrons. The coefficient of the second summation is the set of coherent states defined by Equation (4.1). The first summation provides a distinct contribution from the second, and we concentrate on the second, introducing a resolution of unity in terms of an integral over a delta function as

$$V_2(\mathbf{r}) = -\frac{e}{4\pi\varepsilon}\int \frac{1}{|\mathbf{r}-\mathbf{r}'|}\sum_j \int d^3\mathbf{r}'' \left|\langle \mathbf{r}' \,|\, \mathbf{p}_j, \mathbf{r}''\rangle\right|^2 \delta(\mathbf{r}''-\mathbf{r}_j) d^3\mathbf{r}' \tag{4.4}$$

The squared magnitude Gaussian is independent of the momentum, and is a function only of the difference (squared) of the two coordinate variables. Therefore, we can interchange these in this factor, at the same time changing the notation on the delta function accordingly, and then the integral can be rearranged to give

$$V_2(\mathbf{r}) = \int d^3\mathbf{r}'' V_{cl}(\mathbf{r}'') \left|\langle \mathbf{r}'' \,|\, 0, \mathbf{r}\rangle\right|^2 \tag{4.5}$$

where V_{cl} is the classical potential determined by the charges having only discrete points in phase space. An arbitrary treatment of the first term in Equation (4.3) in this fashion leads us to the result that the non-zero extent of the phase space wave packet of the carriers can be easily moved onto the potential, appearing as a smoothing of the potential by the Gaussian function[46] However, we reiterate that this approach fails when the various eigenvalues of the quantization begin to be resolved. Nevertheless, comparisons with exact solutions of the Schrödinger-Poisson equations for an inversion layer show excellent agreement for those cases in which the approximations are valid.[46,53]

1.4.1.2 Statistical Considerations

From the earliest days of quantum mechanics, there has been an interest in methods that allow the reduction of quantum calculations to classical ones, through the introduction of a suitable *effective potential*. In this regard, one would like to replace the potential in the partition function

$$n \sim \exp\left(-\frac{eV}{k_B T}\right) \tag{4.6}$$

with a modified potential which will describe the density as determined by the quantum wave function. The earliest known approach was provided by Wigner,[54] where he introduced an expansion of the classical potential in powers of $\hbar$ and $\beta = 1/k_B T$, which led to

$$V_{eff}(x) \sim V(x) + \frac{\hbar^2}{8mk_B T}\frac{\partial^2 V}{\partial x^2} + \ldots \tag{4.7}$$

This series led to the well-known Wigner-Kirkwood expansion of the potential that is often used in solutions for the Wigner distribution function. However, the series has convergence problems below the Debye temperature and in cases with sharp potentials, such as the Si-SiO_2 interface. Feynman[55] found a similar result, but with the factor 8 replaced by 24. Feynman also introduced a different approach in which an effective potential is introduced through the free energy; that is, through Equation (4.6) with the classical potential averaged over a Gaussian smoothing function, as in Equation (4.5). For the case of a free particle, he shows that an exact variational minimization leads to a Gaussian weighting of the potential around the classical path, and this automatically includes quantum effects into the trajectory. Indeed, Feynman found that the smoothing parameter σ should have the value

$$\sigma^2 = \frac{\hbar^2}{12mk_B T} = \frac{\lambda_D^2}{24\pi} \tag{4.8}$$

where λ_D is the thermal de Broglie wavelength.

Many people have extended the Feynman approach to the case of bound particles[56–60] and particles at interfaces.[61] The effective potential approach has been recently reviewed by Cuccoli et al.[62] These approaches use the fact that the most-likely trajectory in the path integral no longer follows the classical path when the electron is bound inside a potential well. The introduction of the effective potential and its effective Hamiltonian is closely connected to the return to a phase-space description, as discussed above. This can be done at present only for Hamiltonians containing a *kinetic* energy quadratic in the momenta and a coordinate-only dependence in the potential energy. That is, it is clear that some modifications will have to be made when nonparabolic energy bands, or a magnetic field, are present. However, the Gaussian approximation is well established as the method for incorporating the purely quantum fluctuations around the resulting path. The key new ingredient for bound states (such as in the potential well at the interface of a MOSFET) is the need to determine variationally the dominant path and hence the "correct" value for the parameter σ. For the case in which the bound states are well defined in the potential, both Feynman and Kleinert[57] and Cuccoli et al.[60] find

$$\sigma^2 = \frac{\hbar^2}{4mk_BT}\left[\frac{\coth(f)}{f} - \frac{1}{f^2}\right] \tag{4.9}$$

where

$$f = \frac{\hbar\omega_0}{2k_BT} \tag{4.10}$$

and $\hbar\omega_0$ is the spacing of the subbands. If we take the high-temperature limit, then we can expand for small f, and

$$\sigma^2 \sim \frac{\hbar^2}{12mk_BT} \tag{4.11}$$

to leading order, which agrees with Equation (4.8). In Si, this gives a value of 0.52 nm for the value to be used in the direction normal to the interface (at room temperature). A different mass would be used for transport along the channel, and this gives a value of 1.14 nm.

Another approach to correcting the classical potential arises from the hydrodynamic version of Schrödinger's equation. If it is assumed that the wave function can be written as[63,64]

$$\psi(\mathbf{r},t) = R(\mathbf{r},t)\exp\left[\frac{i}{\hbar}S(\mathbf{r},t)\right] \tag{4.12}$$

the Schrödinger's equation can be separated into two equations for the real and imaginary parts. This gives

$$\frac{\partial S}{\partial t} + \frac{1}{2m}\left(\nabla S\right)^2 + V - \frac{\hbar^2}{2Rm}\nabla^2 R = 0 \tag{4.13}$$

$$\frac{\partial R}{\partial t} + \nabla \cdot \left(\frac{R^2}{m}\nabla S\right) = 0 \tag{4.14}$$

The last equation is a form of the continuity equation, whereas the first is a form of the Euler equation, and in this equation we identify the correction term as the *quantum potential*

$$V_Q(\mathbf{r},t) = -\frac{\hbar^2}{2mR(\mathbf{r},t)}\nabla^2 R(\mathbf{r},t) \tag{4.15}$$

Since the density is identified with the square magnitude of the wave function, Equation (4.15) has become known as the *density gradient* correction to the classical potential. The exact form of the quantum potential can take a variety of shapes, depending upon various approximations for the Wigner function, which have been discussed by Iafrate, Grubin, and Ferry,[65] but Equation (4.15) represents the most common form that has been used in device modeling.[44]

It is important to note that all of these various forms are related to one another. For example, the density-gradient potential is easily derived as a low-order expansion to the actual effective potential. We can expand the effective potential of Equation (4.5) when it is a slowly varying function of position. That is, we use a Taylor series expansion as

$$\begin{aligned} W(x) &= \frac{1}{\sqrt{2\pi}\sigma}\int_{-\infty}^{\infty} V(x+\xi)e^{-\xi^2/2\sigma^2}\,d\xi \\ &\cong \frac{1}{\sqrt{2\pi}\sigma}\int_{-\infty}^{\infty}\left[V(x)+\xi\frac{\partial V}{\partial x}+\frac{\xi^2}{2}\frac{\partial^2 V}{\partial x^2}+\ldots\right]e^{-\xi^2/2\alpha^2}\,d\xi. \end{aligned} \tag{4.16}$$

The first term allows us to bring the potential outside the integral, while the second term vanishes due to the symmetry of the Gaussian. The third term becomes the leading correction term, which gives us

$$V_{eff}(x) = V(x) + \sigma^2\frac{\partial^2 V}{\partial x^2} + \ldots \tag{4.17}$$

We note that this result gives the Wigner form. A value for the smoothing parameter may be found, if we compare with the results of Equation (4.7), to be

$$\sigma^2 = \frac{\hbar^2}{8mk_BT} = \frac{\lambda_D^2}{16\pi} \tag{4.18}$$

which is a factor of 1.5 larger than the Feynman result of Equation (4.11). Asenov et al.[66] have compared the density gradient approach and the effective potential approach and obtained similar results, which is to be expected. This is because the

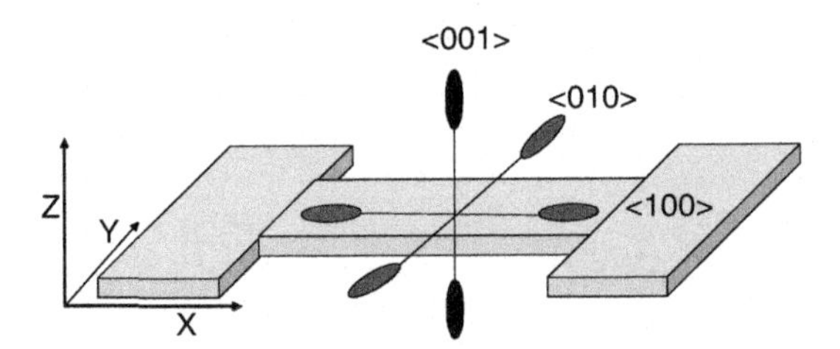

FIGURE 1.3 Crystal orientation of the SOI MOSFET for the quantum simulation (the directions are not to scale). The overlay shows how the six conduction band valleys of Si line up with the coordinate axes. This is discussed further in the text.

approximations on both begin to fail when the quantum corrections become comparable to the classical potential.[49]

1.4.2. Quantum Simulations

There have been many suggestions for different quantum methods to model ultra-small semiconductor devices.[67–69] However, in each of these approaches, the length and the depth are modeled rigorously, and the third dimension (width) is usually included through the assumption that there is no interesting physics in this dimension (lateral homogeneity). Moreover, it is assumed that the mode does not change shape as it propagates from the source of the device to the drain of the device. Other simulation proposals have simply assumed that only one subband in the orthogonal direction is occupied, therefore making higher-dimensional transport considerations unnecessary. These may not be valid assumptions, especially as we approach devices whose width is comparable to the channel length, both of which may be less than 10 nm.

It is important to consider all the modes that may be excited in the source (or drain) region, as this may be responsible for some of the interesting physics that we wish to capture. In the source, the modes that are excited are three dimensional (3D) in nature, even in a thin SOI device. These modes are then propagated from the source to the channel, and the coupling among the various modes will be dependent upon the details of the total confining potential at each point along the channel. Moreover, as the doping and the Fermi level in short-channel MOSFETs increase, we can no longer assume that there is only one occupied subband. In an effort to provide a more complete simulation method, we present a full 3D quantum simulation, based on the use of recursive scattering matrices, which is being used in our group to simulate short-channel, fully depleted SOI MOSFET devices.[70,71]

1.4.2.1 The Device Structure

The device under consideration is a fully-depleted SOI MOSFET structure, shown schematically in Figure 1.3. We orient the x and z directions in order to correspond to the length and the height (thickness of the SOI layer) of the device, respectively. In the x direction, the source and the drain contact regions are 10 nm in length and 18 nm in width (lateral direction, the y axis). In an actual device, the length of the source and the drain of a MOSFET would be much longer, but this length captures the important energy relaxation length. We implement open boundary conditions at the ends of the structure and on the sidewalls. The gate length of this device is 11 nm corresponding to a dimension that will allow the gate to fully control the channel of the device. The actual channel length of the device used in these simulations is 9 nm. The channel itself is 9 nm in width, so that the Si layer is a wide-narrow-wide structure as shown in the figure. The entire structure is on a silicon layer that is taken to be only 6 nm thick, with a 10-nm buried oxide (BOX) layer below this layer. The gate oxide is taken to be 2 nm thick.

An important point relates to the crystal orientation of the device, as indicated in Figure 1.3. As is normal, we assume that the device is fabricated on a [100] surface of the Si crystal, and we then orient the channel so that the current will flow along the <100> direction. This direction is chosen so that all of the principal axes of the conduction band valleys line up with the coordinate axes. By this, we mean that the <010> direction lines up along the y direction and the <001> direction lines up with the z direction, and the six equivalent ellipsoids are oriented along the Cartesian coordinate axes. This is important so that the resulting quantization will split these ellipsoids into three pairs. Moreover, the choice of axes is most useful as the resulting Hamiltonian matrix will be diagonal. In contrast, if we had chosen the <110> direction to lie along the channel, the six ellipsoids would have split into a twofold pair (those normal to the [100] plane) and a fourfold pair, but the Hamiltonian would not be diagonal since the current axis makes an angle with each ellipsoid of the fourfold pair. Using our orientation complicates the wave function, as we will see, but allows for simplicity in terms of the amount of memory needed to store the Hamiltonian and to construct the various scattering matrices (as well as the amount of computational time that is required).

1.4.2.2 The Wave Function and Technique

We can now write a total wave function that is composed of three major parts, one for each of the three sets of valleys. That is, we can write the wave function as a vector

$$\Psi_T = \begin{bmatrix} \Psi^{(x)} \\ \Psi^{(y)} \\ \Psi^{(z)} \end{bmatrix} \tag{4.19}$$

where the superscript refers to the coordinate axis along which the principal axis of the ellipsoid lies (the longitudinal mass direction). Thus, $\Psi^{(x)}$ refers to the two ellipsoids oriented along the x axis in Figure 1.3 (the <100> ellipsoids). Each of these three component wave functions is a complicated wave function on its own. Consider the Schrödinger equation for one of these sets of valleys (i corresponds to x, y, or z valleys):

$$\frac{-\hbar^2}{2}\left(\frac{1}{m_x}\frac{d^2}{dx^2} + \frac{1}{m_y}\frac{d^2}{dy^2} + \frac{1}{m_z}\frac{d^2}{dz^2}\right)\Psi^{(i)} + V(x,y,z)\Psi^{(i)} = E\Psi^{(i)} \tag{4.20}$$

Here, it is assumed that the mass is constant, in order to simplify the equations (for nonparabolic bands, the reciprocal mass enters between the partial derivatives). We have labeled the mass corresponding to the principal coordinate axes, and these take on the values of m_L and m_T as appropriate. We then choose to implement this on a finite difference grid with uniform spacing a. Therefore, we replace the derivatives appearing in the discrete Schrödinger equation with finite difference representations of the derivatives. The Schrödinger equation then reads

$$\begin{aligned} &-t_x\left(\psi_{i+1,j,k}+\psi_{i-1,j,k}\right) \\ &-t_y\left(\psi_{i,j+1,k}+\psi_{i,j-1,k}\right)-t_z\left(\psi_{i,j,k+1}+\psi_{i,j,k-1}\right) \\ &+\left(V_{i,j,k}+2t_x+2t_y+2t_z\right)\psi_{i,j,k}=E\psi_{i,j,k}\ , \end{aligned} \tag{4.21}$$

where t_x, t_y and t_z are the hopping energies

$$\begin{aligned} t_x &= \frac{\hbar^2}{2m_x a^2}, \\ t_y &= \frac{\hbar^2}{2m_y a^2}, \\ t_z &= \frac{\hbar^2}{2m_z a^2}\ . \end{aligned} \tag{4.22}$$

Each hopping energy corresponds with a specific direction in the silicon crystal. The fact that we are now dealing with three sets of hopping energies is quite important.

There are other important points that relate to the hopping energy. The discretization of the Schrödinger equation introduces an artificial band structure, due to the periodicity that this discretization introduces. As a result, the band structure in any one direction has a cosinusoidal variation with momentum eigenvalue (or mode index), and the total width of this band is $4t$. Hence, if we are to properly simulate the real band behavior, which is quadratic in momentum, we need to keep the energies of interest below a value where the cosinusoidal variation deviates significantly from the parabolic behavior desired. For practical purposes, this means that $E_{max} < t$. The smallest value of t corresponds to the longitudinal mass, and if we desire energies of the order of the source-drain bias ~ 1 V, then we must have $a < 0.2$ nm. That is, we must take the grid size to be comparable to the Si lattice spacing!

With the discrete form of the Schrödinger equation defined, we now seek to obtain the transfer matrices relating adjacent slices in our solution space. For this, we will develop the method in terms of slices, and follow a procedure first put forward by Usuki et al.[72,73] and used extensively by our group.[74] This is modified here by the two dimensions in the transverse plane. We begin first by noting that the transverse plane has $N_y \times N_z$ grid points. Normally, this would produce a second-rank tensor (matrix) for the wave function, and it would propagate via a fourth-rank tensor. However, we can reorder the coefficients into a $N_yN_z \times 1$ first-rank tensor (vector), so that the propagation is handled by a simpler matrix multiplication. Since the smaller dimension is the z direction, we use N_z for the expansion, and write the vector wave function as

$$\Psi^{(i)} = \begin{bmatrix} \psi^{(i)}_{1,Ny} \\ \psi^{(i)}_{2,Ny} \\ \cdots \\ \psi^{(i)}_{Nz,Ny} \end{bmatrix} \tag{4.23}$$

Now, Equation (4.21) can be rewritten as a matrix equation as, with s an index of the distance along the x direction,

$$H^{(i)}\Psi^{(i)}(s) - T_x^{(i)}\Psi^{(i)}(s-1) - T_x^{(i)}\Psi^{(i)}(s+1) = EI\Psi^{(i)}(s) \tag{4.24}$$

Here, I is the unit matrix, E is the energy to be found from the eigenvalue equation, and

$$H^{(i)} = \begin{bmatrix} H_0^{(i)}(\mathbf{r}) & \tilde{t}_z^{(i)} & \cdots & 0 \\ \tilde{t}_z^{(i)} & H_0^{(i)}(\mathbf{r}) & \cdots & \cdots \\ \cdots & \cdots & \cdots & \tilde{t}_z^{(i)} \\ 0 & \cdots & \tilde{t}_z^{(i)} & H_0^{(i)}(\mathbf{r}) \end{bmatrix} \tag{4.25}$$

$$T_x^{(i)} = \begin{bmatrix} \tilde{t}_x^{(i)} & 0 & \cdots & 0 \\ 0 & \tilde{t}_x^{(i)} & \cdots & 0 \\ \cdots & \cdots & \cdots & \cdots \\ 0 & 0 & \cdots & \tilde{t}_x^{(i)} \end{bmatrix} \tag{4.26}$$

The dimension of these two supermatrices is $N_z \times N_z$, while the basic Hamiltonian terms of (4.25) have dimension of $N_y \times N_y$, so that the total dimension of these two matrices is $N_yN_z \times N_yN_z$. In general, if we take k and j as indices along y, and η and ν as indices along z, then

$$\left(\tilde{t}_z^{(i)}\right)_{\eta\nu} = t_z^{(i)}\delta_{\eta\nu}, \quad \left(\tilde{t}_y^{(i)}\right)_{kj} = t_y^{(i)}\delta_{kj}, \quad \left(\tilde{t}_x^{(i)}\right)_{ss'} = t_z^{(i)}\delta_{ss'} \tag{4.27}$$

and

$$H_0^{(i)}(\mathbf{r}) = \begin{bmatrix} V(s,1,\eta)+W & t_y^{(i)} & \dots & 0 \\ t_y^{(i)} & V(s,2,\eta)+W & \dots & 0 \\ \dots & \dots & \dots & t_y^{(i)} \\ 0 & 0 & t_y^{(i)} & V(s,N_y,\eta)+W \end{bmatrix} \quad (4.28)$$

The quantity W is $2(t_x^{(i)} + t_y^{(i)} + t_z^{(i)})$ and is, therefore, independent of the valley index.

With this setup of the matrices, the general procedure follows that laid out in the previous work.[72–74] One first solves the eigenvalue problem on slice 0 at the end of the source (away from the channel), which determines the propagating and evanescent modes for a given Fermi energy in this region. The wave function is thus written in a mode basis, but this is immediately transformed to the site basis, and one propagates from the drain end, using the scattering matrix iteration

$$\begin{aligned} \begin{bmatrix} C_1^{(i)}(s+1) & C_2^{(i)}(s+1) \\ 0 & 1 \end{bmatrix} &= \begin{bmatrix} 0 & 1 \\ -1 & (T_x^{(i)})^{-1}\left(EI - H^{(i)}\right) \end{bmatrix} \\ &\times \begin{bmatrix} C_1^{(i)}(s) & C_2^{(i)}(s) \\ 0 & 1 \end{bmatrix} \begin{bmatrix} 1 & 0 \\ P_1^{(i)}(s) & P_2^{(i)}(s) \end{bmatrix}. \end{aligned} \quad (4.29)$$

The dimension of these matrices is $2N_yN_z \times 2N_yN_z$, but the effective propagation is handled by submatrix computations, through the fact that the second row of this equation sets the iteration conditions

$$\begin{aligned} C_2^{(i)}(s+1) &= P_2^{(i)}(s) = \left[-C_2^{(i)}(s) + (T_x^{(i)})^{-1}\left(EI - H^{(i)}\right)\right]^{-1}, \\ C_1^{(i)}(s+1) &= P_1^{(i)}(s) = P_2^{(i)}(s)C_1^{(i)}(s). \end{aligned} \quad (4.30)$$

At the source end, $C_1(0) = 1$, and $C_2(0) = 0$ are used as the initial conditions. These are now propagated to the N_x slice, which is the end of the active region, and then onto the $N + 1$ slice. At this point, the inverse of the mode-to-site transformation matrix is applied to bring the solution back to the mode representation, so that the transmission coefficients of each mode can be computed. These are then summed to give the total transmission and this is used in a version of Equation (2.17) to compute the current through the device (there is no integration over the transverse modes, only over the longitudinal density of states and energy).

If we are to incorporate a self-consistent potential within the device, we must now solve Poisson's equation. Here, the density at each point in the device is determined from the wave function squared magnitude at that point, and this is used to drive Poisson's equation. Our solution for $C_1(N_x + 2)$ is the wave function at this point, and this is back-propagated using the recursion algorithm

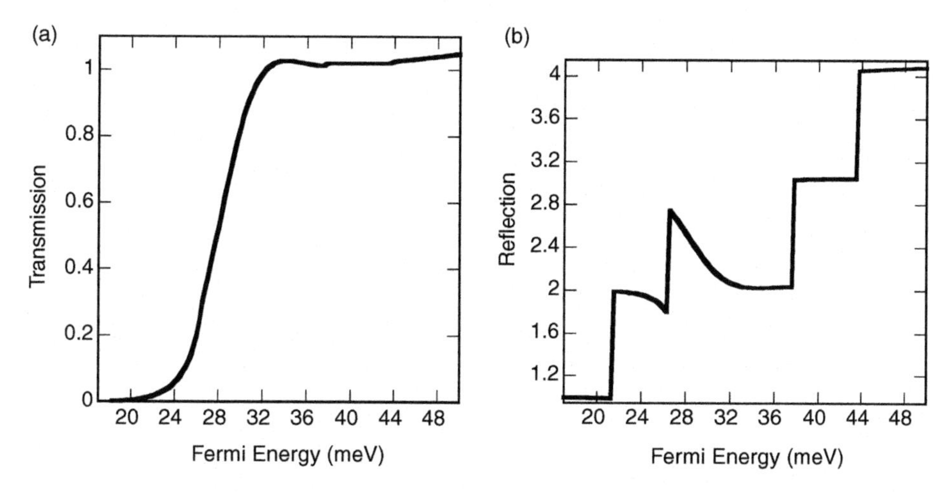

FIGURE 1.4 (a) Transmission and (b) reflection versus Fermi energy for a 9-nm-channel-length SOI MOSFET using hardwall potentials.

$$\Phi_\xi^{(N_x+2,s,i)}(j,\eta) = P_1^{(i)}(s) + P_2^{(i)}\Phi_\xi^{(N_x+2,s+1,i)}(j,\eta) \tag{4.31}$$

Here, as before, the superscript "i" denotes the valley, while j and η denote the transverse position. Here, we are in the mode representation, and ξ is the mode index. The density at any site (s,j,η) is found by taking the sum over ξ of the occupied modes at that site, as

$$n(s,j,\eta) = \sum_\xi \left|\Phi_\xi^{(N_x+2,s,i)}(j,\eta)\right|^2 \tag{4.32}$$

1.4.2.3 Results

All of the equations used so far are written in the absence of a magnetic field, so that the Hamiltonian is symmetric. The various terms become more complicated if a magnetic field is present, as one may want for a study of spin transport, but this is beyond our present interest. Moreover, if there is no intervalley scattering (ballistic transport), then the equations for the three pairs of valleys are uncoupled. If intervalley scattering were to be present, then off-diagonal terms appear in the total Hamiltonian between valleys, and the iteration procedure of (4.29) and (4.30) becomes much more difficult, with the matrices each being a factor of three increased in span. In the following, we will assume that no scattering is present, so that valleys which are unoccupied in the source will remain so throughout the device.

In Figure 1.4(a), we plot the results of the transmission of incident modes as the Fermi energy is varied from 0 to 50 meV for a device at 300 K. Here, hard wall boundary conditions have been used (no self-consistent potential). In this method, we have taken into account the possibility of having the in-plane valleys contributing

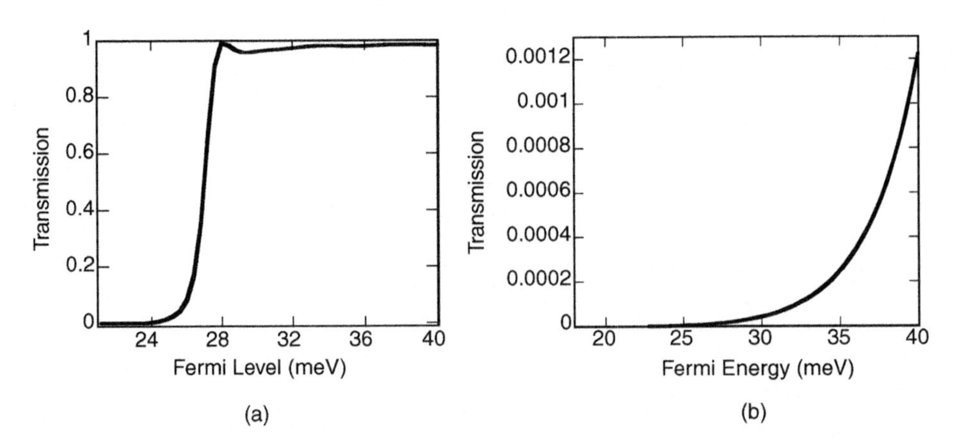

FIGURE 1.5 (a) Transmission for the surface-normal valleys versus Fermi energy for the 9nm-channel-length SOI MOSFET using a self-consistent potential with a gate voltage of 1 V. (b) Transmission for the in-plane (upper) valleys versus Fermi energy for this device.

to the overall conductance of the device. Nevertheless, in the hardwall case, only the two surface normal (<001>) valleys contribute to the conductance. This can be attributed to the fact that the surface normal valleys have the larger effective mass normal to the primary quantization direction (z direction) and, therefore, modes excited in these valleys will be the first to contribute. Further, we see that as the Fermi energy of the system is increased, the number of excited modes in the source of the device grows, but the transmission of these modes through the channel remains constant. This is confirmed in Figure 1.4(b), where the reflection coefficient is plotted against increasing Fermi energy. Clearly, the number of modes increases, but the vast majority of these are reflected at the source channel interface. At approximately 24 meV, we see a decrease in the reflection coefficient followed by a sharp rise and subsequent decline. This behavior is expected as the onset of this decrease marks the point where the MOSFET begins to conduct. As we progress in energy, we see the sharp increase as another mode begins to propagate in the source of the device. This is followed by the exponential decrease back to 2 meV as the channel saturates with a full mode now propagating.

We now compare the hardwall results with results obtained using a self-consistent potential, found from solving Poisson's equation. The n^+ source and drain have been doped $1\times10^{20}cm^{-3}$ whereas the p-type channel of the device has been doped at $1\times10^{18}cm^{-3}$. In Figure 1.5, we plot the transmission resulting from varying the Fermi energy from 0 – 40 meV for all of the valleys, for a gate voltage of 1 V. In the case of the self-consistent potential, the final Fermi energy has been reduced to keep the energies within the artificial band structure. In Figure 1.5, we see that the turn-on energy for the transmission in both the in-plane and perpendicular valleys is very close to that of the hard wall case. In Figure 1.5(a), we see that the self-consistent potentials reduce the contribution from the surface normal valleys. This is because the self-consistent potential squeezes the channel in the lateral y-direction. This greatly raises these valleys due to quantization in this direction, while two of the in-plane valleys are lowered in energy with respect to this first set. In Figure 1.5(b), we also see that the upper valleys have begun to conduct. This can be attributed to the fact that with the self-consistent potential we see a softer variation in the potential. The potential allows for more leakage and higher-order contributions. The in-plane valleys now contribute to the current flow. Although there are more modes excited in the surface normal valleys, most of the modes are reflected at the source-channel constriction.

1.5 QUANTUM DOT SINGLE-ELECTRON DEVICES

Single-electron devices are of great interest, in particular for possible device application in integrated circuits.[75] The ability to control electron charging of a capacitive node by individual electrons makes these devices suitable candidates for memory applications.[76,77] As there are several chapters in this volume devoted to single-electron devices, we mention here only planar devices which are lithographically defined.

A major difficulty in fabricating planar single-electron transistors arises from the lithographic limits required in making small tunnel junctions in which the charging energy of the junction capacitor $e^2/2C >> k_BT$. For room temperature operation, this requirement dictates lithographic control below 10 nm. In general, a single-electron transistor is made of a small "dot" isolated from the source and drain by two small tunnel junctions. For VLSI, it is preferable to work with devices fabricated in a semiconductor system, and quite novel ones have recently been fabricated using sidewall depletion gates[78] Quantum confinement becomes relevant in silicon, and one may be able to observe quantum confinement effects and Coulomb blockade simultaneously in electrical measurements. Recently, single-electron dots have been created within a MOSFET.[78–82] In these structures, the dot is formed in the inversion layer created by a top gate (which is referred to as the inversion gate), with the lateral definition of the dots being provided by side gates (these gates provide the depletion of the dot, and are referred to as the depletion gates) embedded within the gate oxide. In essence, this is a multiple oxide system with stacked gates. The early work on this has recently been reviewed,[81] and the recent work using sidewall depletion gates appears quite promising. The major issue at this point is technological—can the devices be fabricated with sufficiently small dimensions to operate at, or near to, room temperature?

1.6 MANY-BODY INTERACTIONS

In simulations of ultrasmall semiconductor devices, a number of important considerations have been either ignored or have been approximated in a manner that is not representative of the actual physical interactions within the devices. Foremost of these is the study of the Coulomb interaction between the electrons and the impurities and between the individual electrons themselves. This Coulomb interaction has two parts: first, the nature of discrete impurities and how this affects device performance, and, secondly, how the Coulomb interaction affects the *transport* of the carriers through the device. The first of these has been discussed above. Here, we want to turn our attention to the carrier-carrier interaction.

Most ensemble Monte Carlo (EMC) simulation of small semiconductor devices does not include the details of the Coulomb interactions between the individual carriers, primarily because of the computational time and resources required. If carrier-carrier scattering is included, it is typically included through a k-space scattering process without much regard for the energy exchange in the process.[83] In such simulations, Ravaioli (U. Ravaioli, personal communication, 2000) has shown that the carriers will go several tens of nm into the drain before relaxing their energy and directed momentum. If this is a real effect, then actual device sizes will be significantly larger than the gate-related lengths in order to account for the actual hot carrier sizes.[84] Hence, it is important to know if the full Coulomb interaction, treated properly in real space (as opposed to approximations in terms of scattering processes), has a significant role in the transport of carriers in ultrasmall MOSFETs. We have previously discussed a full three-dimensional model of an ultrasmall MOSFET, in which the transport is treated by a coupled EMC and molecular dynamics (MD) procedure to treat the Coulomb interaction in real space.[85,86] Impurities within the device, including the source and drain regions, are treated as discrete charges and are randomly sited according to the nominal doping density of each region. We find that the inclusion of the proper Coulomb interaction significantly affects both the energy and momentum relaxation processes, but also has a dramatic effect on the characteristic curves of the device. Relaxation occurs in the drain over a few nanometers, and the Coulomb "scattering" causes a significant shift in threshold voltage as well as a reduction in actual drain current. These effects are moderated somewhat in an SOI device due to the limited thickness of the Si layer and the small size of the drain.[87]

The inherent real-space tracking of particle positions in the EMC allows us a more exact treatment of the Coulomb interaction between charged particles (particle-ion and particle-particle interactions). This is accomplished through the addition of an MD loop.[29,88–90] This coupled EMC-MD scheme has been shown to give simulation mobility results in excellent agreement with the experimental data for bulk samples with high substrate doping levels. It has also been corrected for both the degeneracy[91] and many-body exchange corrections to the ground state energy of the system.[92] Problems with this EMC-MD approach arise from the fact that both the *e-e* and *e-i* interactions are already included in the self-consistent potential via the solution of the Poisson equation (this is in the Hartree term). The magnitude of the resulting so-called mesh force depends upon the volume of the cell and, for commonly employed mesh sizes in device simulations, usually leads to double counting of the force if a separate Coulomb interaction is added to the EMC transport kernel.[85] Hence, careful treatment of the short-range particle-particle interactions is needed to avoid the double counting of the force.

One brute force way to overcome this difficulty is to identify the correction terms necessary for the inelastic interaction between charge centers within an overall self-consistent particle-based device simulation, thus avoiding the problem of the doublecounting of the force. Briefly, we estimate the smoothed self-consistent potential on the grid points, as determined by the solutions to Poisson equation, and then determine the short-range corrected Coulomb interaction to be used in our MD routine. This scheme has proven to be quite successful in explaining the doping dependence of the low-field electron mobility in highly doped resistors.[85,86] It also gives us confidence that this approach can be successfully used to accurately describe the fluctuations in various device parameters due to the atomistic nature and different distribution of the impurity atoms in the active region of the device. Although ad hoc, it can be based on a more fundamental principle. Quite generally, we can replace the localized carrier by a function that is related strongly to the Gaussian wave packet discussed in Section IV.A.1. This divides the Coulomb potential into a short-range part and a long-range part. For example, we can write

$$\delta(r) \rightarrow \delta(r) + f(r) - f(r) \tag{6.1}$$

so that the Coulomb potential goes into something like

$$\frac{1}{r} \rightarrow \left[1 - erf(r)\right]\frac{1}{r} + erf(r)\frac{1}{r} \tag{6.2}$$

The first term on the left-hand side is a short-range function that vanishes as the magnitude of r increases. On the other hand, the second term is a long-range function that vanishes at short distances. This is, of course, the principle of the potential splitting discussed above, in which the long-range term is found from the solutions of Poisson's equation. Such a cutoff was introduced by Kelbg[93] in order to treat molecular dynamics in plasma problems without incurring the very short-range attraction between ions and electrons. This approach has been shown to be particularly useful in quantum many-body problems.[94] A similar split has been suggested by Kohn et al.[95] for electronic structure calculations, where the short-range potential is kept within the density-functional approach and the long-range potential is used for perturbation theory or configuration-interaction refinements of the results.[96] It is clear that the separation of the Coulomb potential into these short-range and long-range parts has a rich history and validates the splitting discussed above.

An alternative approach is to do the direct elastic Coulomb scattering by the traditional momentum space approach, but then add inelastic plasmon scattering.[97,98] This approach has also been used recently by Fischetti[99] to study the interaction of channel electrons with electrons in the gate. One advantage of this is the separation of the scattering from the role of the impurities in the self-consistent potential. On the other hand, the separation is rather artificial, and it is difficult to account for the density variation in the plasmon description. Moreover, one needs to take a non-equilibrium plasmon distribution function (the Bose-Einstein distribution must be taken at the electron temperature, which is quite difficult to evaluate, especially as a function of position). As a consequence, the best approach is the MD coupled Monte Carlo approach.

In Figure 1.6, we show the energy decay of the channel electrons as they move into the drain region (100) of a 50-nm-gate-length SOI MOSFET. It is clear that the inclusion of the electron-electron interaction causes a more rapid decay, which is

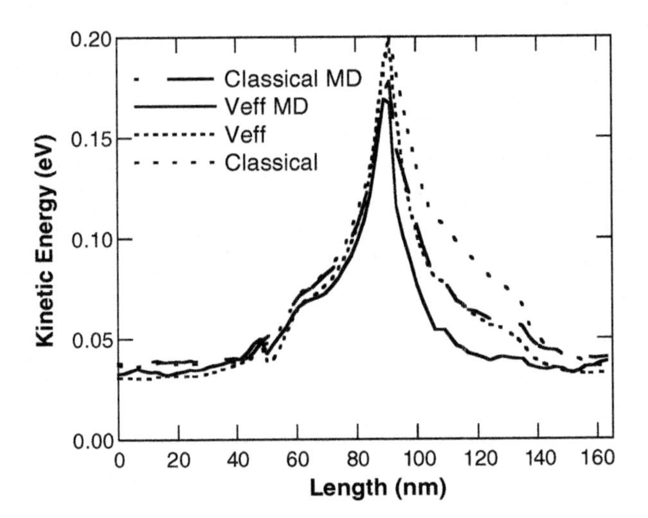

FIGURE 1.6 The variation of the energy along the channel of an SOI MOSFET with a 4-nm-thick film and a 50-nm gate length. Here, the transport is simulated with an ensemble Monte Carlo technique using a molecular dynamics routine to include the carrier-carrier scattering. The solid curve is the result including the carrier-carrier scattering, and the dashed curve is obtained without this scattering.

indicative of plasmon emission being the major loss mechanism (although plasmons do not exist explicitly, this is the energy loss mechanism that explains such a rapid decay). This simulation is for a 4-nm-thick SOI layer, with a drain extension region. For thinner SOI layers, or in the absence of the extension region, the small number of electrons available as a whole really cuts down the effectiveness of this energy relaxation process. This could be a major problem in very small SOI devices in the future.

1.7 ACKNOWLEDGMENTS

The authors have enjoyed many helpful discussions with J. R. Barker, J. P. Bird, S. M. Goodnick, C. Jacoboni, I. Knezevic, S. Milicic, and D. Vasileska, that have aided the flow of this work. The work itself has been funded in part by the Office of Naval Research and the Semiconductor Research Corporation.

REFERENCES

1. Moore, G.E., Cramming more components onto integrated circuits. *Electronics,* 38, 114–117, 1965.
2. Chau, R., 30 nm and 20 nm physical gate length CMOS transistors, 2001 Silicon Nanoelectronics Workshop, Kyoto, Japan, June 10–11, 2001.
3. Yu, B., Wong, H., Joshi, A., Xiang, Q., Ibok, E., and Lin, M-R., 15 nm gate length CMOS transistor, *2001 International Electron Device Meeting Technical Digest,* 937–939, IEEE, New York, 2001.

4. Doris, B., Ieong, M., Kanarsky, T., Zhang, Y., Roy, R.A., Dokumaci, O., Ren, Z., Jamin, F.-F., Shi, L., Natzle, W., Huang, H.-J., Mezzapelle, J., Mocuta, A., Womack, S., Gribelyuk, M., Jones, E.C., Miller, R.J., Wong, H.-SP., and Haensch, W. Extreme scaling with ultrathin Si channel MOSFETs. *2002 International Electron Device Meeting Technical Digest*, 267–270, IEEE, New York, 2002.
5. Barker, J.R. and Ferry, D.K., On the physics and modeling of small semiconductor devices—II: the very small device. *Solid-State Electronics,* 23, 531–544, 1980.
6. Fischetti, M., Theory of electron transport in small semiconductor devices using the Pauli master equation. *J Appl Phys,* 83, 270–291, 1998.
7. Likharev, K., Sub-20 nm electron devices, in Morkoç, H., Ed., *Advanced Semiconductor and Organic Nano-Technique,* Academic Press, New York, 2002.
8. Venugopal, R., Ren, Z., Datta, S., Lundstrom, M.S., and Jovanovic, D., Simulating quantum transport in nanotransistors: real versus mode-space approaches. *J Appl Phys,* 92, 3730–3739, 2002.
9. Natori, K., Ballistic metal-oxide-semiconductor field effect transistor. *J Appl Phys,* 76, 4879–4890,1994.
10. Gilbert, M.J., Akis, R., and Ferry, D.K., Modeling of fully-depleted SOI MOSFETs in 3D using recursive scattering matrices. *J Comp Electron*, 2, 329–334, 2003.
11. Lundstrom, M., Elementary scattering theory of the silicon MOSFET. *IEEE Electron Dev Lett,* 18, 361–363, 1997.
12. Lilienfeld, J.E., Method and Apparatus for Controlling Electric Currents, U. S. Patent 1,745,175, 1930.
13. Ferry, D.K., Hess, H., and Vogl, P., Submicron IGFETs, II. in: Einspruch, N.G., Ed., *VLSI Electronics: Microstructure Science*, 2, 67–103, Academic Press, New York, 1981.
14. Ando, T., Fowler, A., and Stern, F., Electronic properties of two dimensional systems. *Rev Mod Phys,* 54, 437–672, 1982.
15. Landauer, R., Spatial variations of currents and fields due to localized scatterers in metallic conductors, *IBM J Res Develop*, 1, 223–231, 1957.
16. Ferry, D.K. and Bird, J.P., *Electronic Materials and Devices*, Academic Press, San Diego, 2001.
17. Price, P., Monte Carlo calculation of electron transport in solids, in: Willardson, R.K. and Beer, A.C., Eds., *Semiconductors and Semimetals*, 14, 249–334, Academic Press, New York, 1979.
18. Svizhenko, A. and Anantram, M.P., Role of scattering in nanotransistors, 2003 Silicon Nanoelectronics Workshop, Kyoto, Japan, June 8-9, 2003, *IEEE Trans Electron Devices,* 50, 1459–1460, 2003.
19. Natori, K., Ballistic MOSFET reproduces current-voltage characteristics of an experimental device, *IEEE Electron Dev Lett,* 23, 655–657, 2002.
20. Duke, C.B., *Tunneling in Solids*, Academic Press, New York, 1969.
21. Ferry, D.K., *Quantum Mechanics,* 2nd ed., Inst Phys Publ, Bristol, 2001.
22. Blakemore, J.S., *Semiconductor Statistics*, Pergamon Press, New York, 1962.
23. Landauer, R., Electrical resistance of disordered one-dimensional lattice, *Philos Mag,* 21, 863–875, 1970.
24. Landauer, R.. Electrical transport in open and closed systems. *Z Phys B,* 68, 217–228, 1987.
25. Gross, W.J., Vasileska, D., and Ferry, D.K., Ultrasmall MOSFETs: the importance of the full Coulomb interaction on device characteristics. *IEEE Trans Electro Dev,* 47, 1831–1837, 2000.
26. Ravaioli, U. and Ferry, D.F., MODFET ensemble Monte Carlo model including the quasi-two-dimensional electron gas. *IEEE Trans Electron Dev,* 33, 677–680, 1986.

27. Keyes, R.W., Physical limits in semiconductor devices. *Science*, 195, 1230–1235, 1977.
28. Boudville, W.J. and McGill, T.C., Ohmic contacts to *n*-type GaAs, *J Vac Sci Technol B*, 3, 1192–1196, 1985.
29. Joshi, R.P. and Ferry, D.K., Effect of multi-ion screening on the electronic transport in doped semiconductors: a molecular dynamics analysis, *Phys Rev B,* 43, 9734–9739, 1991.
30. Wong, H.-S. and Taur, Y., Three dimensional "atomistic" simulation of discrete dopant distribution effect in sub-0.1 μm MOSFETs, *1993 International Electron Device Meeting Technical Digest*, 705–708, IEEE, New York, 1993.
31. Zhou, J.-R. and Ferry, D.K., Three dimensional simulation of the effect of random impurity distributions on conductance for deep submicron devices. *Proc. 3rd International Workshop on Computational Electronics*, 74-77, Portland, OR, 1994.
32. Zhou, J.-R. and Ferry, D.K., 3D simulation of deep submicron devices: how impurity atoms affect conductance, *IEEE Comp Sci Engr,* 2, 30–37, 1995.
33. Zhou, J.-R. and Ferry, D.K., 3D discrete dopant effects on small semiconductor device physics, in: Hess, K., Leburton, J.P., and Ravioli, U., Eds., *Hot Carriers in Semiconductors*, 491-496, Plenum Press, New York, 1996.
34. Vasileska, D., Gross, W.J., and Ferry, D.K., Modeling of deep submicron MOSFETs: random impurity effects, threshold voltage shifts, and gate capacitance attenuation, *Proc. 6th International Workshop on Computational Electronics*, 259–262, IEEE Press, New York, 1998.
35. Asenov, A., Efficient 3D "atomistic" simulation technique for studying of random dopant induced threshold voltage lowering and fluctuations in decanano MOSFETs, *Proc. 6th International Workshop on Computational Electronics*, 263–266, IEEE Press, New York, 1998.
36. Linton, T.D., Yu, S., and Shaheed, R., Modeling 3D fluctuation effects in highly scaled VLSI devices. *VLSI Design*, 13, 103–110, 2001.
37. Linton, T., Shadhok, M., Rice, B.J., and Schrom, G., Determination of the line edge roughness specification for 34 nm devices, *2002 International Electron Device Meeting Technical Digest*, 303–306, IEEE, New York, 2002.
38. Rack, M.J., Vasileska, D., Ferry, D.K., and Sidorov, M., Surface roughness of SiO_2 from a remote microwave plasma enhanced chemical vapor deposition process, *J Vac Sci Technol B,* 16, 2165–2170, 1998.
39. Asenov, A., Kaya, S., and Davies, J.H., Intrinsic threshold voltage fluctuations in decanano MOSFETs due to local oxide thickness variations, *IEEE Trans Electron Dev*, 49, 112–119, 2002.
40. Goodnick, S.M., Ferry, D.K., Wilmsen, C.W., Lilienthal, Z., Fathy, D., and Krivanek, O.L., Surface roughness at the Si(100)-SiO_2 interface, *Phys Rev B*, 32, 8171–8186, 1985.
41. Brown, A.R., Lema, F.A., and Asenov, A., Intrinsic parameter fluctuations in nanometre scale thin-body SOI devices introduced by interface roughness, *Superlattices and Microstructures,* 34, 283–291, 2003.
42. Vasileska, D., Schroder, D., and Ferry, D.K., Scaled silicon MOSFETs: degradation of the total gate capacitance, *IEEE Trans Electron Dev*, 44, 584–587, 1997.
43. Ferry, D.K., Shifren, L., Ramey, S., and Akis, R., The effective potential in device modeling: the good, the bad, and the ugly, *J Comp Electron*, 1, 59–65, 2002.
44. Zhou, J.-R. and Ferry, D.K., Simulation of ultra-small GaAs MESFET with quantum moment equations, *IEEE Trans Electron Dev*, 39, 473–478, 1992.
45. Ferry, D.K. and Barker, J.R., Open problems in quantum simulation of ultra-submicron devices, *VLSI Design*, 8, 165–172, 1998.

46. Ferry, D.K., The onset of quantization in ultra-submicron semiconductor devices, *Superlatt Microstruc*, 27, 61–66, 2000.
47. Ferry, D.K. and Grubin, H.L., Electrons in semiconductors: how big are they?, *Proc. 6th International Workshop on Computational Electronics*, 84–87, IEEE Press, New York, 1998.
48. Skodje, R.T., Rohrs, H.W., and van Buskirk, J., Flux analysis, the correspondence principle, and the structure of quantum phase space. *Phys Rev A*, 40, 2894–2916, and references therein, 1989.
49. Zurek, W.H., Decoherence, einselection, and the quantum origins of the classical, *Rev Mod Phys,* 75, 715–775, 2003.
50. Glauber, R.J., Coherent and incoherent states of the radiation field. *Phys Rev*, 131, 2766–2788, 1963.
51. Klauder, J.R. and Sudarshan, E.C.G., Fundamentals of Quantum Optics, Benjamin, New York, 1968.
52. Reitz, J.R. and Milford, F.J., Foundations of Electromagnetic Theory, Addison-Wesley, Reading, MA, 1960.
53. Ahmed, S.S., Akis, R., and Vasileska, D., Quantum effects in SOI devices, *Tech Proc 5th Intern Conf Modeling and Simulation of Microsystems*, 518–521, Computational Publications, Boston, 2002.
54. Wigner, E., On the quantum correction for thermodynamic equilibrium, *Phys Rev,* 40, 749–759, 1932.
55. Feynman, R.P. and Hibbs, A.R., Quantum Mechanics and Path Integrals, McGraw-Hill, New York, 1965.
56. Giachetti, R. and Tognetti, V., Variational approach to quantum statistical mechanics of nonlinear systems with applications to Sine-Gordon chains. *Phys Rev Lett*, 55, 912–915, 1985.
57. Feyman, R.P. and Kleinert, H., Effective classical partition functions. *Phys Rev A*, 34, 5080–5084, 1986.
58. Cao, J. and Berne, B.J., Low temperature variational approximation for the Feynman quantum propagator and its application to the simulation of quantum systems, *J Chem Phys,* 92, 7531–7539, 1990.
59. Voth, G.A., On the use of Feynman-Hibbs effective potentials to calculate quantum mechanical free energies of activation. *J Chem Phys*, 94, 4095–4096, 1991.
60. Cuccoli, A., Macchi, A., Neumann, M., Tognetti, V., and Vaia, R., Quantum thermodynamics of solids by means of an effective potential, *Phys Rev B,* 45, 2088–2096, 1992.
61. Kriman, A. and Ferry, D.K., Statistical properties of hard-wall potentials, *Phys Lett A, 138, 8–12,* 1989.
62. Cuccoli, A., Giachetti, R., Tognetti, V., Vaia, R., and Verrucchi, P., The effective potential and effective Hamiltonian in quantum statistical mechanics, *J Phys Cond Matter*, 7, 7891–7938, 1995.
63. Madelung, E., Quantentheorie in hydrodynamischer form. *Z Phys,* 40, 322–328, 1926.
64. Bohm, D., A suggested interpretation of the quantum theory in terms of "hidden" variables, *Phys Rev,* 85, 166–179, 1952.
65. Iafrate, G.J., Grubin, H.L., and Ferry, D.K., Utilization of quantum distribution functions for ultra-submicron device transport, *J Physique*, Suppl 10, C7, 307–312, 1981.
66. Asenov, A., Watling, J.R., Brown, A.R., and Ferry, D.K., The use of quantum potentials for confinement and tunneling in semiconductor devices. *J Comp Electronics,* 1, 503–513, 2002.

67. Pikus, F.G. and Likharev, K.K., Nanoscale field-effect transistors: an ultimate size analysis, *Appl Phys Lett,* 71, 3661–3663, 1997.
68. Datta, S., Nanoscale device modeling: the Green's function method, *Superlatt Microstuct,* 28, 253–278, 2000.
69. Knoch, J., Lengeler, B., and Appenzeller, J., Quantum simulations of an ultrashort channel single-gated n-MOSFET on SOI, *IEEE Trans Elec Dev,* 49, 1212–1218, 2002.
70. Gilbert, M.J., Milicic, S.N., Akis, R., and Ferry, D.K., Modeling fully depleted SOI MOSFETs in 3D using recursive scattering matrices, *J Comp Electron*, 2, 329, 2003.
71. Gilbert, M.J., Akis, R., and Ferry, D.K. Resonant-tunneling behavior and discrete dopant effects in narrow ultrashort ballistic silicon-on-insulator metal-oxide-semiconductor field-effect transistors, *J Vac Sci Technol,* 8, 22, 2039–2044, 2004.
72. Usuki, T., Takatsu, M., Kiehl, R.A., and Yokoyama, N., Numerical analysis of electron-wave detection by a wedge-shaped point contact, *Phys Rev B,* 50, 7615–7625, 1994.
73. Usuki, T., Saito, M., Takatsu, M., Kiehl, R.A., and Yokoyama, N., Numerical analysis of ballistic-electron transport in magnetic fields by using a quantum point contact and a quantum wire, *Phys Rev B,* 52, 8244–8255, 1995.
74. Akis, R., Ferry, D.K., and Bird, J.P., Magnetotransport fluctuations in regular semiconductor ballistic quantum dots. *Phys Rev B*, 54, 17705–17715, 1996.
75. Grabert, H. and Devoret, M.H., Eds., *Single Charge Tunneling*, Plenum Press, New York, 1991.
76. Yano, K., Ishii, T., Hashimoto, T., Kobayashi, T., Murai, F., and Seki, K., Room-temperature single-electron memory, *IEEE Trans Electron Dev,* 41, 1628–1638, 1994.
77. Welser, J.J., Tiwari, S., Rishton, S., Lee, K.Y., and Lee, Y., Room temperature operation of a quantum-dot flash memory, *IEEE Electron Dev Lett,* 18, 278–280, 1997.
78. Kim, D.H., Sung, S.-K., Kim, K.R., Lee, J.D., Park, B.-G., Choi, B.H., Hwang, S.W., and Ahn, D., Silicon single-electron transistors with sidewall depletion gates and their application to dynamic single-electron transistor logic. *IEEE Trans Electron Dev,* 49, 627–635, 2002.
79. Khoury, M., Gunther, A., Pivin, D.P., Jr., Rack, M.J., and Ferry, D.K., Silicon quantum dot in a metal-oxide-semiconductor field effect transistor (MOSFET) structure, *Jpn J Appl Phys,* 38, 469–472, 1999.
80. Khoury, M., Rack, M.J., Gunther, A., and Ferry, D.K., Spectroscopy of a Si quantum dot, *Appl Phys Lett,* 74, 1576–1578, 1999.
81. Khoury, M., Gunther, A., Milicic, S., Rack, M.J., Goodnick, S.M., Vasileska, D., Thornton, T.J., and Ferry, D.K., Single-electron quantum dots in silicon MOS structures. *Appl Phys A,* 71, 415–421, 2000.
82. Simmel, F., Abusch-Magder, D., Wharam, D.A., Kastner, M.A., and Kotthaus, J.P., Statistics of the Coulomb-blockade peak spacings of a silicon quantum dot, *Phys Rev B,* 59, R10441–R10444, 1999.
83. Takenaka, N., Inoue, M., and Inuishi, Y., Influence of inter-carrier scattering on hot electron distribution function in GaAs, *J Phys Soc Japan,* 47, 861–868, 1979.
84. Ferry, D.K. and Barker, J.R., Issues in general quantum transport with complex potentials, *Appl Phys Lett,* 74, 582–584, 1999.
85. Gross, W.J., Vasileska, D., and Ferry, D.K., A novel approach for introducing the electron-electron and electron-impurity interactions in particle-based simulations, *IEEE Electron Dev Lett,* 20, 463–465, 1999.
86. Gross, W.J., Vasileska, D., and Ferry, D.K., Ultra-small MOSFETs: the importance of the full Coulomb interaction on device characteristics, *IEEE Trans Electron Dev,* 47, 1831–1818, 2000.

87. Ramey, S.M. and Ferry, D.K., A New Model for including discrete dopant ions into Monte Carlo simulations, *IEEE Trans Nanotechnol*, 2, 193–197, 2003.
88. Jacoboni, C., Recent developments in the hot electron problem, in *Proc Intern Conf Phys Semicond*, 1195–1199, Tipograf, Marves, Rome, 1974.
89. Bosi, S. and Jacoboni, C., Monte Carlo high-field transport in degenerate GaAs. *J Phys C: Sol State Phys,* 9, 315–321, 1976.
90. Lugli, P. and Ferry, D.K., Dynamical screening of hot carriers in semiconductors from a coupled molecular-dynamics and ensemble Monte Carlo simulation. *Phys Rev Lett,* 56, 1295–1297, 1986.
91. Lugli, P. and Ferry, D.K., Degeneracy in the ensemble Monte Carlo method for high-field transport in semiconductors, *IEEE Trans Electron Dev,* 32, 2431–2437, 1985.
92. Kriman, A.M., Kann, M.J., Ferry, D.K., and Joshi, R., Role of the exchange interaction in the short-time relaxation of a high-density electron plasma, *Phys Rev Lett,* 65, 1619–1622, 1990.
93. Kelbg, G., Zur theorie des mikrofeldes im plasma, *Annalen der Physik,* 13, 385–394, Berlin, 1964.
94. Morawetz, K., Relation between classical and quantum particle systems, *Phys Rev E,* 66, 022103-1–022103-4, 2002.
95. Kohn, W., Meir, Y., Makarov, D.E., and Van der Waals, Energies in density functional theory, *Phys Rev Lett,* 80, 4153–4156, 1998.
96. Pollet, R., Savin, A., Leininger, T., and Stoll, H., Combining multideterminantal wave functions with density functionals to handle near degeneracy in atoms and molecules, *J Chem Phys,* 116, 1250–1258, 2002.
97. Lugli, L. and Ferry, D.F., Electron-electron interaction and high field transport in Si. *Appl Phys Lett,* 46, 594–596, 1985.
98. Lugli, L. and Ferry, D.F., Investigation of plasmon-induced losses in ballistic transport, *IEEE Electron Dev Lett,* 6, 25–27, 1985.
99. Fischetti, M.V., Long-range Coulomb interactions in small Si devices, Part II, effective electron mobility in thin oxide structures, *J Appl Phys,* 89, 1232–1250, 2001.
100. Ramey, S.M. and Ferry, D.K., Quantum modeling of particle-particle interactions in SOI MOSFETs, *Semicond Sci Technolog*, 19, S238–S240, 2004.

2 Practical CMOS Scaling

David J. Frank

2.1 INTRODUCTION

The basic concept of the Field-Effect Transistor (FET) was invented in 1930,[1] and was first reduced to practice in Si/SiO_2 in 1960, by Kahng and Attala.[2] Since the late 1960's the development has been very rapid. The Si metal oxide semiconductor FET (MOSFET) was incorporated into integrated circuits in the early 1970s, and progress since then has followed an exponential behavior that has come to be known as Moore's Law.[3] Device dimensions have been steadily shrinking at a rate of ~2x/6 years, and circuit complexity and industry revenues have been similarly growing exponentially.

This rapid scaling has brought Si technology to the point where various fundamental physical phenomena are beginning to impede the path to further progress. These effects include tunneling through the gate insulator, tunneling through the bandgap, quantum confinement issues, interface scattering, discrete atomistic effects in the doping and at interfaces, and thermal problems associated with very high power densities. Because of these difficulties, many new and interesting changes to the basic MOSFET technology are being explored, including new gate geometries and multiple gates, the use of strain to increase mobility, and the use of extremely thin, fully depleted silicon layers for the FET channel. Unfortunately, even for these new devices, power dissipation and the temperature rise associated with very high power densities remain problems, making it more and more apparent that each technology design point must be optimized for maximum performance subject to application-dependent power constraints.[4]

The organization of this chapter is as follows. Section 2 describes present state-of-the-art complementary MOSFET (CMOS) technology and industry targets for future progress. The third section explains the basic principles of CMOS scaling, and the fourth section describes some of the promising device innovations being considered for future technology generations. The underlying physical issues that limit continued scaling are considered in the fifth section, followed by a more detailed look at limitations imposed by power dissipation, which appears to be the most serious constraint.

2.2 CMOS TECHNOLOGY OVERVIEW

2.2.1 Current CMOS Device Technology

Figure 2.1 illustrates most of the important features of state-of-the-art conventional bulk CMOS technology.[5,6,7] The gates are n- and p-type polysilicon, and are topped with a metal silicide, which lowers the gate series resistance. The gates are patterned down to minimum dimensions that are 50% or more below the general lithographic feature size by means of special lithographic and lateral etching techniques. The

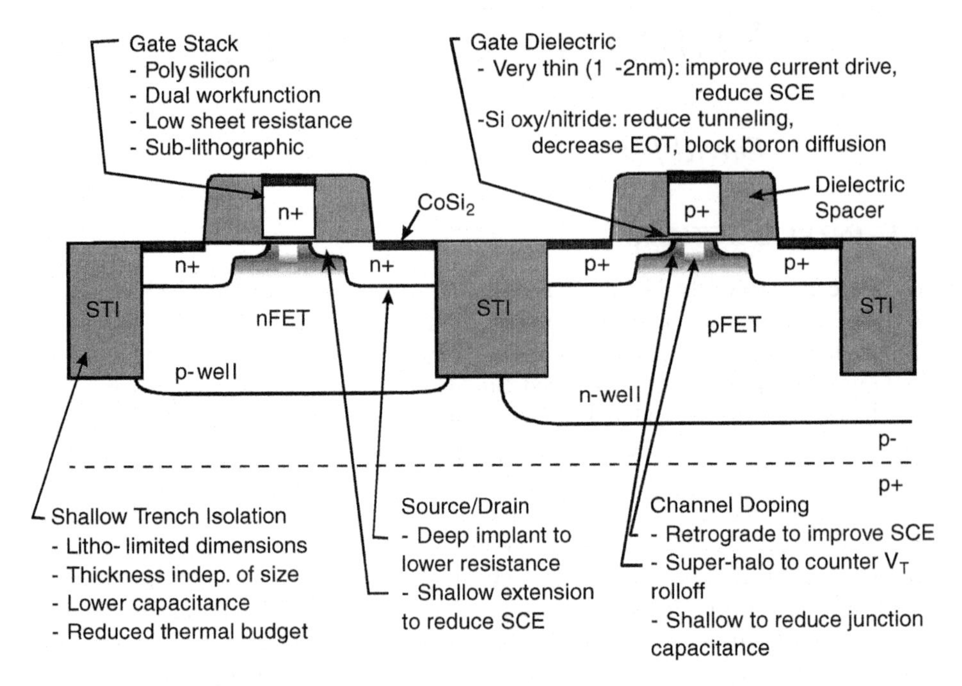

FIGURE 2.1. Conventional Bulk CMOS cross-sectional diagram (from Ref. 7, reproduced with permission. Copyright 2002, Kluwer.)

gate dielectrics can be as thin as 1.0 nm of Si oxynitride for high performance logic. Scaling demands these thin gate insulators in order to keep short channel effects under control and to maximize performance, but tunneling leakage current through these thin insulators has become a major concern for many applications.

Shallow trench isolation (STI) is used to separate the FETs, resulting in very high circuit density. A combination of deep and shallow implants are used for the source and drains, and these must be carefully engineered to reduce short channel effects, prevent gate insulator degradation due to hot electrons and provide low contact resistance between the FET channel and the silicide contacts. The doping profiles in the channel are also very important. Shallow angled ion implants are used to create so-called "halo" doping profiles that are higher near the source and drain and lower in the middle of the channel. Since the halos are defined relative to the edges of the gates, the average doping in the halo overlap region increases when the gate length shrinks. This doping increase tends to compensate for the natural decrease of threshold voltage (V_T) that occurs in very short MOSFETs, enabling the use of FETs with shorter gate lengths than would otherwise be possible.

Although the wiring is not shown in Figure 2.1, it is clearly essential for creating large integrated circuits. Today most of the wire is copper because of its low resistivity and reduced electromigration. Wire-to-wire capacitance is reduced by the use of fluorinated silicate glass (FSG) or organosilicate glass (OSG) for the insulator, with permittivity (k) ranging from 3.7 down to 3.0, and even lower k materials may be in use soon.[8] To keep wire delay under control, a hierarchy of wiring sizes is

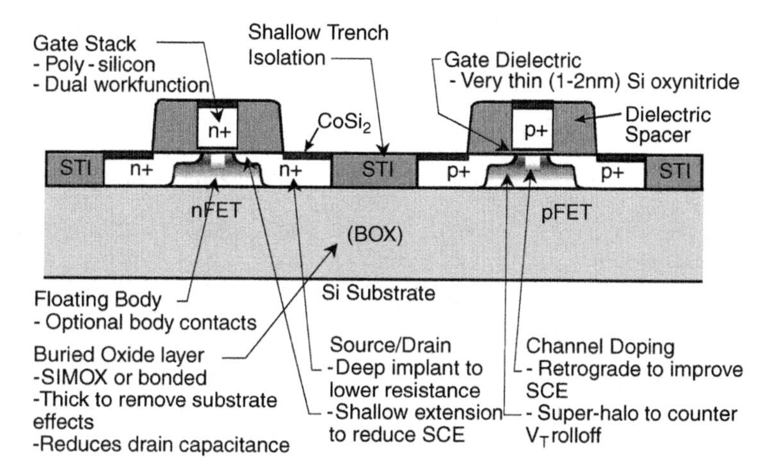

FIGURE 2.2 PD-SOI CMOS technology cross-sectional diagram (from Ref. 7, reproduced with permission. Copyright 2002, Kluwer.)

usually used, from very fine wires at minimum lithographic dimension on the bottom to large "fat" wires on the top.[9]

Partially depleted silicon-on-insulator (PD-SOI) CMOS, shown in Figure 2.2, is also available.[10] It is very similar to bulk CMOS, the main difference being that PD-SOI MOSFETs are fabricated in a thin layer of Si, ~150 nm thick, on top of an insulating SiO_2 layer. In partially depleted SOI, the depletion region in the FET channel is thinner than the silicon layer, leaving some undepleted silicon which acts as a floating body for the FET. The buried oxide (BOX) layer is typically 150 to 250 nm thick and completely insulates the device layer from the substrate. This construction results in source- and drain-to-body junction capacitances that are significantly reduced, which can increase digital switching speed.[11] The floating body eliminates the usual bulk MOSFET body-effect dependencies on source-to-substrate voltage, enhancing some types of circuitry. On the other hand, there are other floating-body effects such as history-dependent body bias and increased output conductance (caused by the injection of majority carriers into the body by impact ionization in the drain region) that may degrade the performance of some circuits.

2.2.2 International Technology Roadmap for Semiconductors (ITRS) Projections

Since 1994 the semiconductor industry has periodically created "roadmaps" showing how it expects CMOS technology to evolve. These roadmaps provide guidance about future device technology and manufacturing capabilities that need to be developed in order to continue the exponential growth of the industry. They are based on device scaling theory and observations about past industry trends (e.g., Moore's Law). The latest of these roadmaps is ITRS'03, which is summarized in Table 2.1.[8] According to these projections, it is hoped that high-performance MOSFETs will reach physical gate lengths of 9 nm by 2016 and be able to achieve local clock speeds of 40 GHz. To do this, some significant changes in device technology are contemplated, including the introduction of new gate insulator materials with high dielectric constants,

TABLE 2.1
2003 ITRS Roadmap Projections*

Year	2003	2004	2005	2007	2010	2013	2016
DRAM ½ pitch (nm)	100	90	80	65	45	32	22
DRAM generation (product)	1G	1G	1G	2G	4G	8G	32G
MPU transistors/chip	153M	193M	243M	386M	773M	1.55G	3.09G
Local clock (GHz)	3.0	4.2	5.2	9.3	15.1	23	40
Number wiring levels	9	10	11	11	12	12	14
Total wire length (km/cm^2)	0.58	0.69	0.91	1.1	1.8	2.5	4.2
Interlayer eff. permittivity	3.3–3.6	3.1–3.6	3.1–3.6	2.7–3.0	2.3–2.6	2.0–2.4	<2.0
High-perf. logic physical gate length (nm)	45	37	32	25	18	13	9
High-perf. logic EOT (nm)	1.3	1.2	1.1	0.9	0.7	0.6	0.5
High-perf. V_{DD} (V)	1.2	1.2	1.1	1.1	1.0	0.9	0.8
High-perf. power (W)	149	158	167	189	218	251	288
Low-power logic physical gate length (nm)	65	53	45	32	22	16	11
Low-power logic EOT (nm)	1.6	1.5	1.4	1.2	0.9	0.8	0.7
Low-power V_{DD} (V)	1.0	0.9	0.9	0.8	0.7	0.6	0.5
Batt-powered power (W)	2.1	2.2	2.3	2.5	2.8	3.0	3.0

* From Ref. 8.

the use of strain and new semiconductors to achieve higher mobility, and the development of alternate FET structures with improved scaling properties. Some of these proposals are discussed in Sections IV and V.

One of the most important difficulties with the roadmap projections is the power dissipation. The trend for power density to increase in each generation of technology is rapidly leading to the point at which the on-chip temperature rise is unacceptable from a reliability point of view. There are many failure mechanisms for Si CMOS that are activated by high temperature, so if the temperature gets too high, the chip fails rather quickly. This temperature constraint translates into a power density constraint that may be mitigated somewhat by cooling innovations, but ultimately requires a tradeoff between circuit density and speed. Maximum speed and maximum density cannot be used simultaneously without endangering the lifetime of the chip. In addition to this problem, the projected technologies have increasing static power dissipation due to higher subthreshold currents associated with the lower V_T's needed to maintain performance and to quantum mechanical tunneling leakage currents. In Section VI it will be shown how these power dissipation considerations lead to optimum scaling limits that vary depending on application.

2.3 SCALING PRINCIPLES

As has been mentioned, continuing progress in CMOS technology is based on the physics of scaling MOSFETs. These scaling principles were originally developed by

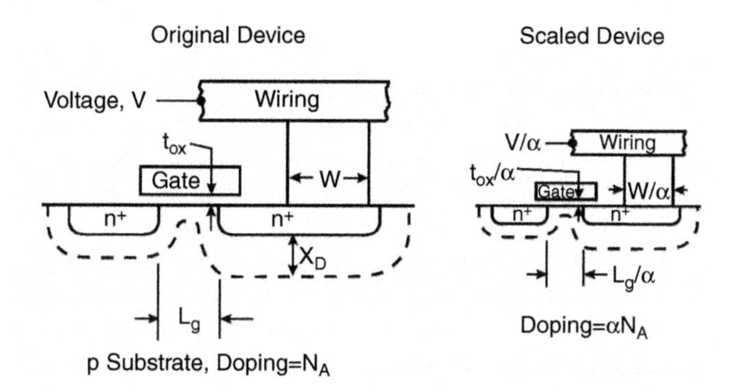

FIGURE 2.3 Schematic illustration of the scaling of Si technology by a factor alpha. (From Ref. 4, reproduced with permission. Copyright 2001 IEEE.)

TABLE 2.2
Technology Scaling Rules for Three Types of Scaling

Physical Parameter	Constant-Electric Field Scaling Factor	Generalized Scaling Factor	Generalized Selective Scaling Factor
Channel length, insulator thickness	$1/\alpha$	$1/\alpha$	$1/\alpha_d$
Wiring width, channel width	$1/\alpha$	$1/\alpha$	$1/\alpha_w$
Electric field in device	1	ε	ε
Voltage	$1/\alpha$	ε/α	ε/α_d
On-current per device	$1/\alpha$	ε/α	ε/α_w
Doping	α	$\varepsilon\alpha$	$\varepsilon\alpha_d$
Area	$1/\alpha^2$	$1/\alpha^2$	$1/\alpha_w^2$
Capacitance	$1/\alpha$	$1/\alpha$	$1/\alpha_w$
Gate delay	$1/\alpha$	$1/\alpha$	$1/\alpha_d$
Power dissipation	$1/\alpha^2$	ε^2/α^2	$\varepsilon^2/\alpha_w\alpha_d$
Power density	1	ε^2	$\varepsilon^2\alpha_w/\alpha_d$

* From Ref. 6.

Dennard in the early 1970s,[12] and have been thoroughly covered in many recent articles and reviews.[4–6,13] This section gives a brief review of these principles and then goes on to discuss the scale length theory for two-dimensional (2-D) effects in MOSFETs.

2.2.1 General Scaling

Figure 2.3 schematically illustrates the basic idea of scaling. It shows how a large FET can be scaled by a factor to yield a smaller FET. Following the principles of electrostatics, if the dimensions, dopings, and voltages are all scaled as shown in the figure, the electric field configuration in the scaled device will be exactly the same as it was in the larger device. These constant electric field scaling relationships are summarized in Table 2.2, in the second column. Within this simple scheme, the speed increases by the factor α while the area and power decrease by $1/\alpha^2$, leaving the power density constant.

In actual practice these ideal rules have not been followed exactly. The first rule "violation" is seen in the voltage, which has been scaled more slowly than dimensions. Initially, this was because of the difficulty of changing industry-wide standard supply voltages and because of a reluctance to give up the extra performance available at higher voltage. More recently, voltage scaling has slowed to a near halt because of the difficulty of lowering the threshold voltage any further. The problem here is that the subthreshold slope does not scale except by lowering the temperature, since it is primarily determined by the thermodynamics of the Boltzmann distribution. Since the subthreshold slope is relatively fixed, the MOSFET off-state leakage current increases exponentially as the threshold voltage is reduced. This puts a limit on how low V_T can be without using up all of the available power for leakage dissipation. To obtain high performance, it is generally necessary for the supply voltage, V_{DD}, to exceed V_T by a ratio of 3 or 4 to 1, so the limit on V_T also tends to limit V_{DD}. Consequently, supply voltage scaling cannot proceed much further. The scaling laws can be adjusted to account for this reduced voltage scaling by introducing an additional scaling factor ε, for the electric field (this ε is greater than 1), and are summarized under "generalized scaling" in the third column of Table 2.2. As can be seen, ε increases the doping and the power dissipation. It also diminishes the long-term reliability and durability of the FET. In fact, this reliability concern has forced the use of lower supply voltages in some previous generations of CMOS, even though it diminished performance somewhat.[13]

In recent technology generations another rule "violation" has arisen, in which the gate length is scaled more than the wiring. This is made possible by new fabrication methods involving specialized optics and masks, trimming the gate photoresist and overetching the gate poly-Si, which enable sublithographic gate length while the wiring remains constrained to the lithographic pitch. The scaling rules for this approach are shown in the final column of Table 2.2 labeled "generalized selective scaling," where there are two spatial dimension scaling parameters, α_d for scaling the gate length and device vertical dimensions, and α_w for scaling the device width and the wiring. Since $\alpha_d > \alpha_w$, this type of scaling enables gate delay to scale faster than in the preceding simpler cases. More details about these scaling approaches can be found in Davari et al.[13]

2.3.2 Characteristic Scale Length

The scaling theory presented in the preceding section tells how to take a known good device design and make it smaller, but how should a good FET be designed in the first place? In a general sense, a "good" MOSFET represents a compromise between long-channel-like and short-channel-like behavior. It needs to behave like a long-channel device by having

- High output resistance,
- High voltage gain and
- Low sensitivity to process variation (e.g., gate length variation).

These characteristics enable one to design robust circuits. On the other hand, it also needs to have certain short-channel behaviors:

- High transconductance,
- High current drive, and
- High switching speed.

These two sets of characteristics are basically in conflict. The gate must be as short as possible, while still being long enough to offer reasonable channel control. The crossover between the long and short channel regimes is determined by the natural length scale, Λ_1, of the FET, which is related to the thickness of the channel. When the channel is much longer than Λ_1 the FET behaves in the classical "long-channel" manner, and the internal electrostatics can be understood from a primarily one-dimensional point of view. When the channel becomes short enough compared to its thickness, two-dimensional (2-D) electrostatic effects occur, and the drain potential can significantly modulate the potential along the channel. When this happens, the long-channel characteristics are degraded.[7]

The extent of the 2-D effects can be estimated by considering the ratio L/Λ_1 for a given FET, where the natural scale length Λ_1 is derived by considering electrostatic solutions of the form $\sinh(n\pi y/\Lambda_n)\sin(n\pi x/\Lambda_n)$ for the potential in the depletion and insulator regions of a MOSFET (where y is in the transport direction and x is in the transverse direction) and applying proper dielectric boundary conditions between the two (14). It is given implicitly as the largest Λ_1 that satisfies[15]

$$0 = k_{Si}\tan\left(\pi\frac{t_I}{\Lambda_1}\right) - k_I\tan\left(\pi\left(1-\frac{t_{Si}}{\Lambda_1}\right)\right) \tag{2.1}$$

for bulk MOSFETs, where t_I is the physical thickness of the insulator, t_{Si} is the thickness of the depletion layer, k_I is the permittivity of the gate insulator, and k_{Si} is the permittivity of Si. This formula is valid for all permittivities and thicknesses, but in the most common regime, where $t_I/\Lambda_1 \ll 1$, it can be approximately solved as[4]

$$\Lambda_1 \approx t_{Si} + \frac{k_{Si}}{k_I}t_I - \frac{\pi^2}{3}\frac{k_{Si}}{k_I}\left(\frac{k_{Si}^2}{k_I^2}-1\right)\left(\frac{t_I}{t_{Si}}\right)^2 t_I\,. \tag{2.2}$$

Figure 2.4 shows how some important FET characteristics depend on the L/Λ_1 ratio for an idealized bulk MOSFET without super halo doping.[4] From this analysis it appears that $L/\Lambda_1 \sim 2.0$ would be a good nominal design point for non-halo bulk MOSFET technologies, since it allows room for gate length tolerances of up to ±30% while still maintaining V_T variation within tolerable bounds. To do better than this, halo or super halo doping profiles are required. As illustrated in Figure 2.1, halo doping, which is implemented by angled ion implants using the gate as a mask, increases the body doping near the source and drain ends of the channel. For short-channel FETs, these halo implants overlap, yielding higher effective channel doping, which tends

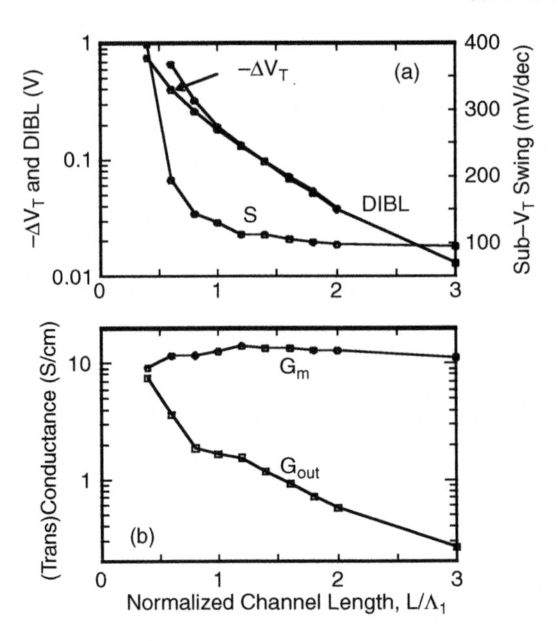

FIGURE 2.4 Dependence of 2-D effects on the L/Λ_1 ratio for an idealized bulk MOSFET: (a) ΔV_T, DIBL and subthreshold swing (S) versus L/Λ_1, and (b) transconductance (G_m) and output conductance (G_{out}) versus L/Λ_1. Based on 2-D FIELDAY simulations with Λ_1 = 13.6 nm (t_{ox} = 1.5 nm, t_{Si} = 10 nm), ΔV_T is determined at V_{DS} = 0.05 V, DIBL is defined as $V_T(V_{DS} = 0.05) - V_T(V_{DS} = 1.0)$, the transconductance is measured at V_{DS} = 1.0 V, $V_G = V_T(V_{DS} = 0.05) + 0.5$ V and the output conductance is measured at the same V_G, and V_{DS} = 0.75 V. (From Ref. 4, reproduced with permission. Copyright 2001 IEEE)

to cancel out some of the threshold voltage rolloff. These profiles also improve the drain-induced barrier-lowering (DIBL) curve by shifting the peak barrier in the channel closer to the source, decreasing the sensitivity of the subthreshold current to drain voltage. As presently practiced, halo doping can lower the L/Λ_1 design point to below 1.5 while still maintaining ±30% gate length tolerance.

One can also achieve a smaller L/Λ_1 design point by improving the processing so that the tolerances become tighter. For example, if tolerance could be improved to <±10%, it might be possible to reduce L/Λ_1 to <1.0.[16]

2.4 EXPLORATORY TECHNOLOGY

As was mentioned in conjunction with the ITRS projections, the increasing difficulties with conventional scaling are driving the semiconductor industry to look at a variety of novel device ideas in the hope of maintaining exponential growth. The difficulties with scaling will be looked at in sections V and VI, but first we will consider some of the innovative new device technologies that are being explored, including new material variations such as strained Si and Si with different crystal orientations, and new device structures using thin Si layers, including fully depleted SOI and multiple-gate MOSFETs.

2.4.1 New Materials

The most promising materials innovations at present are the efforts to obtain higher drive currents from essentially the same Si FETs by increasing the mobility of the silicon. This can (in principle, at least) be done in several ways: the Si channel can be strained, either uniaxially or biaxially, or the crystal orientation can be changed. For the conduction band of Si with a standard (001) surface, biaxial tensile strain causes the sixfold valley degeneracy of Si to split into two lower energy valleys and four higher valleys. The lower energy valleys have low effective mass in the in-plane transport direction, and this significantly increases the electron mobility compared to unstrained Si. The strain also splits the degeneracy of the valence band, yielding higher in-plane mobility for holes, too. For inversion layers in n-channel MOSFETs, mobility improvements in excess of 70% have been observed for biaxially strained Si, as shown in Figure 2.5(a). This higher mobility translates into higher drive current, as shown in the experimental IV curves in Figure 2.5(b).[17] Recently, it has been shown that uniaxially strained Si can show even more mobility enhancement than the biaxial case.[18] For nFETs it has been shown that tensile uniaxial strain in the standard $\langle 110 \rangle$ direction of transport can increase the on-current by up to 10%,[19] and for pFETs compressive uniaxial strain in the $\langle 110 \rangle$ transport direction has yielded up to 35% increase in on-current.[18,20] Finally, it has also been shown recently that although the usual $\langle 110 \rangle$ transport on the (001) surface is the best orientation for electron transport, the mobility of holes can be more than doubled by switching to the (110) surface and the $\langle 1\bar{1}0 \rangle$ transport direction, due to the complex shape of the valence band.[21]

The techniques being developed to implement these different silicon perturbations are quite varied. Biaxial strain is generally accomplished by epitaxially growing the strained Si on an unstrained SiGe layer, as illustrated in Figure 2.6(a). The strain of the Si layer increases in proportion to the Ge mole fraction, which may be ~10 to 35%. The SiGe layer is grown epitaxially on the Si substrate, and there is some fairly significant art to choosing the Ge composition, grading and thickness so that the strain is entirely relaxed and the dislocations are confined deep within the layer.[22] Uniaxial tensile strain for nFETs has been created by using strained dielectric capping layers over the FET, and uniaxial compressive strain for pFETs has been accomplished by regrowing deep SiGe source/drain regions.[19] To take advantage of the crystal orientation effects, one needs to make the nFETs and pFETs in regions with different orientations on the same substrate, necessitating so-called hybrid-orientation ('HOT') substrates. Approaches to making such substrates involve bonded SOI in which the bonded layer has a different orientation than the underlying substrate. FETs can then be built in either the top SOI layer or in epitaxial regions that are grown in holes down to the underlying substrate, thus achieving the two different crystal orientations.[21]

Perhaps the most difficult materials effort underway is the work to replace the SiO_2 gate insulator with an entirely different insulator, such as HfO_2, that would have a higher dielectric constant. This is discussed in Section V.A in connection with tunneling through the gate insulator.

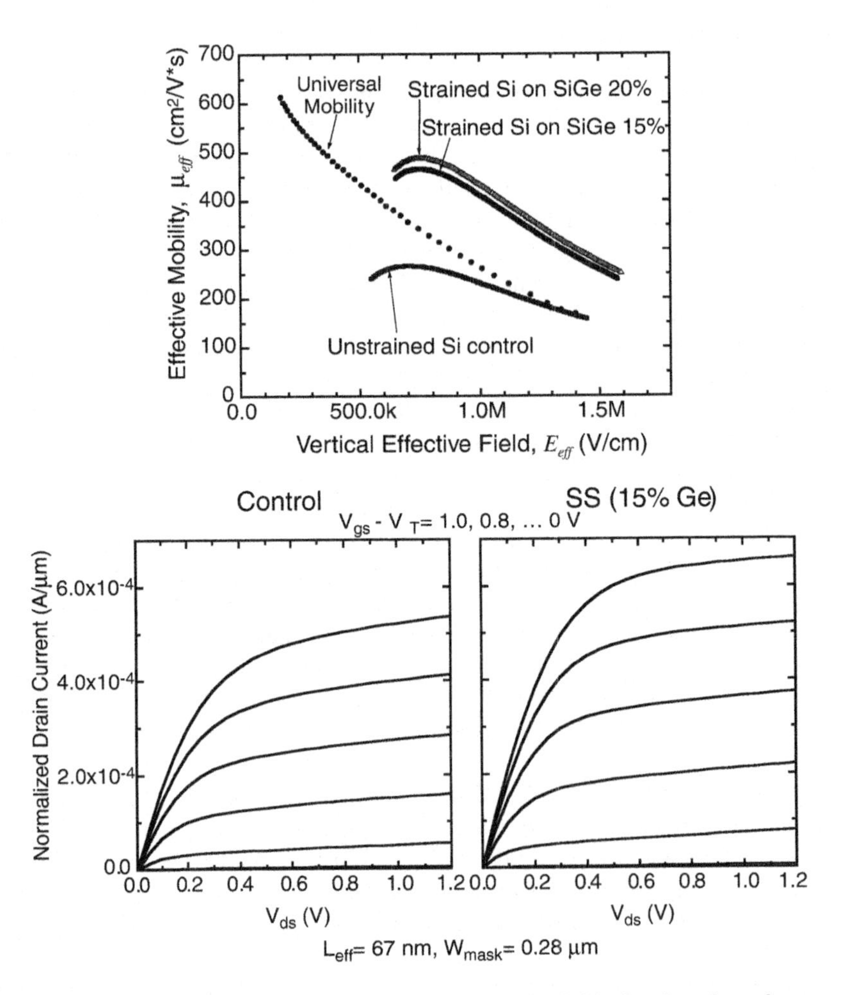

FIGURE 2.5 (a) Mobility versus effective surface electric field, showing the enhancement for strained Si.[17] (b) Experimental I_{DS} versus V_{DS} curves for an unstrained (control) MOSFET and for a strained-Si MOSFET, showing the enhancement resulting from strain. (From Ref. 7, reproduced with permission. Copyright 2002, Kluwer.)

2.4.2 Fully Depleted SOI

The next more complex exploratory device approach is the fully depleted SOI (FD-SOI) MOSFET, illustrated in Figure 2.6(b). Comparing to Figure 2.2 reveals that FD-SOI is very similar to PD-SOI except that the Si layer is much thinner. To guarantee that the layer remains fully depleted[6] and to maintain adequate short-channel control,[23,24] the Si layer should be less than about half the depletion depth of a corresponding bulk device. Advantages of FD-SOI over PD-SOI include elimination of the floating body effect, making circuit design easier, and reduction of the drain capacitance because the depth of the drain-to-body junction is greatly

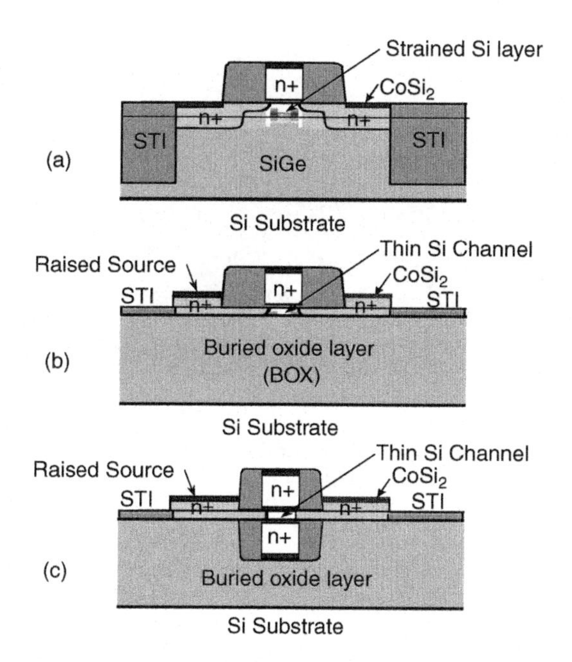

FIGURE 2.6 Schematic illustrations of novel device structures. (a) Biaxially strained-Si MOSFET. (b) Fully depleted SOI MOSFET. (c) Double-gate MOSFET. (From Ref. 7, reproduced with permission. Copyright 2002, Kluwer.)

reduced. The subthreshold slope is also improved, making possible further scaling of V_T and V_{DD}. For example, experiments on 50-nm-gate-length FD-SOI devices showed a subthreshold swing of 75 mV/decade (versus 85 to 90 mV/decade for bulk control samples).[25] Recent experiments on ultrathin SOI layers have probed the limits of FD-SOI scaling by fabricating both nFETs and pFETs with gate lengths below 10 nm, and with SOI thickness down to 4 nm.[26,27] As an example, the 6-nm pFET results are shown in Figure 2.7. Although the IV curves are degraded somewhat due to short channel effects and source/drain resistance, these extremely short devices are still demonstrating good FET behavior.

Problems associated with FD-SOI include the difficulty of controlling the thickness of the very thin Si layers, leading to difficulty in controlling the threshold voltage (which depends on the Si thickness), and the difficulty of achieving low resistance source and drain contacts to such thin Si layers. This latter problem may be overcome through a raised source/drain process, as shown in the excellent results of Chau et al.[25] Another concern, which carries over from PD-SOI, is self-heating because of the low thermal conductivity of SiO_2. Heat generated in the drain of the FET can cause the device to become so hot that its mobility is significantly degraded.

2.4.3 Double-Gate and Multiple-Gate FET Structures

A double-gate FET (DG-FET) is a MOSFET with two gates, one on either side of the channel,[28–31] as shown schematically in Figure 2.6(c). The basic idea is that a second gate can screen the drain field just as well as the body of a bulk MOSFET,

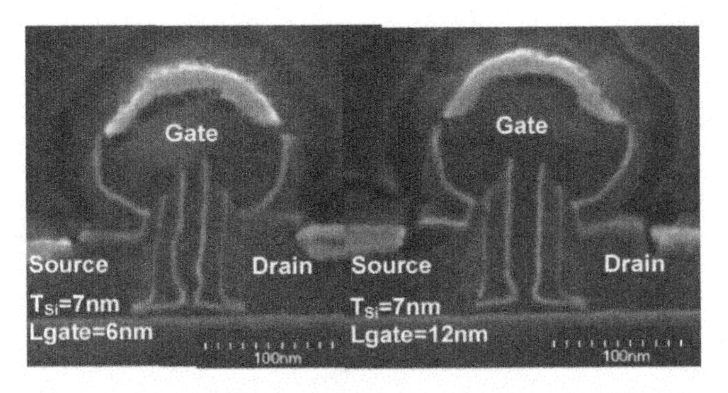

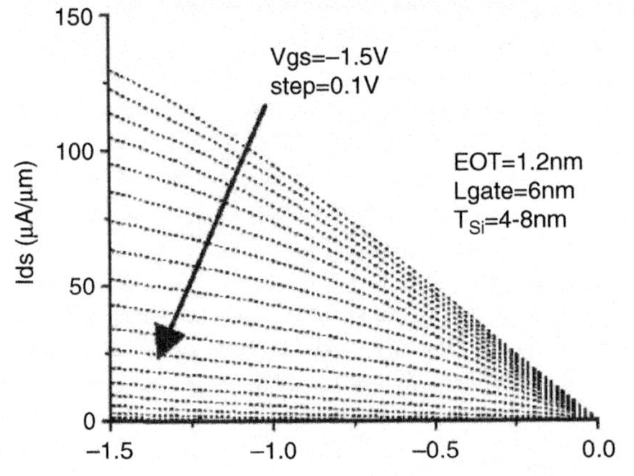

FIGURE 2.7 Ultrasmall FD-SOI pFETs. (a) SEM micrographs of FET crosssections showing 6- and 12-nm gate lengths on 7-nm SOI. (b) I_{DS} versus V_{DS} curves for the 6-nm gate length pFET. (From Ref. 27, reproduced with permission. Copyright IEEE 2002.)

but without the drawbacks of a p-n junction. The two gates keep short channel effects well under control, but unlike the body of a bulk device, the second gate can be switched in conjunction with the first gate to effectively double the switching current of the FET under many circumstances. It is expected that it will be possible to scale DG-FETs further than bulk devices because one can make the Si layer much thinner than the depletion layer of a bulk device, thus enabling the gate length to be scaled smaller, too. Theoretical analyses show that high quality FET characteristics can probably be obtained even for devices with gate length below 10 nm.[32,33]

The DG-FET comes in both a symmetric form, having the same insulator and same gate workfunction on both sides,[31] and various asymmetric forms, involving different gate insulator thicknesses and different gate workfunctions,[34] with the symmetric DG-FETs giving the highest performance. In both cases one can derive a scale length analogous to Equation (2.1), which can be used to characterize the 2-D short channel effects.[14] For the symmetric DG-FET the result can be expressed in a form very similar to Equation (2.1):

$$0 = k_{Si} \tan\left(\pi \frac{t_I}{\Lambda_{1DG}}\right) - k_I \tan\left(\frac{\pi}{2}\left(1 - \frac{t_{Si}}{\Lambda_{1DG}}\right)\right). \tag{2.3}$$

When both gates of a symmetric DG-FET are switched together, there is no capacitor divider effect such as occurs in bulk devices between gate and body. The ideality is near unity, and the subthreshold swing can be nearly ideal, perhaps <70 mV/decade at room temperature. This is highly desirable for low power operation, since it allows the V_T and V_{DD} to be reduced. Another potential scaling advantage of DG-FETs is that it may not be necessary to dope the Si channel if a gate material with suitable workfunction can be found, since the workfunction could set the V_T. This would reduce V_T variation due to discrete doping fluctuations, also enabling lower voltages.[30] Unfortunately, the V_T does become quite sensitive to the Si thickness when the thickness is below ~5 nm, because of the quantization energy associated with confining an electron to such a thin layer.[31]

Geometrically, there are three basic orientations for the DG-FET, depending on the plane and direction of current flow relative to the substrate: planar, vertical, and fin-like, as shown in Figure 2.8(a). Within these orientations there are different forms, depending on the cross section in the plane perpendicular to the transport direction, as illustrated in Figure 2.8(b). Basically, there can be two, three, four, and even cylindrical gates, giving rise to the general description of these devices as "multigate" FETs. Since the electrostatic control of short channel effects is best for the cylindrical case,[35] the more closely the multigate FET resembles a cylinder, the better its scaling properties tend to be.

Since each of the multigate structures that have been proposed has a different fabrication procedure, they each have different advantages and disadvantages with regard to manufacturability, but all of them are presently much more difficult to make than conventional MOSFETs. In a general sense, it has been suggested that planar devices seem to offer the best possibility of process control, since the Si thickness, which is the smallest dimension, is controlled as a layer thickness, whereas the other dimensions are defined by lithography, as in bulk technology. There have been several successful efforts to build planar DG-FETs,[34,36–38] but the fabrication processes for creating a properly aligned lower gate have proven difficult. Vertical DG-FETs have also been fabricated,[39,40] but the method is not very compatible with conventional CMOS processing.

Experimentally, it turns out that FinFETs (Type III in Figure 2.8(a)) have been the easiest to make.[41–44] In these devices the current flows in narrow fins etched into a conventional SOI layer, resulting in a device width determined by the SOI layer thickness, while t_{Si} and the gate length are determined by lithographic processing. It is not yet clear whether this approach can be controlled well enough for manufacturing, but the idea is quite compatible with conventional CMOS processing. One can start with a conventional FET layout, and then modify only the "active area" mask in the gate area so as to create many parallel minimum dimension lines running from source to drain. These lines define the fins. After etching, the gate insulator is grown on the sides of the fins and the gate is deposited conformally, after which it

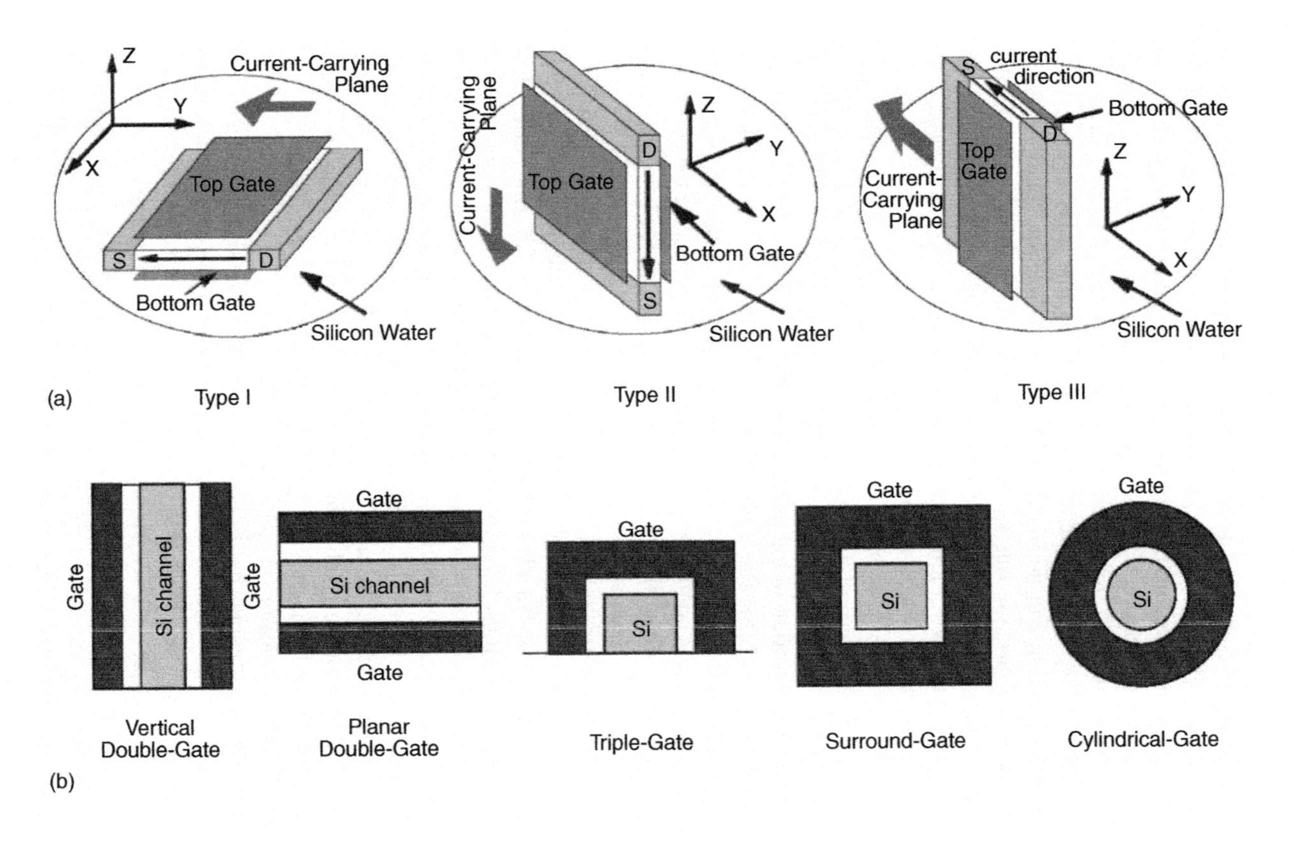

FIGURE 2.8 Schematic illustrations of multi-gate FETs. (a) The three basic orientations for double-gate FETs. (b) Five arrangements of the gate(s) around the channel. These crosssections are in the plane perpendicular to the current transport direction. (Part [a] is from Ref. 6, reproduced with permission. Copyright 1999, IEEE.)

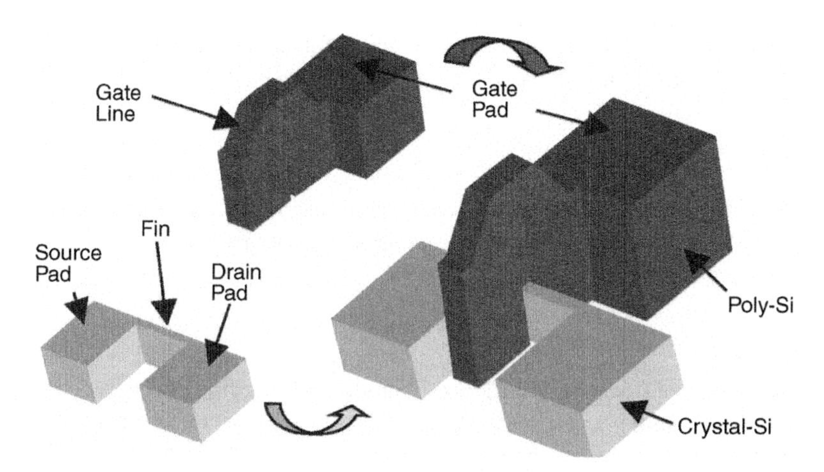

FIGURE 2.9 Pictorial diagram illustrating FinFET geometry. Light grey regions are single crystal Si, while dark regions represent polySi. (From Ref. 43, reprinted by permission. Copyright 2001 IEEE.)

is patterned lithographically and etched in essentially the same way as for conventional processing, yielding a gate that wraps around both sides of the fin (Figure 2.9). It should be possible to design the height of FinFETs so that it exceeds half their pitch, yielding devices with more drive current per unit layout width than conventional MOSFETs.

A conceptually different approach to DG-FETs is obtained by starting with FD-SOI and adding a second gate below the channel. This second gate is farther from the channel than the top gate and is separately biased, but it still acts as a ground plane to improve the short channel behavior of the FD-SOI MOSFET, and at the same time it can be used to tune the V_T. The back gate insulator is expected to be in the ~3- to 10-nm range, based on a compromise between decreasing the parasitic capacitance created by the back gate and increasing the control offered by the back gate. V_T tuning ranges of up to 1 V have been observed in experimental devices.[45] Such a back gate is very interesting for low power use for several reasons. Its close proximity to the device would allow much finer-grained V_T control than is possible with the deep n-well in triple-well bulk technology, making it very convenient to use on a block-by-block basis to reduce the leakage current of blocks that have been turned off.[46] The threshold tuning property can be used in a feedback loop to optimize threshold voltages for low supply voltages.[47] It also allows V_T to be adjusted after manufacture so that chip-to-chip and run-to-run variations in V_T can be removed, thus improving chip yield.[48–50]

There are still many unknowns about the future of DG-FETs. Discrete doping issues and band-to-band tunneling may well prohibit bulk designs below 20 nm, which greatly increases the DG-FETs' apparent advantage. On the other hand, to make competitive DG-FET design points below 20-nm gate length probably requires some sort of halo-like V_T rolloff compensation and metal gates with suitable workfunctions, and neither of these are known processes. It is also still unknown to what extent DG-FET currents will be degraded by self-heating effects. Furthermore, the increased density of gate capacitance per unit chip area that may be available in DG-FETs will tend to adversely impact power density.

2.5 LIMITS TO SCALING

There are many effects that tend to limit scaling, but for the sake of discussion we can divide them into four categories: quantum mechanical, atomistic, thermodynamic, and practical.[7]

2.5.1 QUANTUM MECHANICS

Quantum mechanical scaling limitations include both confinement effects and tunneling effects. Confinement effects occur when electron or hole wavefunctions are squeezed into narrow spaces between barriers. In FETs this primarily happens in the channel, where the charges are squeezed between the gate insulator on one side and the built-in field of the body on the other side. Quantum confinement in this approximately triangular well raises the ground state energy of the electrons or holes, which increases the threshold voltage, and shifts the mean position of the carriers a little farther from the Si-SiO_2 interface. This shift weakens the effect of gate insulator scaling by adding 0.5 to 1.5 nm to the effective oxide thickness, depending inversely on inversion charge density and built-in field. The quantum mechanical ground state energy rise is particularly of concern for potential future SOI FETs with extremely thin bodies (e.g., <5 nm). In such devices the ground state, and hence the V_T, varies inversely with the square of the Si layer thickness. Uncertainty in the layer thickness is expected to translate into large uncertainty in V_T.

Quantum mechanical tunneling is generally more detrimental to scaling than the confinement effects. When electrons or holes tunnel through the barriers of the FET, it causes leakage current. As scaling continues, this ultimately causes unacceptable increases in power dissipation. The leakage may also cause some types of dynamic logic circuits to lose their logic state, but the former problem usually seems to arise first.

There are primarily two forms of tunneling leakage: tunneling current through the gate insulator, and tunneling current through the drain-to-body junction. The first of these is the most prominent and well-known problem. Figure 2.10 illustrates the dependence of these currents on voltage and oxide thickness.[51] In nFETs this current is primarily due to the tunneling of electrons from the channel into the gate. In pFETs with very thin insulators (<1.5 nm) the tunneling current at low voltages can be caused by hole tunneling from channel to gate, but at higher bias it is usually due to the tunneling of electrons from the valence band of the gate into the conduction band of the body. The processes differ because the conduction band barrier is only ~3.5 eV, while the valence band barrier height is ~4.5 eV.

In accordance with the simple Wentzel-Kramers-Brillouin (WKB) approximation,[52] tunneling current generally varies exponentially as

$$J_{tunnel} \sim \exp\left(-(t_I/d_0)\sqrt{m^* E_{eff}}\right), \qquad (2.4)$$

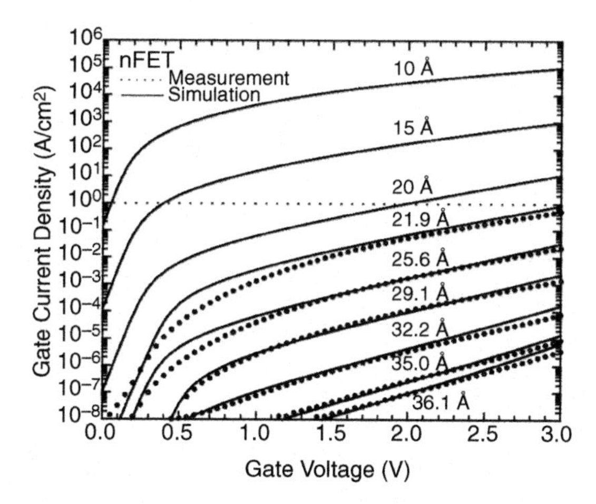

FIGURE 2.10 Calculated (lines) and measured (dots) results for tunnel currents from inversion layers through thin oxides. (Adapted from Lo et al.[51] From Ref. 6, reproduced with permission. Copyright 1999 IEEE.)

where $d_0 \approx 0.1$ nm, m^* is the effective mass in the barrier in units of the electron mass, and E_{eff} is the bias-dependent effective barrier height in eV. Clearly, the insulator thickness cannot be scaled too thin without causing excessive tunneling current. To circumvent this limit on scaling, the industry hopes to use a higher dielectric constant ("high-*k*") insulator instead of using a thinner insulator. This does not result in perfect scaling, but as can be seen from the scale length derived in Equation (2.1) and Equation (2.2), it may work well enough to be useful. Unfortunately, higher permittivity insulators tend to have lower bandgaps and lower barrier heights.[53,22] What is needed is a material with fairly high permittivity and not too low a bandgap, so that one can obtain lower leakage current at the same capacitance per unit area. High-*k* replacements for SiO_2 must also satisfy many other requirements, including thermal stability relative to Si at the necessary high processing temperatures, low diffusion constants, sufficiently matched thermal expansion, and low interface trap density. A significant problem for most of the promising high-*k* insulators for FETs is that they generally cause the channel mobility to be degraded.[54] It is not yet clear how this can be overcome. To date, the only successful "high-*k*" insulators are Si oxynitride composites, with $k \sim 5$ to 6, but HfO_2 ($k \sim 20$) and Hf silicates ($k \sim 10$) are presently considered promising [55]. The use of metal gates in conjunction with high-*k* insulators presents interesting possibilities. It allows increased capacitance by eliminating depletion of the polySi gate, while at the same time tending to result in higher threshold voltages due to the tendency for most metals to display roughly midgap workfunctions.

Band-to-band tunneling between the body and drain of an FET is the second important source of tunneling leakage current. This occurs when the FET is in the "off" state, with high drain voltage and low gate voltage. Since the gate potential can significantly modulate this current, it is often referred to as GIDL (gate-induced drain leakage). Recent measurements have shown that band-to-band current varies as $e^{-d_{B2B}/\lambda}$, where d_{B2B} is the minimum physical distance from a point in the conduction band to a point in the valence band at the same energy, and $\lambda = 0.38$ nm is the characteristic length scale for the tunneling.[56] These experiments suggest that this form of tunneling will become a significant cause of power dissipation when the body doping reaches the 10^{19} cm^{-3} regime. Since direct band-to-band tunneling depends on conduction band states being lined up with valence band states, it may be avoided by arranging the body and drain potentials so that the bands never line up. In bulk MOSFETs this might be accomplished by forward biasing the body, which might be an interesting option at low temperature,[4] but it is unlikely that it would be applied to anything except very high-performance computing.

2.5.2 Atomistic Effects

The atomistic effects that cause limitations to scaling are those in which the discreteness of matter gives rise to large statistical variations in small devices. These statistical variations occur because the atoms or molecules tend to display Poisson statistics in their number or position, and the Poisson distribution for small numbers can become very wide. Wide distributions are the exact opposite of what is needed to successfully manufacture extremely large-scale integrated circuits. There are at least three major concerns in this area: the discreteness of dopant atoms, interface roughness at the Si-SiO_2 interface, and line edge roughness (LER), which is also partly related to the discreteness of energy (photons).

The discrete doping problem has probably received the most attention. The difficulty is that although the average concentration of doping is quite well controlled by the usual ion implantation and annealing processes, these processes do not control the exact placement of each dopant. This results in randomness at the atomic scale, which causes spatial fluctuations in the local doping concentration, resulting in device-to-device variation in MOSFET threshold voltages. The uncertainty in the number of dopants, N, in any given device is expected to vary as $\sqrt{N}$, in keeping with Poisson statistics, so that the fractional uncertainty and, hence, the threshold variation, σ_{VT}, may become quite large. In addition, the uncertainty in the placement of the dopants creates additional V_T uncertainty of a similar magnitude.[57] Furthermore, since σ_{VT} varies as $1/\sqrt{width}$, narrow devices such as those in static random access memory (SRAM) cells are most affected by this effect. Current generation SRAM devices are already approaching the point of having only a few hundred dopant atoms controlling the threshold voltage, and the resulting σ_{VT} is expected to make the design of robust SRAMs increasingly difficult.[58] This problem is compounded by the large number of devices in an SRAM, which can cause the statistical tail to extend to more than 5 sigma.

The effects of doping fluctuations on MOSFETs have been investigated by many workers, both experimentally and via modeling.[57,59,60] The most quantitatively accurate modeling results use randomly placed dopants in full three-dimensional (3-D) MOSFET simulations to fully resolve the effects of dopant number and placement.[61–63,57] An example of such a calculation is shown in Figure 2.11, which reveals the wide variation in subthreshold behavior that is expected in an aggressively scaled 11-nm bulk MOSFET due to random dopant placement.[15] Although these variations

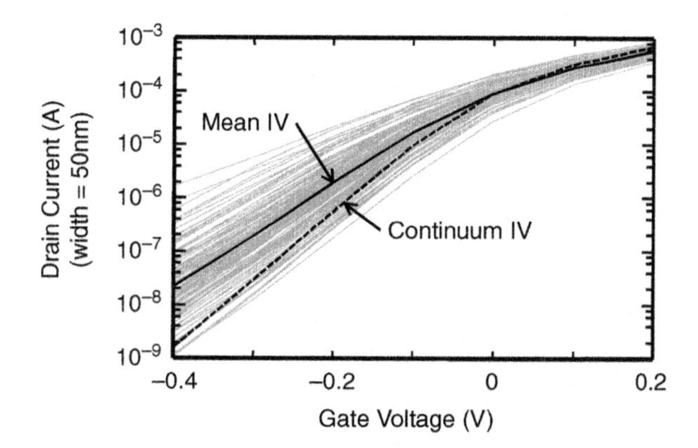

FIGURE 2.11 Simulated IV curves for 100 different 11-nm channel length bulk MOSFETs with discretely placed dopants. Each grey curve corresponds to a different random placement of the dopants, derived in a Monte Carlo manner from the designed average doping profiles. The solid black line is the geometric average of the 100 curves, and the dashed line is the IV curve expected from continuum doping profiles. (From Ref. 15, reproduced with permission. Copyright 2002 IBM.)

are quite large, there are several methods that might be used to reduce the variation. The most straightforward approach for bulk devices is to move the dopants in the body back away from the surface using highly retrograde channel doping profiles. Simulations show that such profiles can lower σ_{VT} up to 2× compared to uniformly doped channels.[63,57] The best way to eliminate these fluctuations is to remove the doping, which may be possible in ultra-thin SOI or double-gate FETs using a metal gate workfunction to set V_T, rather than doping.[30,64]

The second atomistic effect of concern regarding scaling is the atomic roughness of the interface between the Si channel and the SiO_2 gate insulator. The somewhat random character of the bonds in the amorphous SiO_2 causes atomic-scale variations in the position of the surface. In addition to causing random V_T variations in small devices,[65] this roughness causes extra scattering of the carriers in an inversion layer, decreasing the mobility and hence the drive current of an FET.[66] The problem is exacerbated at the higher oxide fields that are reached in more highly scaled devices. When the silicon layer is made extremely thin (<5 nm), with oxide layers on both sides, these atomic variations combine with the quantum confinement effect to cause significant scattering and a rapid loss of mobility.[67] Finally, high-resolution 3-D numerical simulations have shown that these atomic thickness variations also cause increased threshold voltage variation.[68]

The third atomistic effect is line edge roughness (LER). Discrete molecular effects in the photoresist exposure and development cause unevenness on the edges of patterns, an example of which is shown in Figure 2.12. When this pattern is then etched (e.g., by reactive ion etching) into an underlying structure, the molecular randomness of the etch may still further increase this roughness. This is particularly problematic for the gates of aggressively scaled MOSFETs, since they are usually overetched in some way to make the gate length smaller than the nominal lithographic

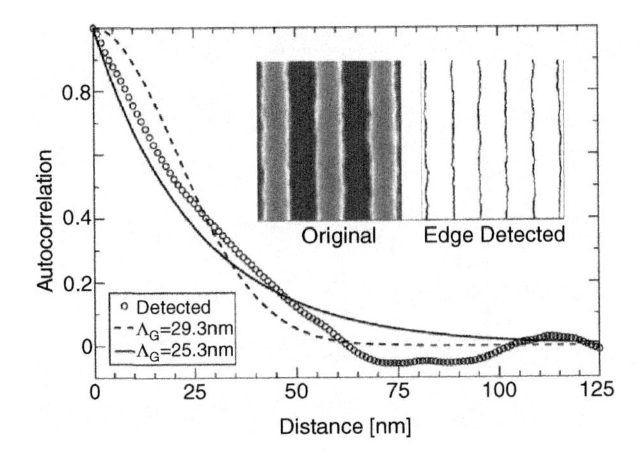

FIGURE 2.12 Example of line edge roughness (LER). Inset shows a micrograph of 100-nm EUV lines and their edges. Main plot shows the autocorrelation function of these edges, along with a Gaussian and an exponential fit. (From Ref. 69, reproduced by permission. Copyright 2003, IEEE.)

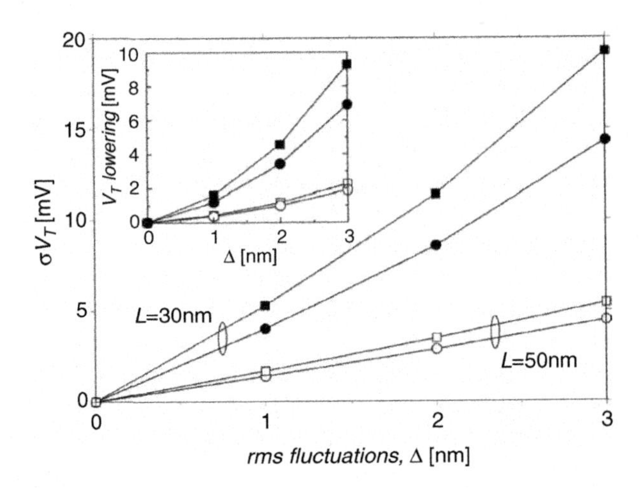

FIGURE 2.13 Threshold voltage uncertainty versus rms magnitude of the line edge roughness for 30 × 50 and 50 × 50 nm MOSFETs at V_{DS} = 1.0 V (squares) and V_{DS} = 0.1 V (circles). (From Ref. 69, reproduced by permission. Copyright 2003, IEEE.)

dimension. This etching of the gate does not, however, reduce LER (but may increase it), with the result that the random variation in gate length of the FET may become quite significant. Some 3-D simulations of this effect have shown that it leads to still further threshold voltage variation,[69] as shown in Figure 2.13. Since the origin of LER is in the discrete molecular composition and interactions of the resist, it is an open research question as to how much it can be reduced. Furthermore, as lithography moves toward shorter-wavelength, higher-energy light sources, the quantization of photons may also come into play, as fewer photons are required to expose minimum size features, leading to more statistical variation in the number of photons actually received.[70]

2.5.3 Thermodynamic Effects

Thermodynamics is just as significant in limiting scaling as the preceding effects. The first way it limits scaling is in its control of the subthreshold behavior of MOSFETs. The subthreshold current of a MOSFET originates in the high-energy tail of the statistical distribution of carriers in its source region. The carriers in the source are governed by Fermi-Dirac statistics, and so the tail of the distribution is essentially Boltzmann. Only the carriers with high enough energy to pass over the channel barrier are eligible to become drain current, and so the subthreshold current varies exponentially with gate voltage as

$$I_{DS} \cong I_{VT} 10^{(V_G - V_T)/S} \tag{2.5}$$

where S is the subthreshold swing and I_{VT} is the current at which V_T is defined. Since $S \cong (\ln 10)\eta k_B T / e$, where η is the ideality (≥ 1), k_B is Boltzmann's constant and T is the temperature, the only way to scale V_T without also changing I_{off} ($=I_{DS}(V_G=0)$) is to scale T. There are different limits on I_{off} for different applications, but for each such limit, thermodynamics clearly sets an accompanying lower limit on V_T, and since V_T is limited, supply voltage scaling is also limited.

The second thermodynamic issue is that all of the energy used in conventional computation is dissipated. It is converted to heat that must be removed. The energy associated with charging and discharging the circuit capacitances is mostly dissipated in the drains of the driving transistors, with some (perhaps 10 to 20%) being dissipated in the wiring. This would be less a problem if the voltage could be fully scaled as in Table 2.2, but thermodynamics prevents that, as already noted. Reversible computing schemes that recycle some of the charging energy (usually via inductors) have been investigated for CMOS,[71,72] but so far they do not appear practical enough for widespread use. In addition to the dynamic energy, all of the static leakage current in CMOS is also dissipative, consuming additional power and generating more heat within the transistors that must be removed. As discussed below, all of this heat creates practical problems that tend to limit scaling.

2.5.4 Practical Considerations

The consumption of energy and removal of heat are not fundamental limits but are constrained by practical limitations. The power source must be practical for a given application, and the removal of the heat must also be practical. For example, many electronic applications are portable. These devices need to be battery powered and light weight to serve their intended purpose. Since lightweight batteries do not contain very much energy, the power level in these applications must be kept low, so that they will run for a reasonable period of time on the available energy. Any heatsinks must also be lightweight. These requirements lead to a wide range of actual power constraints, from ~10 W for a laptop computer that must run for an hour or two on fairly heavy rechargeable batteries, to ~1 μW for a wristwatch that must run for a year on a very small cell.

For a home computer running on "wall" power, the total power limit may perhaps be as high as ~1 KW, based on limitations to how much power one can draw from a single outlet and on the undesirability of dumping so much heat into one's living space. Consumers would probably prefer the present power levels of ~100 W. For the high-end computer market, one of the important considerations is the cost of the cooling technology. The cost of cooling technology generally increases with increasing power density, and as power densities climb above ~100 W/cm^2, the cost increases very rapidly.

Another practical consideration is the cost of power or energy.[73] The cost over the lifetime of a product depends very greatly on how the product is used. For high-end servers that are running all the time, the cost of power from the utility may be as low as 5 $/W (which can come to a lot of money for a 1-MW server farm). For a home computer that spends most of its life in standby, the cost may be well below 1 $/W. On the other hand, for devices powered by disposable batteries the cost can be very high. At the extreme end, wristwatch power costs ~30 M$/W (10 yrs × 1 battery/yr × $3/battery ÷ 1 μW). For a lower case, a PDA whose two batteries must be replaced every two weeks probably costs ~300 $/W (2 yrs × 50 batteries/yr × $1/battery ÷ 300 mW). For applications powered by rechargeable batteries, the cost of power tends to be intermediate between "wall" power and disposable battery power because of the limited number of charging cycles obtainable with present-day rechargeable batteries. As will be discussed further in the next section, reasonable engineering and economic balance requires the power level to vary roughly inversely with the cost of the power, imposing a further practical power constraint on scaling.

2.6 POWER-CONSTRAINED SCALING LIMITS

Since power dissipation is such an important constraint on scaling, much recent circuit and technology research has focused on this problem. The circuit work is beyond the scope of this chapter, but on the technology side the need for lower power dissipation can be addressed by seeking to optimize technology design parameters in such a way as to minimize the power. Optimization of the threshold and supply voltage has received particular attention.[74–76] One can carry the optimization approach much further, however. For any given device structure and material set, everything about the technology (gate length, oxide thickness, threshold voltage, supply voltage, wiring density, etc.) can be optimized to maximize the performance of a given application subject to its own particular power constraint. In this way, the end of scaling becomes an optimization problem. This naturally leads to the result that different applications have different optimal end-of-scaling device designs.[4] An example of such optimizations is discussed below.[73]

The first step in carrying out such optimizations is to choose what to optimize. In recent work[73] two different types of optimizations have been considered. In the first approach, the total logic transition rate (*LTR*, the total number of transitions by individual logic gates per second) of a processor core is maximized while subject

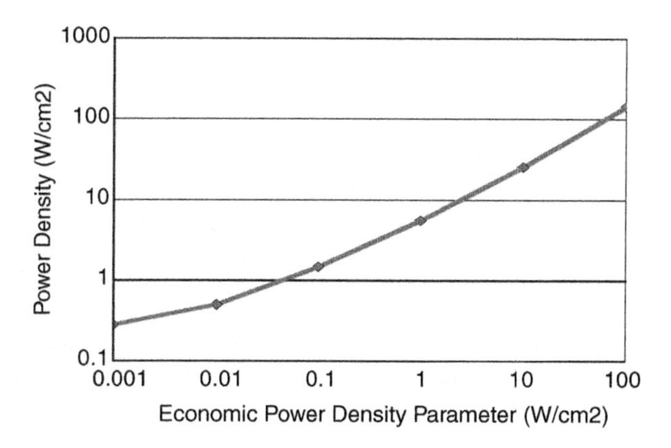

FIGURE 2.14 Optimum power density corresponding to different values of P_{econ}, the economic power density parameter. (From Ref. 73, reproduced by permission. Copyright 2002, IEEE.)

to a fixed total power constraint. The second approach adopts an economics oriented view, and seeks to maximize the Return on Investment (*ROI*), defined as

$$ROI = \frac{LTR \cdot t_{Life}}{Area \cdot C_A + Power \cdot C_P}, \tag{2.6}$$

where t_{Life} is the expected power-on lifetime of the product, C_A is cost per unit area of Si and C_P is the cost/Watt of energy over the power-on lifetime of the product that was discussed in the previous section. *ROI* attempts to capture the customer's desire to get the most computation per dollar spent, including the cost of the energy. It merges performance, power, and area into a single variable and does not require a power constraint. Furthermore, by rearranging the equation a little, one obtains

$$ROI = \frac{LTR \cdot t_{Life}}{Area \cdot C_A \left(1 + P_{Dens} / P_{econ}\right)} \tag{2.7}$$

where P_{Dens} is the average electrical power density, and $P_{econ} = C_A / C_P$ is a purely economic ratio with units of power density that can be used to characterize different applications. For example, if a high-end processor costs ~\$100/cm^2 and power costs ~\$5/W, then $P_{econ} = 100/5 = 20$ W/cm^2. A low power PDA running on disposable batteries might cost ~\$10/cm^2 and ~\$300/W, giving $P_{econ} = 0.03$ W/cm^2. Not surprisingly, when optimizations over technology variables are run to maximize *ROI* at different values of P_{econ}, the optimum electrical power density is found to correlate quite well with the economic power density parameter, as shown in Figure 2.14.

For either type of optimization, one must develop a set of material, device, circuit, and system level models[77] that, when taken together, enable power and delay to be defined as functions of the underlying technology parameters.[17,78] For example,

$$Power(V_{DD}, V_T, t_I, t_{Si}, L_G, \langle w \rangle, w_{hp}, d_R, w_R)\ , \tag{2.8}$$

and

$$Delay(V_{DD}, V_T, t_I, t_{Si}, L_G, \langle w \rangle, w_{hp}, d_R, w_R)\ , \tag{2.9}$$

where $L_G = L_G(t_I, t_{Si})$ is the gate length, and $\langle w \rangle, w_{hp}, d_R$, and w_R are the mean FET width for logic gates, the wiring half-pitch, the repeater spacing (for long wires) and the repeater width, respectively.[73] Then, one must impose constraints, and optimize over the remaining variables. Note that these functions also depend on various system, circuit, device, wiring, and technology assumptions, which must be specified.

The optimizations considered here use a variety of underlying models and assumptions that are detailed in Ref. 73. The system that is optimized is modeled as a 10^7 NAND gate processor core (which could form the basis of larger multiprocessor systems). The "standard" conditions used here assume an oxynitride gate insulator, k = 2.5 wiring dielectric, a junction temperature of 65°C, a moderately low activity factor for the logic, and a clock distribution and driver network that uses fully half of the total power available to the processor.

Optimization results for the limits of bulk CMOS scaling using the two different methods described above are shown in Figure 2.15. These are the results of optimizing over the eight independent variables in Equation (2.8) and Equation (2.9) simultaneously. As can be seen, the optimization results are essentially the same for both methods, indicating that these scaling limits are quite robust, and not strongly dependent on the approach. Of particular interest in these results are the optimal FET dimensions, which are in the 20 to 30-nm range, depending on power level. This means that smaller FETs will have *less* system performance, contrary to what one would expect from scaling. This loss of performance occurs when the increasing leakage of smaller FETs takes away from the power available for useful switching and the increasing device performance isn't sufficient to compensate. Eventually, if one scales too far, all of the power budget will be consumed in leakage dissipation, and none will be available for doing computation. The optimizations show that larger devices with thicker oxides are required for lower power applications, as would be expected in order to reduce tunneling leakage to the levels necessary for low power operation. At the high-power end of these optimizations the power of the CPU core is only 1 W, which may seem low, but it is already leading to power density in excess of 100 W/cm^2. Keeping the circuits cool becomes problematic at power levels much above this. Some of the novel devices proposed for the ITRS roadmap may enable scaling to somewhat smaller dimensions, but this type of power-constrained optimization suggests that the end of useful scaling is in sight.

One of the important uses of this type of technology optimization is in evaluating the potential impact of various proposed technology changes on system-level performance. Figure 2.16 shows an example of this, in which the optimum system performance of various technology enhancements is compared by normalizing to

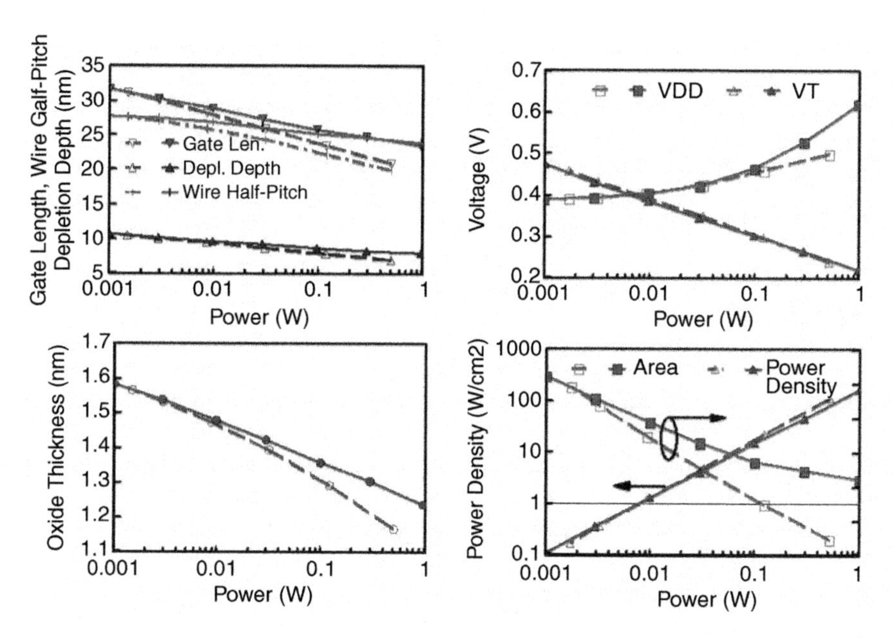

FIGURE 2.15 Power-constrained device and technology optimization results, comparing the optimization method. Solid lines and solid symbols are for maximizing *LTR* subject to fixed power constraints. Dashed lines and open symbols are for maximizing *ROI* at various values of P_{econ}. (Partly from Ref. 73, reproduced by permission. Copyright 2002, IEEE.)

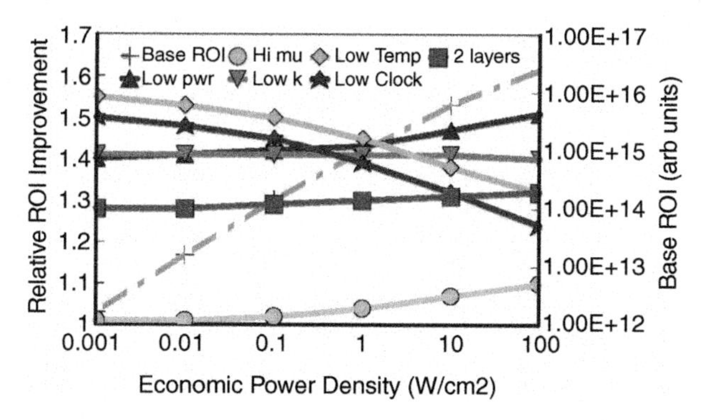

FIGURE 2.16 Relative performance improvements for various possible technology enhancements, as a function of P_{econ}, the economic power density parameter. (From Ref. 73, reproduced by permission. Copyright 2002, IEEE.)

the base optimization case in which there are no enhancements.[73] By accumulating many such one-time changes, the industry can hope to continue increasing the performance of computational electronics, even though it will no longer be "scaling" in the conventional sense. The "low pwr" and "low clock" cases refer to circuit-oriented innovations that reduce the amount of static and clock power, increasing the power available for useful computation. On the technology side, reduced operating temperature appears promising, as does the "low k" case with reduced wiring permittivity (1.5 vs. 2.5), and the "2 layers" case in which 3-D integration is envisioned as enabling the use of two layers of circuitry on top of each other. In the "high mu" case strained Si is imagined to increase the FET current 1.3× relative to the base case, resulting in some advantage for high performance applications. Hopefully these types of analyses can shed light on the most fruitful directions for future device research.

2.7 SUMMARY

Progress in CMOS technology over the past 30 years has been based on the well-understood principles of scaling and has yielded exponential growth in the electronics and computer industries. Unfortunately this scaling is starting to run into fundamental limits, due to various physical effects that do not scale properly, including quantum mechanical tunneling, the discreteness of dopants, thermodynamics, and practical issues associated with power dissipation. Various novel materials and device structures are being investigated in the hope of extending the long march of progress in CMOS, including strained Si devices, FD-SOI, high-*k* gate insulators, and multiple-gate FETs. Many of these ideas hold promise, but in the end the limiting effects will still be there, and the industry will probably eventually turn to complex technology optimization procedures in order to produce the best possible products.

ACKNOWLEDGMENTS

The author is indebted to many co-workers for valuable discussions, including Asen Asenov, Bob Dennard, Wilfried Haensch, Jakub Kedzierski, Steve Kosonocky, Tak Ning, Ed Nowak, Philip Restle, Ken Rim, Paul Solomon, Yuan Taur, and H.-S. Philip Wong.

REFERENCES

1. Lilienfeld, J.E., Method and Apparatus for Controlling Electric Currents, U.S. Patent 1,745,175, 1930.
2. Kahng, D. and Atalla, M.M., Silicon–Silicon Dioxide Field Induced Surface Devices, presented at IRE Solid-State Device Res. Conf., Pittsburgh, PA, June 1960.
3. Bondy, P.K., Moore's Law Governs the Silicon Revolution, *Proc. IEEE,* 86, 78–81, 1998.
4. Frank, D.J., Dennard, R. H., Nowak, E., Solomon, P. M., Taur, Y., and Wong, H.-S. P., Device Scaling Limits of Si MOSFETs and their Application Dependencies, *Proc. IEEE*, 89, 259–288, 2001.
5. Taur, Y., Buchanan, D., Chen, W., Frank, D., Ismail, K., Lo, S.-H., Sai-Halasz, G., Viswanathan, R., Wann, H.-J. C., Wind, S., and Wong, H.-S., CMOS scaling into the nanometer regime, *Proc. IEEE,* 85, 486–504, 1997.
6. Wong, H.-S. P., Frank, D. J., Solomon, P. M., Wann, H.-J., and Welser, J., Nanoscale CMOS, *Proc. IEEE*, 87, 537–570, 1999.

7. Frank, D. J., CMOS device technology trends for power-constrained applications, in *Power Aware Design Methodologies*, Ed. by M. Pedram and J. M. Rabaey, Kluwer Academic Publishers, Boston, 2002.
8. Semiconductor Industry Association (SIA), *International Technology Roadmap for Semiconductors, 2003 Edition*, SEMATECH, USA, Austin, TX, retrieved from http://public.itrs.net
9. Sai-Halasz, G. Performance trends in high-end processors, *Proc. IEEE*, 83, 20, 1995.
10. Sleight, J. W., Varekamp, P. R., Lustig, N., Adkisson,J., Allen, A., Bula, O., Chen, X., Chou, T., Chu, W., Fitzsimmons, J., Gabor, A., Gates, S., Jamison, P., Khare, M., Lai, L., Lee, J., Narasimha, S., Ellis-Monaghan, J., Peterson, K., Rauch, S., Shukla, S., Smeys, P., Su, T.-C., Quinlan, J., Vayshenker, A., Ward, B., Womack, S., Barth, E., Blery, G., Davis, C., Ferguson, R., Goldblatt, R., Leobandung, E., Welser, J., Yang, I., and Agnello, P., A high performance 0.13 µm SOI CMOS technology with a 70 nm silicon film and with a second generation low-k Cu BEOL, in *IEDM Tech. Dig.*, 245–248, 2001.
11. Puri, R. and Chuang, C. T., SOI digital circuits: Design issues, in Thirteenth Int. Conf. VLSI Design, 474–479, 2000.
12. Dennard, R.H., Gaensslen, F.H., Yu, H.N., Rideout, V.L., Bassous, E., and LeBlanc, A.R., Design of ion-implanted MOSFET's with very small physical dimensions, *Jour. Solid St. Circuits,* SC-9, 256–268, 1974.
13. Davari, B., Dennard, R. H., and Shahidi, G. G., CMOS scaling, the next ten years, *Proc. IEEE*, 89, 595–606, 1995.
14. Frank, D.J., Taur, Y., and Wong, H.-S. P., Generalized scale length for two-dimensional effects in MOSFET's, *IEEE Elec. Dev. Lett.*, 19, 385–387, 1998.
15. Frank, D.J., Power-constrained CMOS scaling limits, *IBM J. Res. Devel.*, 46(2/3), March/May 2002.
16. Solomon, P. M. and Djomehri, I. J., Overscaling, Design for the future, IBM Research Report, RC22379, Jan. 2002.
17. Rim, K., Koester, S., Hargrove, M., Chu, J., Mooney, P.M., Ott, J., Kanarsky, T., Ronsheim, P., Ieong, M., Grill, A., and Wong, H.-S.P., Strained Si NMOSFETs for high performance CMOS technology, in *Symp. VLSI Technology Dig.*, 59–60, 2001.
18. Mistry, K., Armstrong, M., Auth, C., Cea, S., Coan, T., Ghani, T., Hoffmann, T., Murthy, A., Sandford, J., Shaheed, R. Zawadzki, K., Zhang, K., Thompson, S., and Bohr, M., Delaying forever: Uniaxial strained silicon transistors in a 90nm CMOS technology, in *Symp. VLSI Technology Dig.*, 50–51, 2004.
19. Ghani, T., Armostong, M., Auth, C., Bost, M., Charvat, P., Glass, G., Hoffman, T., Johnson, K., Kenyon, C., Klaus, J., McIntyre, B., Mistry, K., Murthy, A., Sandford, J., Silberstein, M., Sivakumar, S., Smith, P., Zawadzki, K., Thompson, S., and Bohn, R., A 90nm high volume manufacturing logic technology featuring novel 45nm gate length strained silicon CMOS transistors, *IEDM Tech. Dig.*, 978–981, 2003.
20. Chidambaram, P.R., Smith, B.A., Hall, L.H., Bu, H., Chakravarthi, S., Kim, Y., Samoilov, A.V., Kim, A.T., Jones, P.J., Irwin, R.B., Kim, M.J., Rotondaro, A.L.P., Machala, C.F., and Grider, D.T., 35% drive current improvement from recessed-SiGe drain extensions on 37 nm gate length PMOS, in *Symp. VLSI Technology Dig.*, 48–49, 2004.
21. Yang, M., Ieong, M., Shi, L., Chan, K., Chan, V., Chou, A., Gusev, E., Jenkins, K., Boyd, D., Ninomiya, Y., Pendleton, D., Surpris, Y., Heenan, D., Ott, J., Guarini, K., D'Emic, C., Cobb, M., Mooney, P., To, B., Rovedo, N., Benedict, J., Mo, R., and Ng, H., High performance CMOS fabricated on hybrid substrate with different crystal orientations, in *IEDM Tech. Dig.*, 453–456, 2003.

22. Wong, H.-S. P., Beyond the conventional transistor, *IBM J. Res. Devel.*, 46, 133–168, 2002.
23. Wann, C., Assaderaghi, F., Shi, L., Chan, K., Cohen, S., Hovel, H., Jenkins, K., Lee, Y., Sadana, D., Viswanathan, R., Wind, S., and Taur, Y., High performance 0.07 μm CMOS with 9.5ps gate delay and 150GHz ft, *IEEE Electron Dev. Lett.*, 18, p. 625, 1997.
24. Su, L.T., Hu, H., Jacobs, J.B., Sherony, M., Wei, A., and Antoniaadis, D.A., Tradeoffs of Current Drive vs. Short Channel Effect in Deep Submicrometer Bulk and SOI MOSFET's, in *IEDM Tech. Dig.*, 649–652, 1994.
25. Chau, R., Kavalieros, J., Doyle, B., Murthy, A., Paulsen, N., Lionberger, D., Barlage, D., Arghavani, R., Roberds, B., and Doezy, M., A 50nm Depleted-Substrate CMOS Transistor (DST), in *IEDM Tech. Dig.*, 621–624, 2001.
26. Doris, B., Ieong, M., Zhu, H., Zhang, Y., Steen, M., Natzle, W., Callegari, S., Narayanan, V., Cai, J., Ku, S.H., Jamison, P., Li, Y., Ren, Z., Ku, V., Boyd, D., Kanarsky, T., D'Emie, C., Newport, M., Dobuzinsky, D., Deshpande, S., Petrus, J., Jammy, R., and Haensch, W., Device Design Considerations for Ultra-Thin SOI MOSFETs, in *IEDM Tech. Dig.*, 631–634, 2003.
27. Doris, B., Ieong, M., Kanarsky, T., Zhang, Y., Roy, R.A., Dokumaci, O., Ren, Z., Jamin, F-F., Shi, L., Natzle, W., Huang, H.-J., Mezzapelle, J., Mocuta, A., Womack, S., Gribelyuk, M., Jones, E.C., Miller, R.J., Wong, H-S P., and Haensch, W., Extreme Scaling with Ultra-Thin Si Channel MOSFETs, in *IEDM Tech. Dig.*, 267–270, 2002.
28. Sekigawa, T. and Hayashi, Y., Calculated threshold-voltage characteristics of an XMOS transistor having an additional bottom gate, *Solid State Electronics*, 27, p. 827, 1984.
29. Fiegna, C., Iwai, H., Wada, T., Saito, T., Sangiorgi, E., and Ricco, B., A New Scaling Methodology for the 0.1 - 0.025 μm MOSFET, in *Symp. VLSI Technol. Dig.*, 33–34, 1992.
30. Frank, D. J., Laux, S. E., and Fischetti, M. V., Monte Carlo Simulation of a 30 nm Dualgate MOSFET: How Far Can Si Go?, in *IEDM Tech. Dig.*, 553–556, 1992.
31. Wong, H.-S. P., Frank, D.J., and Solomon, P.M., Device Design Considerations for Double-Gate, Ground-Plane, and Single-Gated Ultra-Thin SOI MOSFETs at the 25 nm Channel Length Generation, in *IEDM Tech. Dig.*, 407–410, 1998.
32. Naveh, Y. and Likharev, K.K., Modeling of 10-nm-scale ballistic MOSFETs, *IEEE Elec. Dev. Lett.*, 21, 242–244, 2000.
33. Pikus, F.G. and Likharev, K.K., Nanoscale field-effect transistors: An ultimate size analysis, *Appl. Phys. Lett.*, 71(25), 3661–3663, Dec. 1997.
34. Tanaka, T., Suzuki, K., Horie, H., and Sugii, T., Ultrafast Low-Power Operation of p+-n+Double-Gate SOI MOSFETs, in *Symp. VLSI Technol. Dig.*, 11–12, June 1994.
35. Wang, J., Solomon, P.M., and Lundstrom, M., A general approach for the performance assessment of nanoscale silicon FETs, *IEEE Trans Electron Dev*, 51, 1366 – 1370, 2004.
36. Wong, H.-S., Chan, K., and Taur, Y., Self-Aligned (Top and Bottom) Double-Gate MOSFET with a 25 nm Thick Silicon Channel, in *IEDM Tech. Dig.*, 427–430, 1997.
37. Lee, J.-H., Tarashi, G., Wei, A., Langdo, T., Fitzgerald, E.A., and Antoniadis, D., Super Self-Aligned Double-Gate (SSDG) MOSFETS Utilizing Oxidation Rate Difference and Selective Epitaxy, in *IEDM Tech. Dig.*, 71–74, Dec. 1999.
38. Guarini, K.W., Solomon, P.M., Zhang, Y., Chan, K.K., Jones, E.C., Cohen, G.M., Krasnoperova, A., Ronay, M., Dokumaci, O., Bucchignano, J.J., Lavoie, C., Cabral, C., Jr., Ku, V., Boyd, D.C., Petrarca, K.S., Babich, I.V., Treichler, J., Kozlowski, P.M., Newbury, J.S., D'Emic, C.P., Sieina, R.M., and Wong, H.-S. Triple-Self-Aligned, Planar Double-Gate MOSFETs: Devices and Circuits, in *IEDM Tech. Dig.*, 425–428, 2001.

39. Hergenrother, J.M., Monroe, D., Klemens, F.P., Komblit, A., Weber, G.R., Mansfield, W.M., Baker, M.R., Baumann, F.H., Bolan, K.J., Bower, J.E., Ciampa, N.A., Cirelli, R.A., Colonell, J.I., Eaglesham, D.J., Frackoviak, J., Gossman, H.J., Green, M.L., Hillenius, S., King, C., Kleiman, R., Lai, W.Y.C., Lee, J.T.-C., Liu, R.-C., Maynard, H., Moris, M., Oh, S.-H., Pai, C.-S., Rafferty, C., Rosamilia, J., Sorsch, T., and Vuong, H.-H., The Vertical Replacement-Gate (VRG) MOSFET: A 50-nm Vertical MOSFET with Lithography Independent Gate Length, in *IEDM Tech. Dig.*, 75–78, 1999.
40. Hergenrother, J.M., Wilk, G.D., Nigam, T., Klemens, F.P., Monroe, D., Silverman, P.J., Sorsch, T.W., Busch, B., Green, M.L., Baker, M.R., Boone, T., Bude, M.K., Ciampa, N.A., Ferry, E.J., Fiory, A.T., Hillenius, S.J., Jacobson, D.C., Johnson, R.W., Kalavade, P., Keller, R.C., King, C.A., Kornblit, A., Krautter, H.W., Lee, J.T.-C., Mansfield, W.M., Miner, J.F., Morris, M.D., Oh, S.-H, Rosamilia, J.M., Sapjeta, B.J., Short, K., Steiner, K., Muller, D.A., Voyles, P.M., Grazul, J.L., Shero, E.J., Givens, M.E., Pomarede, C., Mazanec M., and Werkhoven, C., 50 nm Vertical Replacement-Gate (VRG) nMOSFETs with ALD HfO_2 and Al_2O_3 Gate Dielectrics, in *IEDM Tech. Dig.*, 51–54, 2001.
41. Huang, X., Lee, W.-C., Ku, C., Hisamoto, D., Chang, L., Kedzierski, J., Anderson, E., Takeuchi, H., Choi, Y.-K., Asano, K., Subramanian, V., King, T.J., Bokor J., and Hu, C., Sub 50-nm FinFET: PMOS, in *IEDM Tech. Dig.*, 67–70, 2001.
42. Lindert, N., Chang, L., Choi, Y.-K., Anderson, E., Lee, W.-C., King, T.-J., Bokor, J., and C. Hu, C., Sub-60-nm quasi-planar FinFETs fabricated using a simplified process, *IEEE Electron Dev. Lett.*, 22(10), 487–489, 2001.
43. Kedzierski, J., Fried, D.M., Nowak, E.J., Kanarsky, T., Rankin, J.H., Hanafi, H., Natzle, W., Boyd, D., Zhang, Y., Roy, R.A., Newbury, J., Yu, C., Yang, Q., Saunders, P., Willets, C.P., Johnson, A., Cole, S.P., Young, H.E., Carpenter, N., Rakowski, D., Rainey, B.A., Cottrell, P.E., Ieong M., and Wong, H.-S. P., High-Performance Symmetric-Gate and CMOS Compatible V_T Asymmetric-Gate FinFET Devices, in *IEDM Tech. Dig.*, 437–440, 2001.
44. Choi, Y.-K., Lindert, N., Xuan, P., Tang, S., Ha, D., Anderson, E., King, T.-J., Boker, J., and Hu, C., Sub-20nm CMOS FinFET Technologies, in *IEDM Tech. Dig.*, 421–424, 2001.
45. Yang, I., Vieri, C., Chandrakasan, A., and Antoniadis, D., Backgated CMOS on SOIAS for dynamic threshold voltage control, *IEEE Trans. Elec. Dev.*, 44, p. 822, 1997.
46. Kosonocky, S.V., Irnmediato, M., Cottrell, P., Hook, T., Mann, R., and Brown, J., Enhanced Multi-Threshold (MTCMOS) Circuits Using Variable Well Bias, in *Int. Symp. Low Power Electronics and Design*, 165–169, 2001.
47. von Kaenel, V.R., Pardoen, M.D., Dijkstra, E., and Vittoz, E. A., Automatic Adjustment of Threshold and Supply Voltages for Minimum Power Consumption in CMOS Digital Circuits, in *IEEE Symp. Low Power Electronics, Digest of Technical Papers*, 78–79, 1994.
48. Narendra, S., Antoniadis, D., and De, V., Impact of Using Adaptive Body Bias to Compensate Die-to-Die V_T Variation on Within-Die V_T Variation, in *Int. Symp. Low Power Electronics and Design*, 229–232, 1999.
49. Narendra, S., Haycock, M., Mooney, R., Govindarajulu, V., Erraguntala, V., Wilson, H., Vangal, S., Pangal, A., Seligman, E., Nair, R., Borkar, N., Hofsheier, J., Menon, S., Bloechel, B., Dermer, G., Borkar, S., and De, V., 1.1V 1GHz Communications Router with On-Chip Body Bias in 150nm CMOS, in *Proc. 2002 ISSCC*, paper 16.4, 2002.
50. Tschanz, J., Kao, J., Narendra, S., Nair, R., Antoniadis, D., Chandrakasan, A., and De, V., Adaptive Body-Bias for Reducing Impacts of Die-to-Die and Within-Die Parameter Variations on Microprocessor Frequency and Leakage, in 2002 ISSCC, paper 25.7, 2002.

51. Lo, S.-H., Buchanan, D.A., Taur, Y., and Wang, W., Quantum-mechanical modeling of electron tunneling current from the inversion layer of ultra-thin-oxide nMOSFET's, *IEEE Electron Dev. Lett.*, 18, 209–211, 1997.
52. Sze, S. M., *Physics of Semiconductor Devices, 2nd Edition.* John Wiley & Sons, New York, 1981.
53. Robertson, J., Band offsets of wide-band-gap oxides and implications for future electronic devices, *J. Vacuum Science and Technology B*, 18(3), 1785–1791, 2000.
54. Fischetti, M., Neumayer, D., and Cartier, E., Effective electron mobility in Si inversion layers in MOS systems with a high-k insulator: The role of remote phonon scattering, *J. Appl. Phys.*, 90(9), p. 4587, 2001.
55. Barlage, D., Arghavani, R., Dewey, G., Doczy, M., Doyle, B., Kavalieros, J., Murthy, A., Roberds, B., Stokley, P., and Chau, R., High-Frequency Response of 100nm Integrated CMOS Transistors with High-k Gate Dielectrics, in *IEDM Tech. Dig.*, 231–234, 2001.
56. Solomon, P.M., Frank, D.J., Jopling, J., D'Emic, C., Dokumaci, O., Ronsheim, P., and Haensch, W.E., Tunnel Current Measurements on P/N Junction Diodes and Implications for Future Device Design, in *IEDM Tech. Dig.*, 233–236, 2003.
57. Asenov, A., Brown, A.R., Davies, J.H., Kaya, S., and Slavcheva, G., Simulation of intrinsic parameter fluctuations in decananometer and nanometer-scale MOSFETs. *IEEE Trans. Elec. Dev.,* 50, 1837–1852, 2003.
58. Bhavnagarwala, A.J., Tang, X., and Meindl, J.D., The impact of intrinsic device fluctuations on CMOS SRAM cell stability, *IEEE J. Solid-St. Circ.,* 36, 658–665, 2001.
59. Mizuno, T., Iwase, M., Niiyama, H., Shibata, T., Fujisaki, K., Nakasugi, T., Toriumi, A., and Ushiku,, U., Performance Fluctuations of 0.1 μm MOSFETs—Limitation of 0.1 μm ULSI's, in *Symp. VLSI Technology Dig.*, 13–14, 1994.
60. Stolk, P.A., Widdershoven, F.P., and Klaassen, D. B. M., Modeling statistical dopant fluctuations in MOS transistors, *IEEE Trans. Elec. Dev.*, 45, 1960–1971, 1998.
61. Wong, H.-S. P. and Taur, Y., Three-Dimensional 'Atomistic' Simulation of Discrete Microscopic Random Dopant Distributions Effects in Sub-0.1 μm MOSFETs, in *IEDM Tech. Dig.*, 705–708, 1993.
62. Wong, H.-S. P., Taur, Y., and Frank, D., Discrete random dopant distribution effects in nanometer-scale MOSFETs, *Microelectronic Reliability*, 38, 1447–1456, 1998.
63. Frank, D.J., Taur, Y., Ieong, M., and Wong, H.-S. P., Monte Carlo Modeling of Threshold Variation due to Dopant Fluctuations, in *Symp. VLSI Technol. Dig.*, 169–170, 1999.
64. Frank, D. J. and Wong, H.-S. P., Simulation of Stochastic Doping Effects in Si MOSFETs, in Proc. Int. Workshop on Computational Electronics, 2–3, May 2000.
65. Asenov, A., Kaya, S., and Davies, J.H., Intrinsic threshold voltage fluctuations in decanano MOSFETs due to local oxide thickness variations, *IEEE Trans. Electron Devices*, 49(1), 112–119, Jan. 2002.
66. Ishihara, T., Matsuzawa, K., Takayanagi, M., and Takagi, S., Comprehensive understanding of electron and hole mobility limited by surface roughness scattering in pure oxides and oxynitrides based on correlation function of surface roughness, *Jpn. J. Appl. Phys.*, 41, 2353–2358, 2002.
67. Uchida, K. and Takagi, S. Carrier scattering induced by thickness fluctuation of silicon on insulator film in ultrathin body metal oxide semiconductor field effect transistors, *Appl. Phys. Lett.*, 82, 2916–2918, 2003.
68. Brown, A.R., Adamu-Lema, F., and A. Asenov, A., Intrinsic parameter fluctuations in nanometre scale thin-body SOI devices introduced by interface roughness, *Superlattices and Microstructures*, 34, 283–291, 2003.

69. Asenov, A., Kaya, S., and Brown, A.R., Intrinsic parameter fluctuations in decananometer MOSFETS introduced by gate line edge roughness, *IEEE Trans. Elec. Dev.* 50, 1254–1260, 2003.
70. Brunner, T.A., Why optical lithography will live forever, *J. Vac. Sci. Technol. B, Microelectron. Nanometer Struct. (USA)*, 21, 2632–2637, 2003.
71. Athas, W., Tzartzanis, N., Mao, W., Peterson, L., Lal, R., Chong, K., Moon, J.-S., Svensson,L., and Bolotski, M., The design and implementation of a low-power clockpowered microprocessor, *IEEE J. Solid-State Circuits*, 35(11), 1561–1570, Nov. 2000.
72. Frank, D. J., Comparison of High Speed Voltage-Scaled Conventional and Adiabatic Circuits, in *Int. Symp. Low Power Electronics and Design (ISLPED), Digest of Tech. Papers,* p. 377, 1996.
73. Frank, D. J., Power Constrained Device and Technology Design for the End of Scaling, in *IEDM Tech. Dig.*, 643–646, 2002.
74. Chen, Z., Burr, J., Shott, J., and Plummer, J.D., Optimization of Quarter Micron MOSFETs for Low Voltage/Low Power Applications, in *IEDM Tech. Dig.*, 63–65, 1995.
75. Frank, D.J., Solomon, P., Reynolds, S., and J. Shin, J., Supply and Threshold Voltage Optimization for Low Power Design, in *Proc. 1997 Int. Symp. Low Power Electronics and Design*, 317–322, 1997.
76. Bhavnagarwala, A.J., Austin, B.L., Bowman, K.A., and Meindl, J.D., A minimum total power methodology for projecting limits on CMOS GSI, *IEEE Trans. VLSI Sys.*, 8, 235–251, 2000.
77. Meindl, J.D., Low power microelectronics: Retrospect and prospect, *Proc. IEEE*, 83, 619–635, 1995.
78. Eble, J.C., A Generic System Simulator with Novel On-Chip Cache and Throughput Models for Gigascale Integration, Ph.D. Thesis, Georgia Inst. Tech., Nov. 1998.

3 The Scaling Limit of MOSFETs due to Direct Source-Drain Tunneling

Hisao Kawaura

3.1 INTRODUCTION

The past 40 years have seen intensive research on the scaling down of metal-oxide-semiconductor field-effect transistors (MOSFETs), which has improved both the operating speed and integration density of silicon logic large scale integrated circuits (LSIs). The International Technology Roadmap for Semiconductors (ITRS)[1] predicts that the scaling will continue, and that in 2016, the gate length in multiple processor units (MPUs) will reach 9 nm (Figure 3.1). Several research groups[2–4] have already reported on the development of 10-nm-level MOSFETs shown in Figure 3.2. It is, therefore, quite likely that the scaling down to sub-10-nm is achievable in the future. However, a number of problems still remain in the scaling of MOSFETs as shown in Figure 3.3. The problems in Figure 3.3 can be categorized into three groups: physical, technical, and economical problems. The physical problems are those that cannot be solved in physical principle. The technical problems are solvable in principle but difficult to solve with current technology. The economical problems are those that are solvable with current technology, but their solution is very costly.

According to the scaling principle,[5] the thickness of a gate dielectric should be scaled in accordance with the scaling of its lateral size to prevent short-channel effects. The ITRS predicts that the thickness for high-performance devices will reach 2 nm at a gate length of 30 nm. Conventionally, SiO_2 films grown on a (100)-oriented silicon substrate have been used as gate dielectrics because of their reliability and the low trap density at the Si-SiO_2 interface. However, the SiO_2 films thinner than 2 nm allow direct tunneling through the film ("gate-dielectric tunneling") and increase the power consumption of LSIs (Figure 3.4). This is a serious problem, but it can be solved by replacing SiO_2 with an appropriate dielectric with a high permittivity (high-κ).[6] Therefore, the gate-dielectric tunneling problem can be reduced to a technical problem.

When the gate length is further reduced, another tunneling effect takes place. In very-short-gate MOSFETs, carriers in the source can quantum-mechanically tunnel through the potential barrier below the gate without the help of thermal activation even under the complete cut-off conditions (i.e., "direct source-drain tunneling")[7] as shown in Figure 3.4. According to the Wentzel-Kramers-Brillouin

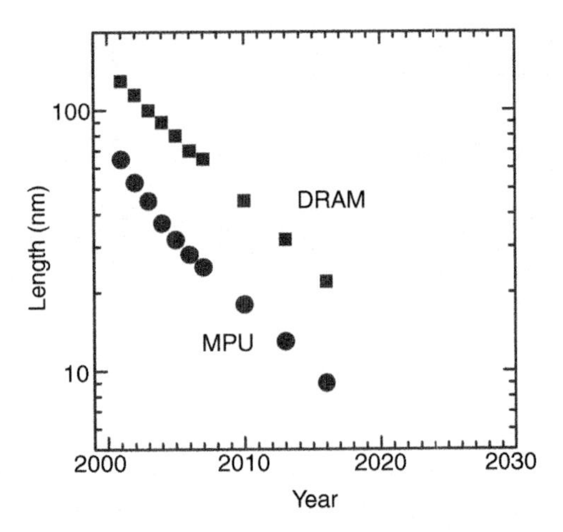

FIGURE 3.1 Trends in the scaling of silicon LSIs. The solid circles and rectangles correspond to the physical gate length of multiple processor units (MPUs) and the half pitch of dynamic random access memories (DRAMs), respectively.

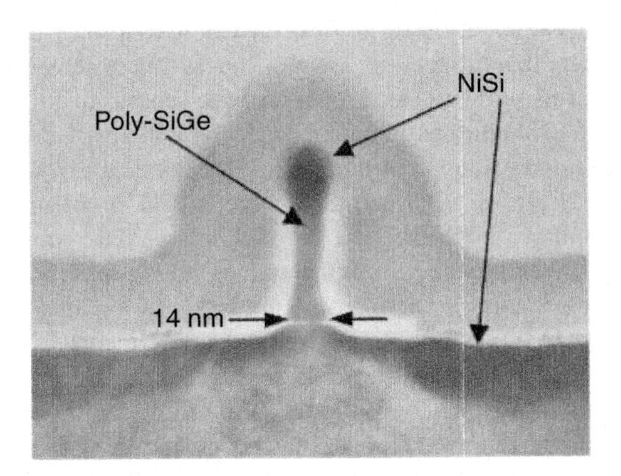

FIGURE 3.2 Cross-sectional TEM image of a 14-nm-gate-length MOSFET (From Ref. 4, © 2002 IEEE).

(WKB) approximation,[8] the transmission probability, T, for a rectangular-shaped potential barrier with width d and height ϕ is given by

$$T \propto \exp\left[-2\left(2m^*/\hbar^2\right)^{1/2}\left(q\phi\right)^{1/2} d\right] \tag{3.1}$$

where m^* and q are the effective mass of silicon and the elementary charge, respectively. Since the tunneling current has weaker ϕ dependence than the thermal current which depends exponentially on ϕ, when direct source-drain tunneling

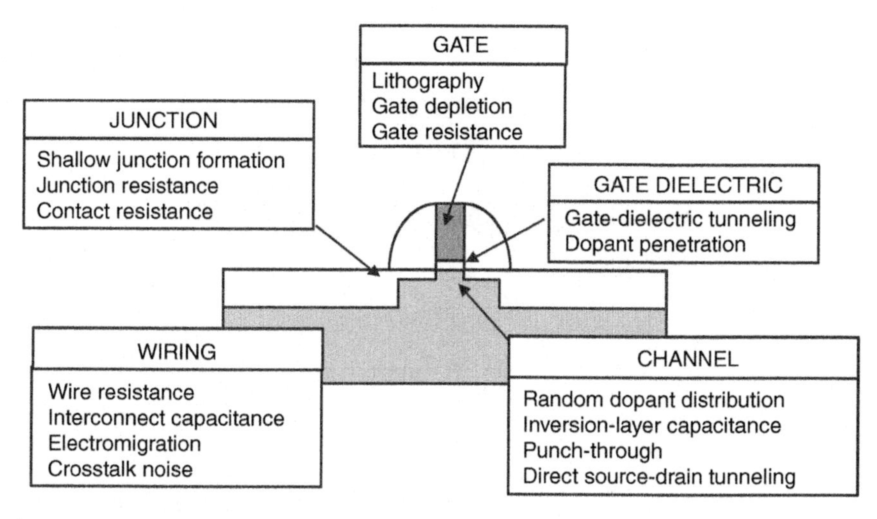

FIGURE 3.3 Problems in the scaling of MOSFETs.

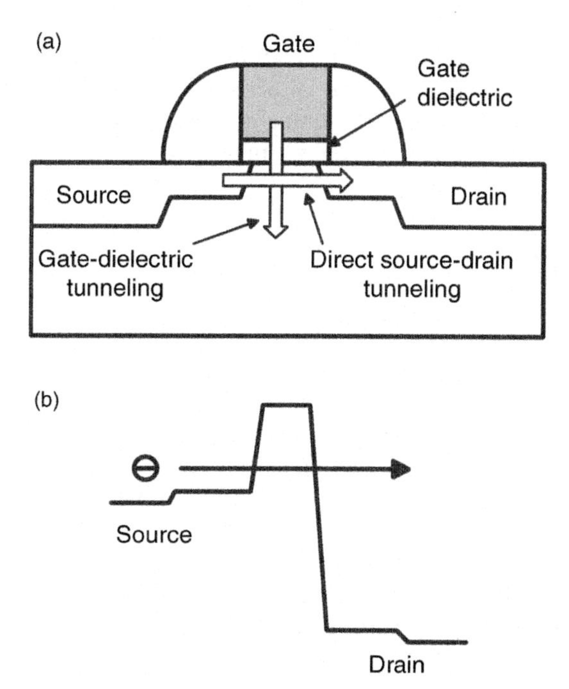

FIGURE 3.4 (a) Two types of tunneling effects: gate-dielectric tunneling and direct source-drain tunneling. (b) band diagram between the source and the drain at cut-off conditions (From Ref. 7).

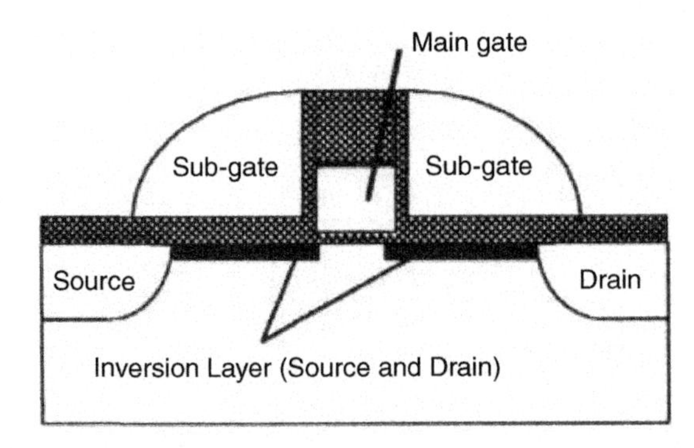

FIGURE 3.5 Schematic crosssection of an ILD-MOSFET using an inversion layer as the source and drain (From Ref. 9, © 1993 IEEE).

occurs, it significantly degrades the subthreshold swing (S-factor) and, consequently, inhibits low-power operation of digital LSIs. From Equation (3.1), the barrier width giving transmission probabilities of 1×10^{-3} and 1×10^{-6} at a barrier height of 100 mV can be estimated to be 10 and 5 nm, respectively. This result indicates that direct source-drain tunneling will occur below 10-nm generation. What is important is that direct source-drain tunneling is not a technical problem but a physical one because T is determined only by the effective mass of silicon except for ϕ and d. Therefore, although direct source-drain tunneling is not as urgent a problem as gate-dielectric tunneling, it is very important to study direct source-drain tunneling in detail.

In this chapter, direct source-drain tunneling is experimentally investigated using nanometer-scale devices, and the scaling limit of MOSFETs is discussed in terms of direct source-drain tunneling.

3.2 EJ-MOSFETS

3.2.1 Concept of EJ-MOSFETs

According to the scaling principle, the vertical size of a device should be scaled in accordance with the scaling of its lateral size. The reduction of the junction depth is one of the key problems in the vertical scaling of devices; another one is the reduction of the thickness of the gate dielectric. According to the ITRS, a junction depth of at least 5 nm is needed for 10-nm generation, but this depth is hard to obtain with current doping technologies.

The idea of electrically induced junctions is attractive for reducing the junction depth. The idea was first implemented in ILD-MOSFETs (MOSFETs using an inversion layer as the source/drain).[9] Figure 3.5 shows a schematic illustration of an ILD-MOSFET. The ILD-MOSFET has conductive subgates adjacent to the main gate to induce junctions. As shown in Figure 3.6, short-channel effects can be significantly suppressed around the main-gate-length of 100 nm. This result is understandable because the electrically induced junction is extremely shallow (typically 5 nm) and can cover even sub-10-nm generation.

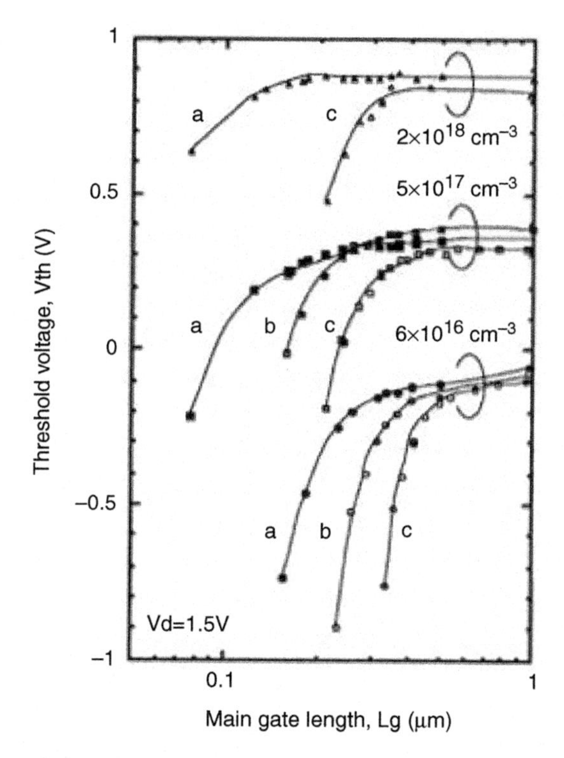

FIGURE 3.6 Dependence of the threshold voltage on the main-gate length for (a) ILD-MOSFET, (b) lightly doped drain MOSFET (LDD-MOSFET), and (c) single-drain MOSFET for various dopant concentrations in the substrate (From Ref. 9, © 1993 IEEE). The junction depth is 50 nm for the LDD-MOSFET (b) and 110 nm for the single-drain MOSFET (c). Since the junction depth for the ILD-MOSFET is extremely small (5 nm), the threshold voltage lowering caused by short-channel effects in the sub-100-nm region is effectively minimized.

In light of this background, an electrically variable shallow junction MOSFET (EJ-MOSFET) was proposed.[10] Figure 3.7 shows a schematic illustration of an EJ-MOSFET. The EJ-MOSFET has two gates: a lower gate and an upper gate. These gates are insulated from each other with an intergate-oxide layer. The upper gate electrically induces the inversion layers that are self-aligned to the lower gate, and the lower gate controls the current between the inversion layers. Since the structure comprising the lower gate and the two inversion layers is electrically equivalent to a MOSFET, we hereafter refer to the inversion layers as "source/drain junctions." Figure 3.8 shows the calculated depth distribution of the carrier concentration in the source/drain junctions. The distribution is obtained by numerically coupling a Shrödinger equation and a Poisson equation. The junction depth, x_j, is extremely shallow (4.5 nm) and can cover the sub-10-nm generation according to ITRS.

Although the EJ-MOSFET is designed to emulate conventional MOSFET operations, there are several differences in operation between EJ-MOSFETs and conventional MOSFETs. First, the carrier distribution in the junctions is affected by

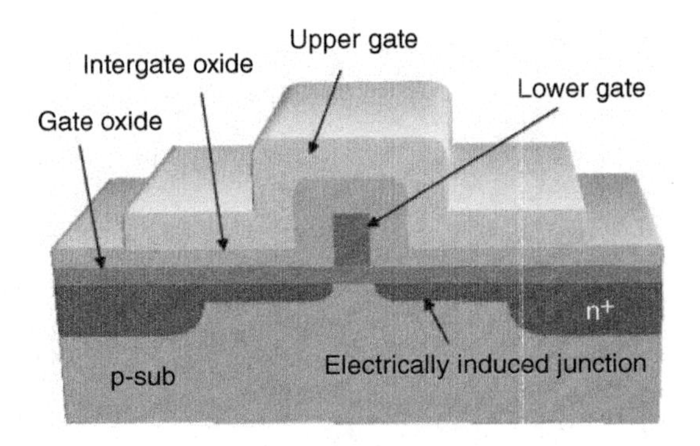

FIGURE 3.7 Schematic illustration of an EJ-MOSFET.

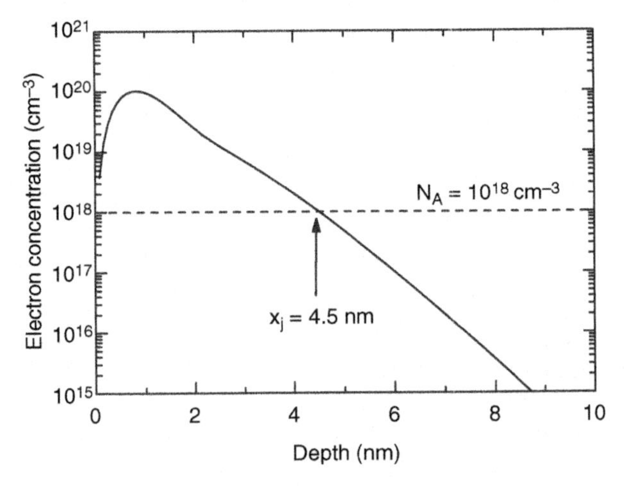

FIGURE 3.8 Calculated depth distribution of the carrier concentration in the source/drain junctions (From Ref. 10). An upper-gate voltage of 18 V, a substrate dopant concentration (N_A) of 10^{18} cm^{-3}, a gate oxide thickness of 2 nm, and an integrate-oxide thickness of 20 nm are assumed. The junction depth (x_j) is as small as 4.5 nm.

lower-gate and drain biases because source/drain regions are not actually doped but electrically induced. In order to reduce these effects, the upper-gate bias should be as high as possible. Second, the EJ-MOSFET has a large coupling capacitance between the lower and upper gates due to the thin intergate oxide layer (~20 nm). The large capacitance limits the high-speed AC operation of the EJ-MOSFETs. Hence, the device should be operated under DC-bias conditions.

3.2.2 Fabrication of the Device Structure

Because direct source-drain tunneling appears below the 10-nm generation, nanometer-scale devices are needed to observe it. Recent progress in electron-beam lithography has enabled the formation of 10-nm patterns. J. Fujita and colleagues have

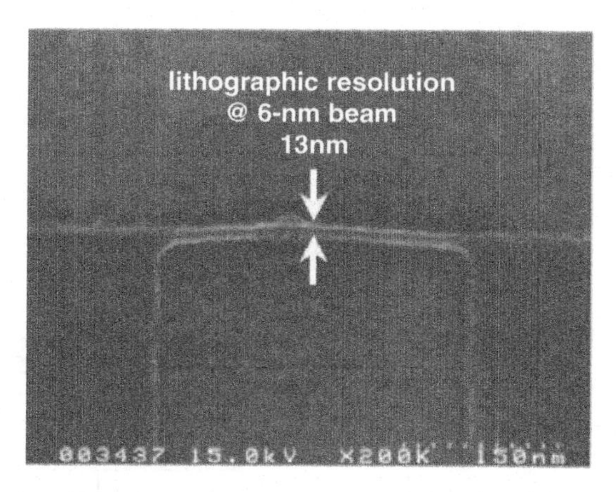

FIGURE 3.9 Scanning-electron-microscope image of a calixarene pattern formed by electron-beam lithography with a beam diameter of 6 nm. The total lithographic resolution is 13 nm.

found that calixarene works as a negative-type electron-beam resist, and it has a surprisingly high resolution (10 nm).[11] Figure 3.9 shows a scanning-electron-microscope (SEM) image of a calixarene pattern formed by an electron-beam writing system with a beam diameter of 6 nm. In this case, the total lithographic resolution is 13 nm. The fabrication sequence of EJ-MOSFETs[12] is shown in Figure 3.10. It is fully compatible with the fabrication sequence of conventional complementary MOS (CMOS) devices except for the use of calixarene resists for lower-gate definition.

Figure 3.11 shows a schematic illustration of an EJ-MOSFET. The polySilicon lower gate consists of three sections: two long sections on each side of the gate and a short section in the center.[13] The long sections help prevent the twisting of resist patterns of the short section during the development of the resist. For the same reason, the width of the short section is limited to 200 nm. The upper gate completely covers the short section and partially overlaps the long sections. The subthreshold current is determined by the short section because the short section has a lower threshold voltage, V_{th}, than the long regions due to the short-channel effects. Because the discussion in this chapter is focused on the subthreshold region, we hereafter refer to the length of the short section as the lower–gate length, L_{LG}. It is worth pointing out that this type of device is not suitable for studying strong inversion characteristics because the highly resistive source/drain regions mask high-current characteristics.

Figure 3.12 is a cross-sectional transmission-electron-microscope (TEM) image of the fabricated EJ-MOSFET.[14] As a result of accurate pattern transfer into the polysilicon film by reactive-ion etching (RIE), the lower-gate length (14 nm) was approximately equal to the lithographic resolution (13 nm). However, the length was not sufficiently short to detect direct source-drain tunneling, which appears at the scale below 10 nm. In order to further shorten the lower gate, side etching was used after the main etching. Figure 3.13 shows an SEM image of a lower gate fabricated using the side-etching technique.[13] The lower gate was uniformly slimmed and its length was 8 nm.

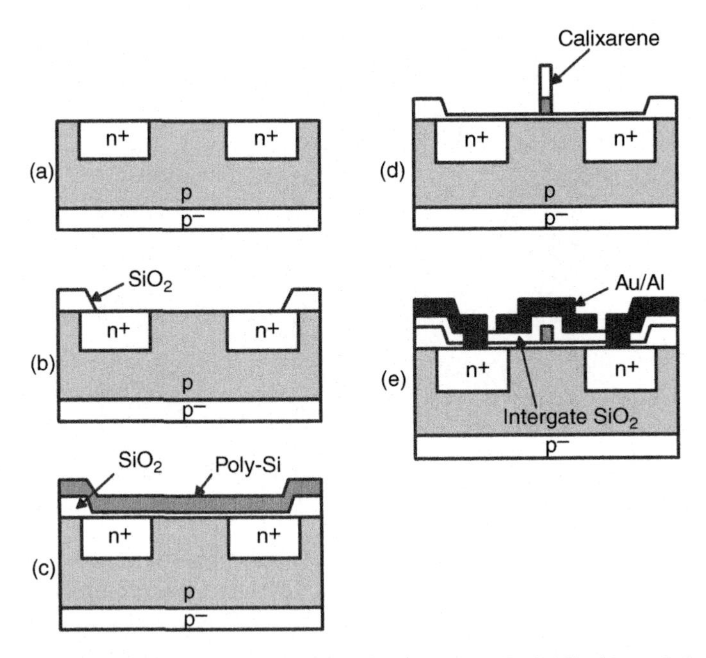

FIGURE 3.10 Fabrication sequence of EJ-MOSFETs (From Ref. 12). (a) p-tab formation by boron doping and n⁺-region formation by arsenic-ion implantation. (b) Formation of a field region. (c) Gate oxide (5 nm) and polySi (40 nm) growth followed by phosphorus doping into the polySi. (d) Patterning of calixarene by electron-beam lithography and polySi etching by RIE. (e) Intergate-oxide growth, contact-window opening and simultaneous formation of the upper gate and electrodes.

3.2.3 Basic Operation

Figure 3.14 shows the current-voltage characteristics of EJ-MOSFETs at the upper-gate voltage V_{UG} of 13 V and room temperature. When the lower-gate voltage, V_{LG}, is higher than 1 V, a parasitic channel is formed below the long sections of the lower gate, increasing the drain current, I_D. At $V_{LG} < 1$ V, I_D is determined by the short section and behaves as in conventional MOSFETs. As L_{LG} decreases, I_D increases through a decrease in V_{th}, and the slope in the saturation current increases. These are caused by the aggravation of the short-channel effects.

Figure 3.15 shows the subthreshold characteristics at room temperature for various L_{LG}s. The drain voltage V_D is set at 0.5 V. As pointed out, an increase in the current at $V_{LG} \sim 1$ V is attributed to the turning on of the parasitic channels. As in the conventional MOSFETs, the current for $V_{LG} < 1$ V exponentially depends on V_{LG} for all L_{LG}s. As L_{LG} decreases, so does V_{th}, and the S-factor increases due to the aggravation of the short-channel effects. Nevertheless, the ON/OFF current ratio exceeds 10^6 even at $L_{LG} = 8$ nm. This indicates that EJ-MOSFETs are suitable for studying direct source-drain tunneling. It is worth pointing out that this device operation is limited to the subthreshold region because of the high parasitic resistance in the source/drain junctions. Nevertheless, we can investigate direct source-drain tunneling because the tunneling effect is detectable even in the low-current region.

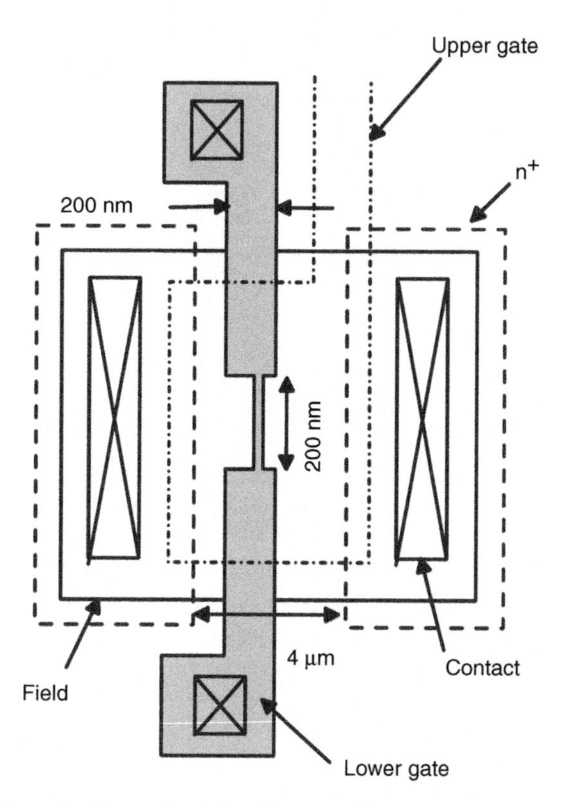

FIGURE 3.11 Schematic illustration of an EJ-MOSFET (From Ref. 13).

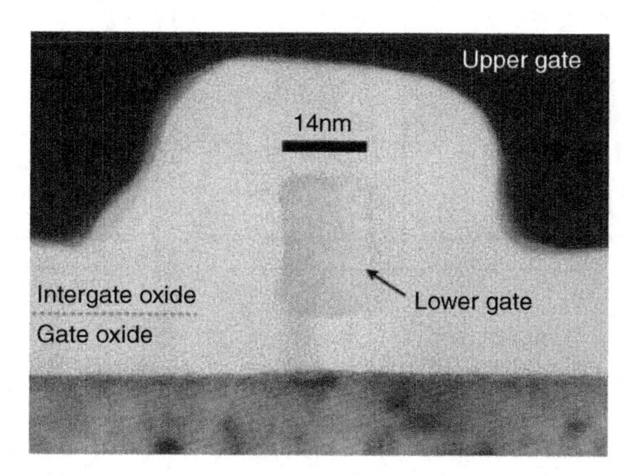

FIGURE 3.12 Cross-sectional TEM image of the central region of an EJ-MOSFET (From Ref. 14, © 2000 IEEE).

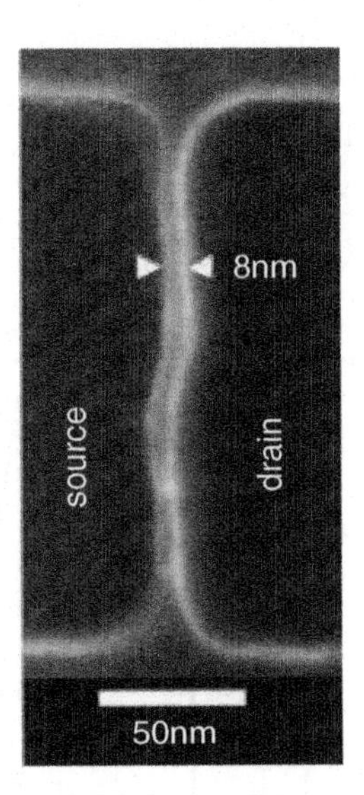

FIGURE 3.13 SEM image of a lower gate fabricated by side etching (From Ref. 13). The lower-gate length is 8 nm. The slight bending in the shape is not intrinsic but was caused by the drift of the SEM image during exposure. All electrical measurements for $L_{LG} = 8$ nm were performed using this device. After the measurements, the upper gate and the intergate oxide were removed by wet etching, and the SEM image was obtained.

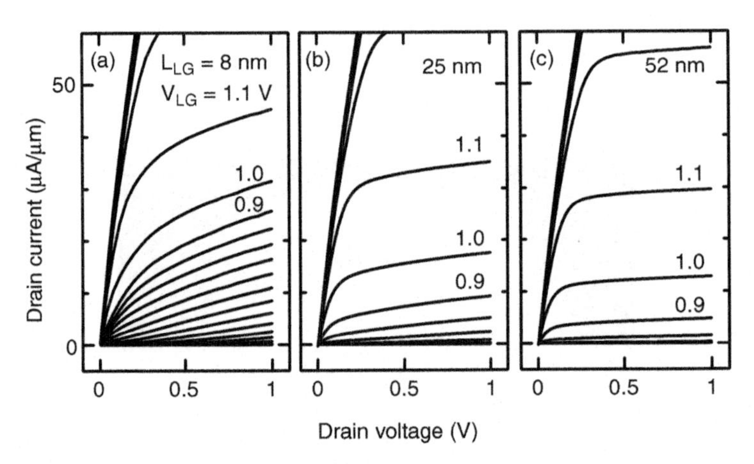

FIGURE 3.14 Current-voltage characteristics of EJ-MOSFETs at room temperature for (a) $L_{LG} = 8$ nm, (b) $L_{LG} = 25$ nm and (c) $L_{LG} = 52$ nm. The upper-gate voltage is 13 V. The lower-gate voltage (V_{LG}) is applied from 0 to 1.5 V in steps of 0.1 V. The currents are normalized by the width of the short section in the lower gate (i.e. 200 nm).

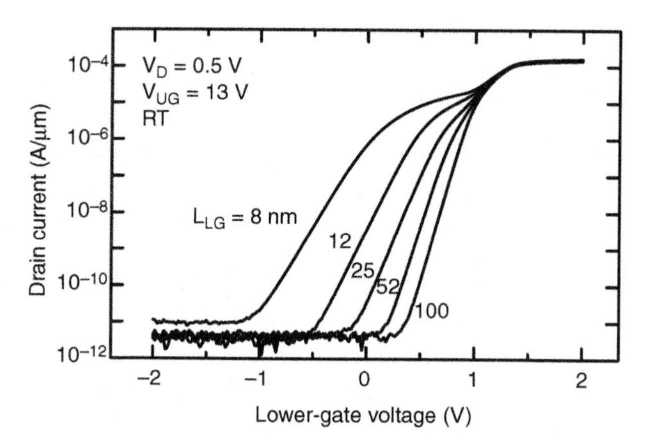

FIGURE 3.15 Subthreshold characteristics for various L_{LG}s at room temperature. The upper gate (V_{UG}) and the drain voltage (V_D) are fixed at 13 V and 0.5 V, respectively. The currents are normalized by the width of the short section in the lower gate (i.e. 200 nm).

3.3 DIRECT SOURCE-DRAIN TUNNELING

3.3.1 Detection of the Tunneling Current

Figure 3.16 shows the temperature dependence of the subthreshold current for various lower-gate lengths.[7] At $L_{LG} = 52$ nm, the current decreases with decreasing temperature (d). As the lower-gate length decreases, the current at 25 K approaches that at 50 K as shown in (b) and (c). When L_{LG} is 8 nm, the current at 25 is almost the same as at 50 K (a). Thus, as L_{LG} decreases, the temperature dependence of the subthreshold current weakens.

In order to make this point clearer, the temperature dependence of the S-factor is plotted for various lower-gate lengths in Figure 3.17. The S-factor behaves differently in the high- and low-temperature regions. In the high-temperature region, the S-factor is proportional to the temperature for all the lower-gate lengths. This is clearly seen at temperatures above 100 K. In the low-temperature region, in contrast, the S-factor tends to remain constant against the temperature. This behavior can be more clearly observed at shorter L_{LG}. Here we define T_C as the temperature at the cross-point of two curves in Figure 3.17. It roughly corresponds to the cross-over temperature between the low- and high-temperature regions.

In the high-temperature region ($T > T_C$), the subthreshold current strongly depends on the temperature. This implies that in the high-temperature region, electrical transport is based on thermal processes. There are two types of thermal conduction: thermal diffusion and thermionic emission for long and short gates, respectively. For both types of conduction, the S-factor is given by the same expression

$$S = \frac{kT}{q} \ln 10 \cdot \left(1 + \frac{C_D}{C_i} \right) \tag{3.2}$$

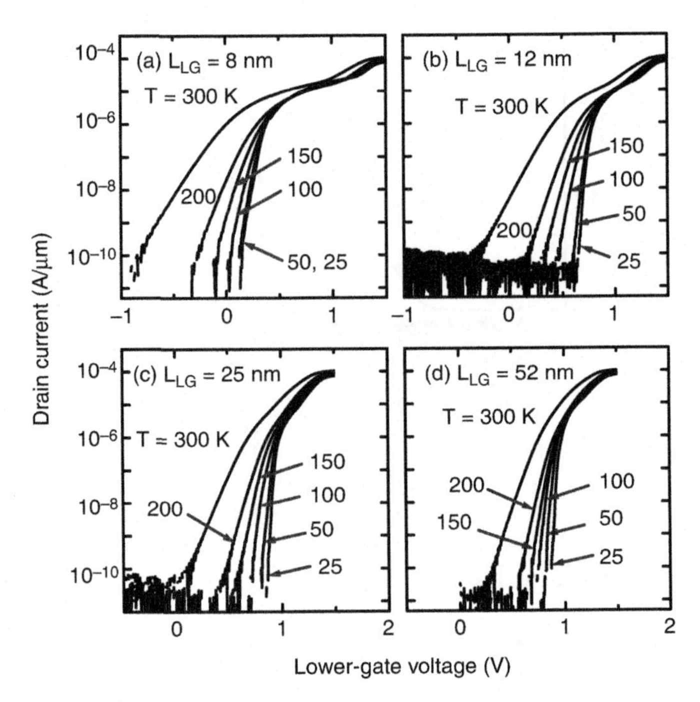

FIGURE 3.16 Temperature dependence of subthreshold currents at V_{UG} = 13 V and V_D = 0.5 V. The lower-gate length is (a) 8, (b) 12, (c) 25, and (d) 52 nm (From Ref. 7). The currents are normalized by the width of the short section in the lower gate (i.e. 200 nm).

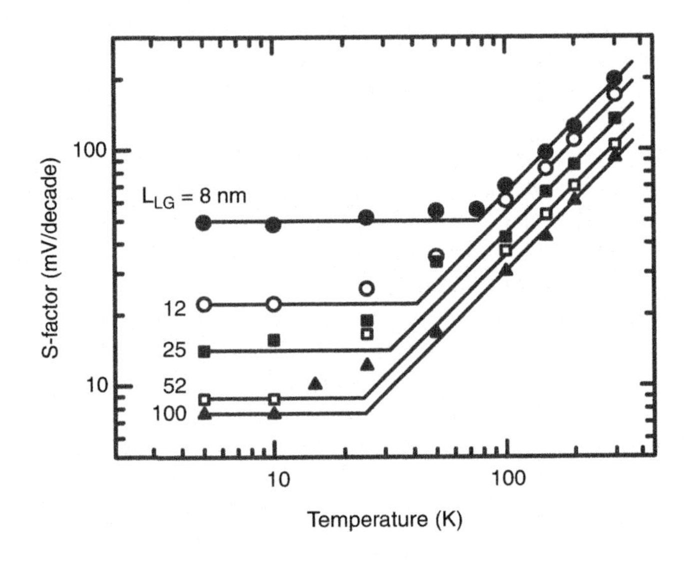

FIGURE 3.17 Temperature dependence of the subthreshold swing (S-factor) for various lower-gate lengths (From Ref. 13). The solid lines indicate extrapolations from the high- and low-temperature region; S is proportional to T at high T, and S is constant at low T.

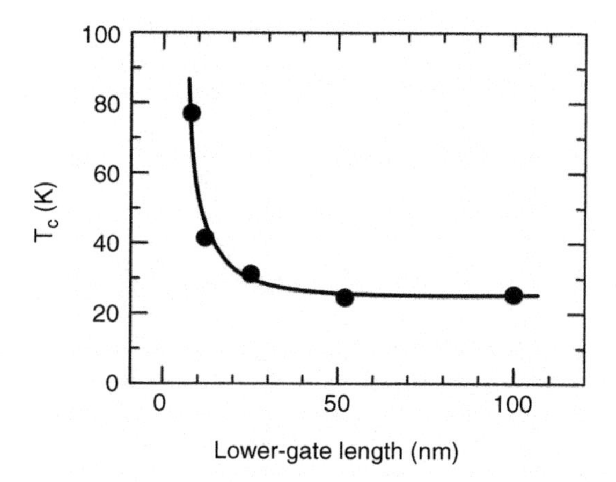

FIGURE 3.18 Dependence of T_C on the lower-gate length.

where C_D and C_i are the capacitances of the depletion layer and the gate dielectric, respectively.[15] As shown in Equation (3.2), the subthreshold current is proportional to T, and its behavior is consistent with that observed at $T > 100$ K.

When the temperature decreases, the Fermi level approaches the valence band edge, and the interface state density at the Fermi level increases. When the effect of the interface states is large, Equation (3.2) should be changed to

$$S = \frac{kT}{q} \ln 10 \cdot \left(1 + \frac{C_D + C_{it}}{C_i}\right) \tag{3.3}$$

where C_{it} is the capacitance associated with the interface state. Since the interface state density at the band edge is higher than in the center of the band gap, Equation (3.3) gives a more gradual temperature dependence than does Equation (3.2). This behavior can be typically seen at 25 nm $< L_{LG} <$ 100 nm and 25K $< T <$ 100 K.

When the temperature is low, the S-factor tends to remain constant against the temperature. As the lower-gate length decreases, the temperature range at which the S-factor remains constant extends towards higher temperatures. As a result, the T_C increases with the decreasing lower-gate length. Figure 3.18 shows the dependence of the T_C on the lower-gate length. When the lower-gate length decreases below 15 nm, the T_C significantly increases and reaches the liquid-nitrogen temperature (77 K) at a lower-gate length of 8 nm. This behavior can be attributed to direct source-drain tunneling as discussed below.

Figure 3.19 shows three possible types of electrical conduction between the source and the drain: thermal, thermally assisted source-drain tunneling and direct source-drain tunneling. In the thermal process, carriers are thermally excited in the source, and they go over the potential barrier beneath the gate. The thermal process, therefore, has a strong temperature dependence. In the thermally assisted source-drain tunneling process, carriers are thermally excited in the source, and then they tunnel slightly beneath the peak of the potential barrier. While this type of conduction

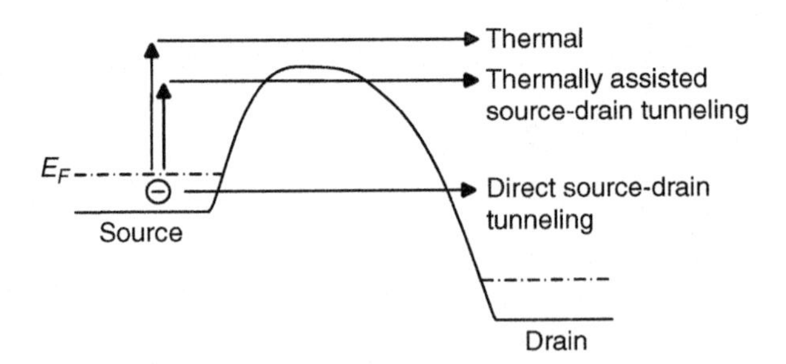

FIGURE 3.19 Three types of conduction between the source and the drain: thermal, thermally assisted source-drain tunneling and direct source-drain tunneling (From Ref. 7).

belongs to the tunneling process, it has a strong temperature dependence because it accompanies the thermal excitation in the source prior to tunneling. In contrast, direct source-drain tunneling does not need any thermal excitation, and, therefore, it has a weak dependence on temperature. The tunneling current (I_{tunnel}), which is the sum of the thermally assisted source-drain tunneling current and the direct source-drain tunneling current, can be written as

$$I_{tunnel} = \int_{0}^{\infty} qv(E)\left[f_S(E,T) - f_D(E,T)\right]n(E)P(E)dE \tag{3.4}$$

where $v(E)$, $f_S(E,T)$, $f_D(E,T)$, $n(E)$ and $P(E)$ are, respectively, the velocity along the channel, the Fermi-Dirac distribution function in the source and the drain, the density of states, and the transmission probability.[16] If the degenerate energy in the source (ΔE_{deg}) is large compared to the thermal energy, $3/2kT$, and the transmission probability is high enough to allow direct source-drain tunneling, Equation (3.4) can be rewritten as a temperature-independent expression

$$I_{tunnel} \approx \int_{E_C}^{E_C+\Delta E_{deg}} qv(E)n(E)P(E)dE \tag{3.5}$$

where E_C is the energy of the conduction band edge in the source. In Equation (3.5), $f_D(E,T)$ is ignored because the drain voltage V_D (= 0.5 V) is sufficiently high compared to $3/2kT$. Since the degenerate energy in actual devices is 68 meV, $\Delta E_{deg} > 3/2kT$ (= 6.7 meV) is satisfied at 77 K. Therefore it can be concluded from Equation (3.5) that the weak temperature dependence of the subthreshold currents observed at low temperatures originates in the direct source-drain tunneling.

3.3.2 Numerical Study of the Tunneling Current

The tunneling current was analyzed numerically. The calculation procedure is as follows. First, the two-dimensional potential distribution within the EJ-MOSFET was calculated using a commercially available device simulator (Athena and Atlas[17]).

Then the one-dimensional surface potential distribution between the source and the drain was approximated with a step function, and the electron wave function, φ_j, in the j-th segment was assumed to be a linear combination of two plain waves

$$\varphi_j = A_j \exp(ik_j x) + B_j \exp(-ik_j x) \tag{3.6}$$

where

$$k_j = \sqrt{2m^*(E_x - U_j)/\hbar^2} \tag{3.7}$$

where E_x, U_j, and m^* are, respectively, the electron energy for motion along the channel, the potential energy of the j-th segment, and the electron effective mass.[18] The global wave function spreading between the source and the drain can be obtained by smoothly connecting each plain wave at the segment edge. Using the obtained global wave function, the transmission probability $T(E_x)$ can be calculated, and then the drain current, $I(E_x)dE_x$, is obtained from the following expression

$$I(E_x)dE_x = \frac{qm^*}{2\pi^2\hbar^3} P(E_x) \int_{E_x}^{\infty} \left[f_S(E,T) - f_D(E,T) \right] dE dE_x \tag{3.8}$$

Then the thermal current, $I_{thermal}$, and the tunneling current, I_{tunnel}, can be obtained as

$$I_{thermal} = \int_{\phi}^{\infty} I(E_x) dE_x \tag{3.9}$$

$$I_{tunnel} = \int_{0}^{\phi} I(E_x) dE_x \tag{3.10}$$

where ϕ is the barrier height seen from the source. In the calculations, dE_x and the inversion layer depth were set at 0.2 meV and 1 nm, respectively. The degenerate energy in the source/drain regions was assumed to be 68 meV which corresponds to an electron concentration of 1×10^{20} cm^{-3} in the source/drain regions. Because the ground state in the electrically induced source/drain regions has a transverse effective mass (m_t) for motion in the (100) plane, the effective mass was set at $m_t = 0.19\ m_0$, where m_0 is the free electron mass. It should be noted that this analysis method is not very precise because the electrical potential is neither treated in a two-dimensional manner nor self-consistently coupled with the carrier concentration. Nevertheless, the model gives semiquantitative information about the direct source-drain tunneling.

Figure 3.20 shows calculated results for the subthreshold currents ($I_{thermal} + I_{tunnel}$) for various temperatures and lower-gate lengths. The calculation successfully reproduces the behavior: the curves at T = 25 and 50 K approach each other as L_{LG} decreases. The thermal ($I_{thermal}$) and tunneling (I_{tunnel}) components are separately

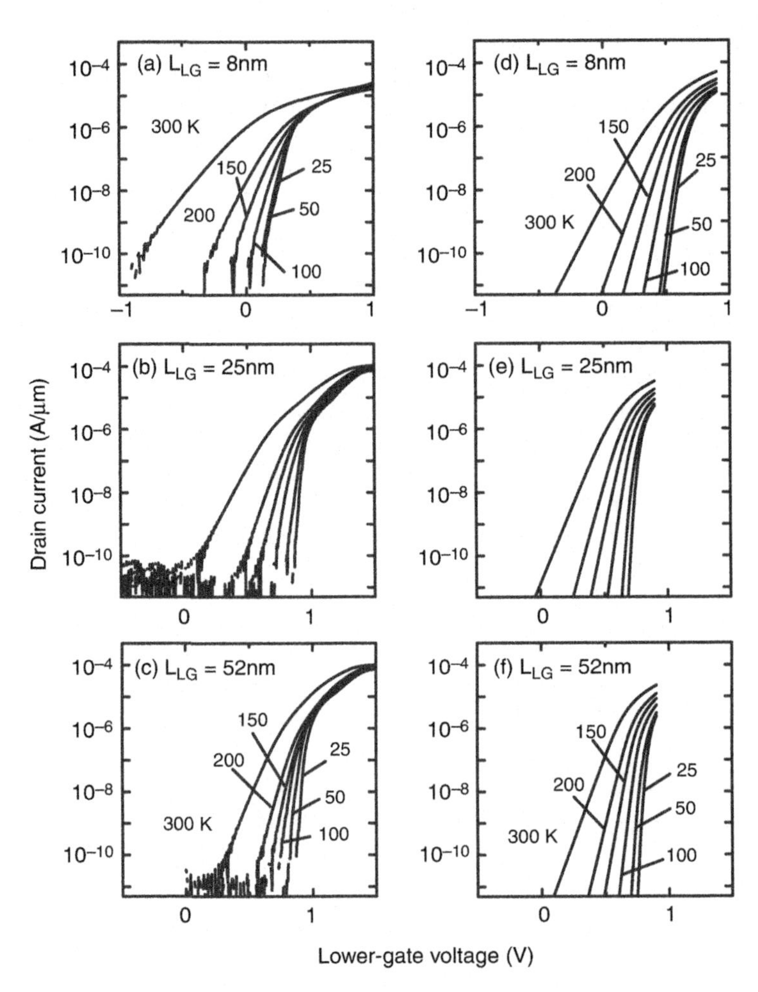

FIGURE 3.20 Calculated subthreshold currents of EJ-MOSFETs at various temperatures ($T = 300$, 200, 150, 100, 50, and 25 K) (From Ref. 7). The lower-gate length is (d) 8, (e) 25, and (f) 52 nm. For comparison, experimental results are re-plotted: The lower-gate length is (a) 8, (b) 25, and (c) 52 nm.

plotted in Figure 3.21. The tunneling components at low temperatures behave differently depending on the L_{LG}. In order to make this point clearer, the temperature dependencies of the threshold voltage for the tunneling component (V_{th}^{tunnel}) and the thermal component ($V_{th}^{thermal}$) are plotted in Figure 3.22, where the threshold voltage is defined as the voltage that gives the current of 1 nA / μm. At high temperatures, V_{th}^{tunnel} is higher than $V_{th}^{thermal}$. As the temperature decreases, V_{th}^{tunnel} approaches $V_{th}^{thermal}$, and it becomes lower than $V_{th}^{thermal}$ at low temperatures. If the crossover temperature (T_C) is defined as a temperature giving $V_{th}^{thermal} = V_{th}^{tunnel}$, T_C can be estimated to be 130, 50, and 25 K at $L_{LG} = 8$, 25, and 52 nm, respectively. As can be seen in Figure 3.22, V_{th}^{tunnel} depends strongly on the temperature in the temperature region above T_C, and its dependence on the temperature weakens at temperatures

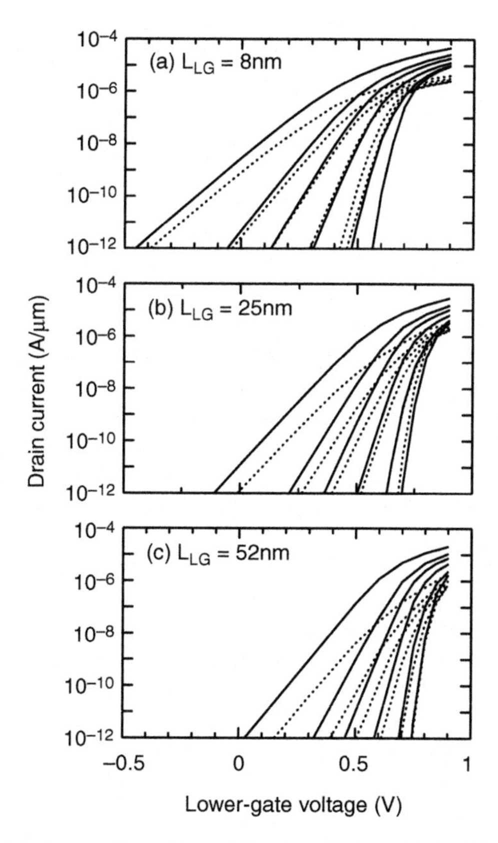

FIGURE 3.21 Calculated tunneling (dotted line) and thermal (solid line) components of currents at various temperatures (T = 300, 200, 150, 100, 50, and 25 K) (From Ref. 7). The lower-gate length is (a) 8, (b) 25, and (c) 52 nm. Each component decreases with decreasing temperature.

below T_C. This indicates that the tunneling process is based on different conduction mechanisms below and above T_C.

The main tunneling process at $T > T_C$ is thermally assisted source-drain tunneling as shown in Figure 3.19. Figure 3.23 shows the calculated tunneling current density as a function of electron energy along the channel for L_{LG} = 8 nm. As can be seen with the curve at T = 300 K, the tunneling current at $T > T_C$ can be attributed to the electrons having a much higher energy than the Fermi energy, E_F. Although this type of conduction belongs to the tunneling process, it has similar temperature- and barrier-height dependencies to the thermal one because it relies on the thermal activation in the source. In addition, whenever the thermally assisted source-drain tunneling occurs, it induces thermal processes. Therefore, the thermally assisted source-drain tunneling current behaves as if it were part of the thermal current, and does not significantly affect the scaling of MOSFETs.

In contrast, the main tunneling process at $T < T_C$ is direct source-drain tunneling. As can be seen with the curve at T = 25 K in Figure 3.23, the tunneling current at

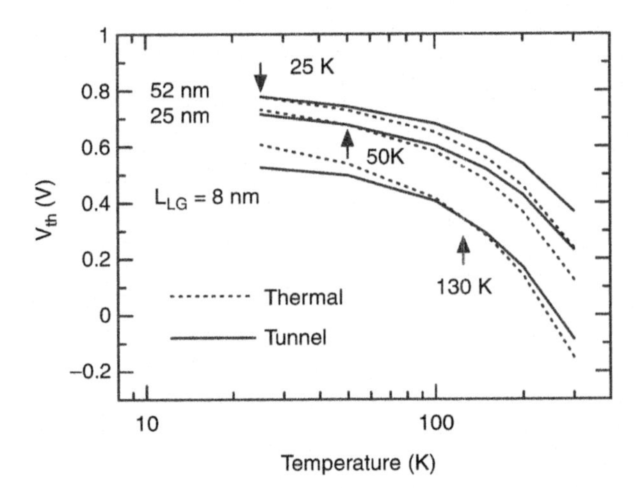

FIGURE 3.22 Calculated threshold voltage of the tunneling (solid line) and thermal (dotted line) components as a function of temperature for L_{LG} = 8, 25, and 52 nm (From Ref. 6). Crossover temperatures (T_Cs) are indicated by the arrows: T_C = 130, 50, and 25 K for L_{LG} = 8, 25, and 52 nm, respectively.

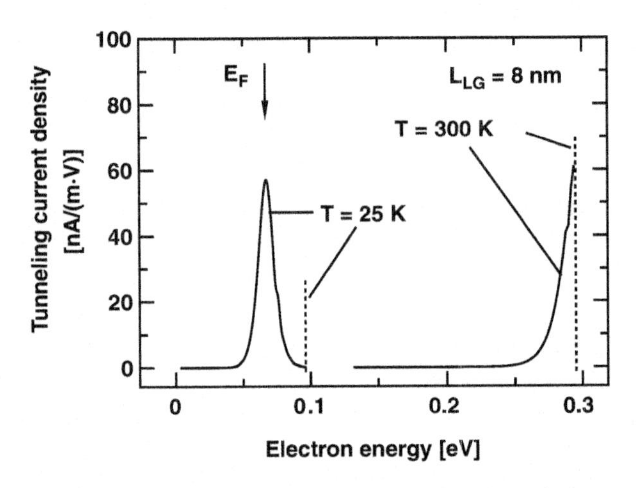

FIGURE 3.23 Calculated tunneling current density as a function of electron energy along the channel at T =25 and 300 K for L_{LG} = 8 nm (From Ref. 7). Lower-gate biases giving a total current of around 1 nA were chosen: V_{LG} =0 and 0.55 V at T = 300 and 25 K, respectively. The origin of the horizontal axis corresponds to the conduction band edge in the source. The peak energies of the potential barrier beneath the lower gate are shown by the dotted lines. The Fermi energy (E_F) in the source is shown by the arrow.

$T < T_C$ is caused by electrons with an energy around E_F. Since the tunneling probability increases exponentially with the decreasing barrier width, the decrease in L_{LG} significantly enhances direct source-drain tunneling and weakens the temperature dependence of the subthreshold current.

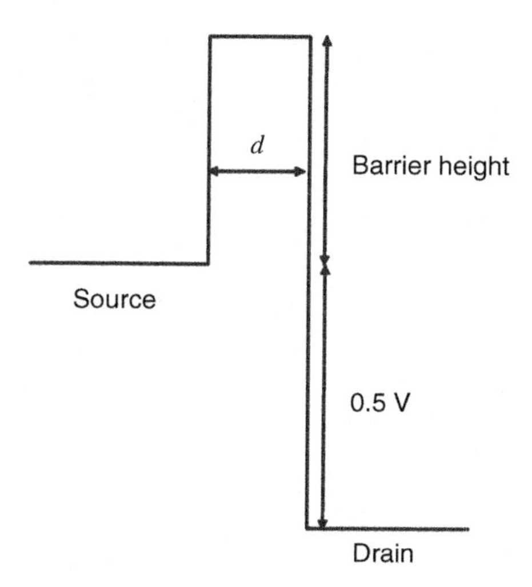

FIGURE 3.24 Schematic of a rectangular-shaped surface potential distribution between the source and the drain with a barrier width, *d*. The drain voltage is set at 0.5 V.

3.4 THE SCALING LIMIT OF MOSFETS

3.4.1 Estimation of Direct Source-Drain Tunneling in MOSFETs

As shown in the previous section, when the gate length decreases, direct source-drain tunneling controls the subthreshold characteristics, and this results in a large S-factor. A large S-factor increases the high power consumption and limits the scaling of MOSFETs. In this section, the scaling limit of MOSFETs is described, which is estimated by calculation. In the calculation, the surface channel potential distribution between the source and the drain is simplified to a rectangular-shaped potential distribution as shown in Figure 3.24. Although the actual potential distribution in MOSFETs is not like that, the simplified potential distribution is used because the actual potential distribution below deep 10-nm generation is still unknown, and the rectangular-shaped distribution has a well-defined barrier width, *d*. Hence, the calculated results give us a rough idea about the characteristics of electrical transport as a function of *d*, which corresponds to the effective channel length rather than the physical gate length of MOSFETs.

The calculation procedure is the same as discussed in the previous section. Because the light effective mass of carriers significantly affects the tunneling process, the effective mass ratios for the electrons and holes, which are isotropic in the (100) plane, were set at 0.19 and 0.16, respectively. For simplicity, all the electrons and holes in the source were assumed to have a light effective mass.

Figure 3.25 shows the barrier-height dependence of the tunneling and thermal components for various barrier widths. The left and right figures correspond to the electron and hole currents, respectively. Since the barrier height is controllable by

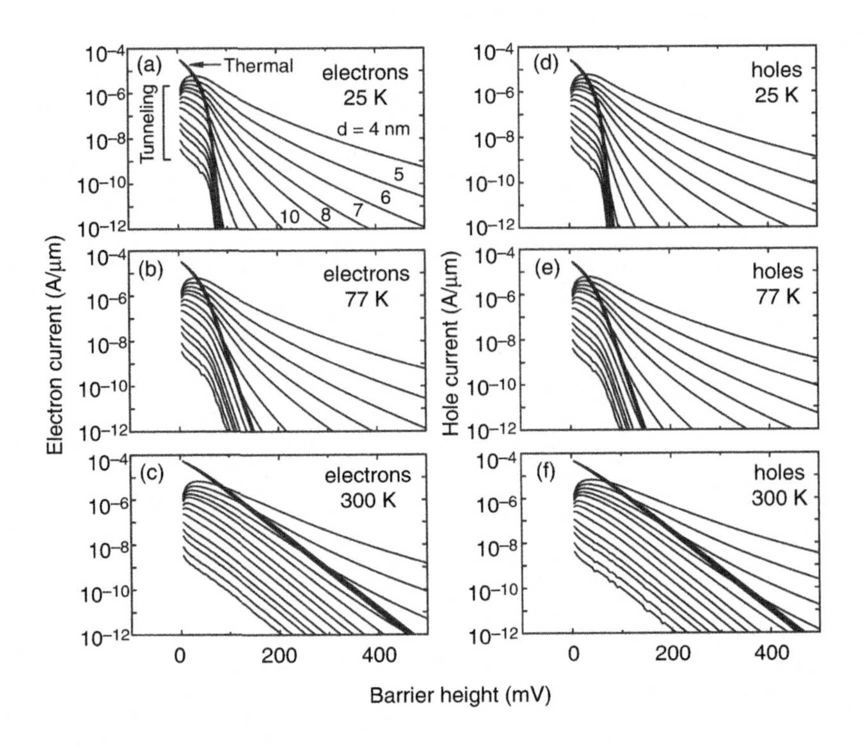

FIGURE 3.25 Calculated tunneling and thermal components as a function of barrier height for the rectangular-shaped channel potential distribution (From Ref. 7). Effective mass ratios of 0.19 and 0.16 were used to calculate the electron currents [(a), (b), and (c)] and the hole currents [(d), (e), and (f)], respectively. The currents were calculated for various barrier widths: $d = 4, 5, 6, 7, 8, 10, 12, 15, 20, 25, 30, 40$, and 50 nm. Since the thermal component is almost independent of d, all the thermal components overlap one another.

the gate voltage, Figure 3.25 can be considered as representing the subthreshold characteristics: (a)-(c) and (d)-(f) correspond to n- and p-MOSFETs, respectively. As shown in Figure 3.25, there is little difference between the electron and hole tunneling currents because the electron and hole light effective masses are approximately the same. Since no scattering is incorporated in this model, that is, ballistic transport is assumed, all the thermal currents are almost identical. In contrast, the tunneling current strongly depends on the barrier width. If we define the crossover barrier width, d_C, as a barrier width at which the tunneling and thermal components are the same (1 nA / μm), d_Cs are estimated as in Table 3.1. Roughly speaking, the d_Cs at $T = 25$, 77, and 300 K are 18, 12, and 6 nm, respectively. When the barrier width is smaller than d_C,, direct source-drain tunneling is dominant and, consequently, leads to a large S-factor. Since a large S-factor is detrimental to low-power operation of digital LSIs, direct source-drain tunneling limits the scaling of MOSFETs. Roughly speaking, this scaling limit is 12 nm at $T = 77$ K and 6 nm at $T = 300$ K.

TABLE 3.1
Crossover Barrier Width (d_C) for Electrons and Holes

T(K)	d_c (nm) Electrons	Holes
25	17.3	18.5
77	11.5	12.1
300	6.0	6.5

3.4.2 Future Trends in Post-6-nm MOSFETs

Until now, the scaling limit of MOSFETs has been studied on the assumption that devices are formed on the surface of the (100)-oriented silicon substrates. Now we would like to discuss the feasibility of further scaling of MOSFETs by eliminating that assumption.

There are two strategies for improving the performance of post-6-nm devices. One is by enhancing the carrier mobility by suppressing the carrier scattering, and the other is by suppressing the tunneling probability by increasing the effective mass of carriers. The first one makes it possible to improve the device performance without scaling down the device itself. The second one enables further scaling of the devices by reducing direct source-drain tunneling. However, the second strategy is not as effective as the first one because an increase in the effective mass reduces the carrier mobility. Hence, only the first strategy is discussed here.

Strained silicon technology is a promising technology for reducing carrier scattering.[19–21] When the silicon is epitaxially grown on a Si_xGe_{1-x} substrate, tensile strain is introduced, and it gives rise to a valley splitting in the conduction and valence bands. As a result, the carrier concentration in the lower energy states (the lower-energy valley in the conduction band and the lower-energy valence band) reduces the carrier scattering and, consequently, enhances the carrier mobility. An enhancement of several tens of percentage points has been demonstrated experimentally. However, since the lower-energy states have a smaller effective mass than the higher-energy states, the carrier concentration in the lower-energy states leads to an increase in the direct source-drain tunneling currents to the tune of several tens of percentage points. To avoid this problem, the gate should be made slightly longer because the tunneling current depends drastically on the barrier width. As shown in Figure 3.25, an increase in the tunneling current of several tens of percentage points can be easily avoided by increasing the barrier width by as little as 0.5 nm, which does not significantly reduce the drain current because the drain current linearly (or more weakly in the ballistic transport regime) depends on the barrier width.

The quantum-wire transistor is another candidate for suppressing carrier scattering.[22] Because carrier scattering is limited to the one-dimensional direction, the scattering rate can be significantly reduced. However, an enhancement in mobility has not been demonstrated yet. This is probably because the fabrication method of quantum wires has not been fully developed. The development of damage-free fabrication techniques for uniform wire structures may improve the carrier mobility.

3.5 CONCLUSION

In order to continue the scaling of MOSFETs, a number of problems have to be resolved. Direct source-drain tunneling is one of the most serious problems because it is not only harmful to low-power operation but also difficult to avoid in ultimately scaled MOSFETs. The tunneling effects were experimentally studied using EJ-MOSFETs fabricated by nanometer-scale lithography. As the gate length decreased, the direct source-drain tunneling current grew and became detectable at higher temperatures. At a gate length of 8 nm, the temperature at which the current was detectable reached 77 K. Based on the results of a one-dimensional simulation incorporating the tunneling effects, it was shown that at an effective channel length (not a physical gate length) below 6 nm, direct source-drain tunneling would appear and degrade the subthreshold swing. Because a large subthreshold swing is detrimental to low-power operation, it is predicted that the scaling limit of silicon digital LSIs will be around 6 nm.

If, instead of the (100)-oriented bulk silicon substrate, a different substrate is used, the device performance can be further improved. Strained silicon MOSFETs and quantum-wire MOSFETs are promising candidates for high performance MOSFETs in the post-6-nm generation.

ACKNOWLEDGMENTS

I would like to acknowledge many fruitful discussions with T. Sakamoto, T. Baba, Y. Ochiai, J. Fujita, S. Matsui and J. Sone. The assistance of N. Iguchi in the device fabrication is gratefully acknowledged.

REFERENCES

1. International Technology Roadmap For Semiconductor, 2001, retrieved from http://public.itrs.net
2. Yu, B., Wang, H., Joshi, A., Xiang, Q., Ibok, E. and Lin, M., 15 nm Gate Length Planar CMOS Transistor, IEEE International Electron Devices Meeting,Washington, DC, Dec 2–5, 2001.
3. Boeuf, F., Skotnicki, T., Monfray, S., Julien, C., Dutartre, D., Martins, J., Mazoyer, P., Palla, R., Travel, B., Ribot, P., Sondergard, E. and Sanquer, M., 16 nm planar NNOSFET Manufacturable Within State-of-the-Art CMOS Process Thanks To Specific Design And Optimization, IEEE International Electron Devices Meeting, Washington, DC, Dec 2–5, 2001.
4. Hokazono, A., Ohuchi, K., Takayanagi, M., Watanabe, Y., Magoshi, S., Kato, Y., Shimizu, T., Mori, S., Oguma, H., Sasaki, T., Yoshimura, H., Miyano, K., Yasutake, N., Suto, A., Dachi, K., Fukui, H., Watanabe, T., Tamaoki, N., Toyoshima, Y. and Ishiuchi, H., 14 nm Gate Length CMOSFETs Utilizing Low Thermal Budget Process with PolySiGe and Ni Salicide, IEEE International Electron Devices Meeting, San Francisco, CA, Dec 8–11, 2002.

5. Dennard, R., Gaensslen, F., Yu, H., Rideout, V., Bassous, E. and LeBlanc, A., Design of ion-implanted MOSFET's with very small physical dimensions, *IEEE J. Solid-State Circuits,* SC-9, 256–268, 1974.
6. Iwai, H. and Ohmi, S., Trend of CMOS downsizing and its reliability, 13th Eur. Symp. Reliability Electron Dev., Failure Physics and Analysis, Rimini, Italy, Oct 7–11, 2002.
7. Kawaura, H. and Baba, T.. Direct tunneling from source to drain in nanometer-scale silicon transistors, *Jpn. J. Appl. Phys.*, 42, 351–357, 2003.
8. Landau, L. D. and Lifshitz, E. M., *Quantum Mechanics*, Addison-Wesley, Boston, 1958.
9. Noda, H., Murai, F. and Kimura, S., Threshold Voltage Controlled 0.1-μm MOSFET Utilizing Inversion Layer as Extremely Shallow Source/Drain, International Electron Devices Meeting, Washington, DC, Dec 5–8, 1993.
10. Kawaura, H., Sakamoto, T., Baba, T., Ochiai, Y., Fujita, J., Matsui, S. and Sone, J. Proposal of pseudo source and drain MOSFETs and evaluation for 10-nm gate MOSFETs, *Jpn. J. Appl. Phys.,* 36, 1569–1573, 1997.
11. Fujita, J., Ohnishi, Y., Ochiai, Y., and Matsui, S., Ultrahigh resolution of calixarene negative resist in electron beam lithography, *Appl. Phys. Lett.,* 68, 1297–1299, 1996.
12. Kawaura, H. and Sakamoto, T., Electrical Transport in Nano-Scale Silicon Devices, *IEICE Trans. Electron*, E84-C, 1037–1042, 2001.
13. Kawaura, H., Sakamoto, T., and Baba, T., Observation of source-to-drain direct tunneling current in 8 nm gate electrically variable shallow junction metal-oxide-semiconductor field-effect transistors, *Appl. Phys. Lett.*, 76, 3810–3812, 2000.
14. Kawaura, H., Sakamoto, T., Baba, T., Ochiai, Y., Fujita, J., and Sone, J., Transistor characteristics of 14-nm-gate-length EJ-MOSFET's, *IEEE Trans. Electron Devices*, 47, 856–860, 2000.
15. Sze, S. M., *Physics of Semiconductor Devices. 2nd ed.*, John Wiley, New York, 1981.
16. Chang, L. and Hu, C., MOSFET scaling into the 10 nm regime, *Superlattice and Microstructures,* 28, 351–355, 2000.
17. Silvaco, Description of Athena (process simulator) and Atlas (device simulator), Silvaco International, http://www.silvaco.com, accessed Mar. 13, 2005.
18. Ando, Y. and Itoh, T., Calculation of transmission tunneling current across arbitrary potential barriers, *J. Appl. Phys.,* 61, 1497–1502, 1987.
19. Basu, P. K. and Paul, S. K., Reduced intervalley scattering in strained Si/SixGe1-x quantum wells and enhancement of electron mobility: A model calculation, *J. Appl. Phys.,* 71, 3617–3619, 1992.
20. Takagi, S., Hoyt, J. L., Welser, J. and Gibbsons, J. F., Comparative study of phonon-limited mobility of two-dimensional electrons in strained and unstrained Si metal-oxide-semiconductor field-effect transistors, *J. Appl. Phys.,* 80, 1567–1577, 1996.
21. Oberhuber, R., Zandler, G. and Vogl, P., Subband structure and mobility of two-dimensional holes in strained Si/SiGe MOSFET's, *Phys. Rev. B*, 58, 9941–9948, 1998.
22. Sakaki, H., Scattering suppression and high-mobility-effect of size-quantized electrons in ultrafine semiconductor wire structures. *Jpn. J. Appl. Phys.*, 19, L735–L738, 1980.

4 Quantum Effects in Silicon Nanodevices

Toshiro Hiramoto

4.1 INTRODUCTION

The size of silicon metal oxide semiconductor field-effect-transistors (MOSFETs) in very large scale integrated ciruits (VLSIs) has been miniaturized for over 30 years in order to attain higher performance and higher integration. The gate length of the state-of-the-art complementary MOSFET (CMOS) devices has reached less than 60 nm, and the most advanced MOSFETs in the research level have the gate length less than 10 nm.[1,2] As the horizontal dimensions such as gate length are scaled down, the vertical dimensions such as gate dielectric thickness and depletion layer thickness have been also rapidly scaled down to suppress the short channel effect. Thus, the carriers in advanced MOSFETs are strongly confined vertically, resulting in the quantum confinement effect.

Although the quantum effects and their device applications have been widely studied in the field of compound semiconductors, it is generally recognized in the field of silicon devices that the quantum effects are not favorable effects. This is mainly because the quantum effects are usually very sensitive to device size, and thus, the small distribution of the device size will result in large variation in the device characteristics when the quantum effect dominantly affects the device performances. Moreover, the tunneling current that is caused by the quantum effect rapidly increases when the thickness of gate dielectrics becomes less than 2 nm, which results in the exponential increase in standby power dissipation. It is also well known that the drive current in MOSFETs with very thin gate insulator will not be improved due to the finite thickness of electron inversion layer that is also induced by the quantum effect. Therefore, suppressing these kinds of quantum effects have been the device design guideline of advanced MOSFETs. However, the device size has now reached the nano-scale regime. The quantum effects will certainly take place and affect the device characteristics and they cannot be considered negligible any more. It is very important to positively utilize the quantum effects and take the quantum effects into consideration in the design of silicon nano-scale devices.

In this chapter, quantum effects in silicon nanodevices are described. First, the basic understanding of the band structure and quantum confinement effects in silicon are described. Next, the researches on the observation and analysis of quantum effects in MOSFETs are reviewed. Then, characteristics of ultranarrow-channel MOSFETs that exhibit stronger quantum effects than usual MOSFETs are described. More focus is placed on the positive utilization of the quantum effects that will enhance the device performance and will possibly break the fundamental scaling limit of silicon MOSFETs.

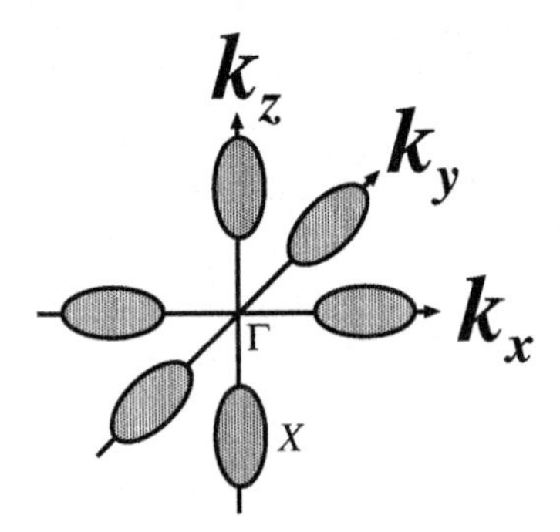

FIGURE 4.1 A schematic of constant energy surfaces of six equivalent X-valleys in a Brillouin zone in silicon.

4.2 QUANTUM EFFECTS IN MOSFETS

4.2.1 Band Structures of Silicon

The lowest points in the conduction band of silicon lie along the Δ-directions in the Brillouin zone. These points are apart from the center of the zone (Γ-point) and called X-points. There are six Δ-directions and, therefore, there are six equivalent minima at X-points in the conduction band of silicon,[3] as shown in Figure 4.1. All the valleys at X-minima are equivalent in terms of energy in bulk silicon. The energy in each valley can be expressed by a parabolic approximation. However, the energy depends on the direction and thus it is anisotropic.

The picture of the valence band is much more complicated than that of the conduction band.[4,5] There are three types of holes at Γ-point in the valence band: the light hole, the heavy hole, and the split-off hole. The light hole and heavy hole are degenerated at the top of the bands, and most of holes are populated in the light and heavy hole hands. The top of the light hole band and the top of heavy hole band are non-parabolic and it is also anisotropic. Figure 4.2 shows a schematic of the light and heavy holes. The energy difference of the light/heavy holes and the split-off hole is 44 meV. The split-off hole band cannot be ignored at room temperature, because the splitting energy is comparable with the thermal energy at room temperature.

4.2.2 Surface Quantization

Figure 4.3 shows a schematic diagram of the conduction band in MOS structure of bulk silicon. The Si-SiO2 interface is inverted by the gate bias and electrons are populated at the interface. When the impurity concentration in the channel is high, which is often the case in advanced MOSFETs, potential slope near the interface is so steep that electrons in the inversion layer are confined by gate oxide and the steep potential slope. The electron energy perpendicular to the interface is now quantized into discrete states (ε_n). Please note that the total energy (E_n) is not quantized because electrons are not quantized in parallel to the interface. Each state (referenced by n)

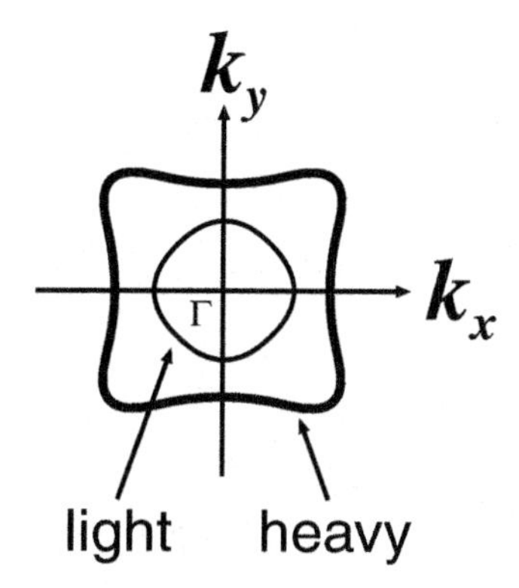

FIGURE 4.2 A schematic of constant energy surfaces of a light hole and a heavy hole in silicon.

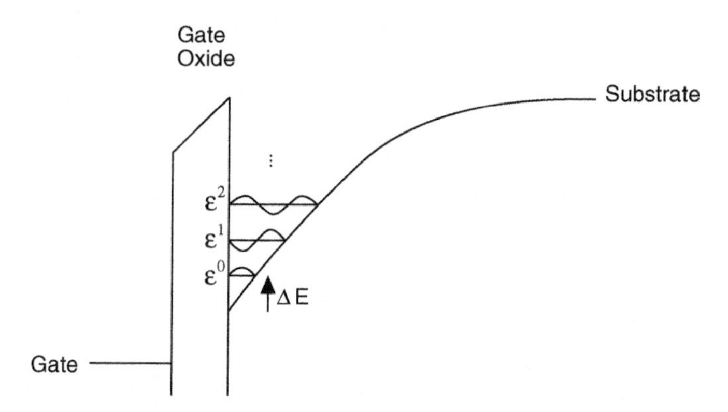

FIGURE 4.3 A schematic of the conduction band in a bulk MOS structure where the gate is positively biased.

forms a subband. The ground state energy rises from the conduction band edge as shown in Figure 4.3. This phenomenon is called surface quantization. Higher gate voltage is required to invert the Si-SiO_2 interface under the surface quantization, which results in the increase in threshold voltage. Therefore, threshold voltage increases as the impurity concentration increases.

The surface quantization also takes place for holes. The fall in the ground state energy in the valence band leads to the decrease in threshold voltage. However, the decrease in threshold voltage corresponds to the increase in threshold voltage in the absolute value in p-type MOSFET. The rise in the ground state in the conduction band and the fall in the ground state in the valence band results in the effective bandgap widening.

The surface quantization is often expressed by a triangular well approximation. Although the triangular well approximation is a simple model, it is highly reliable. This is because the potential near the interface is almost a triangular shape, and the

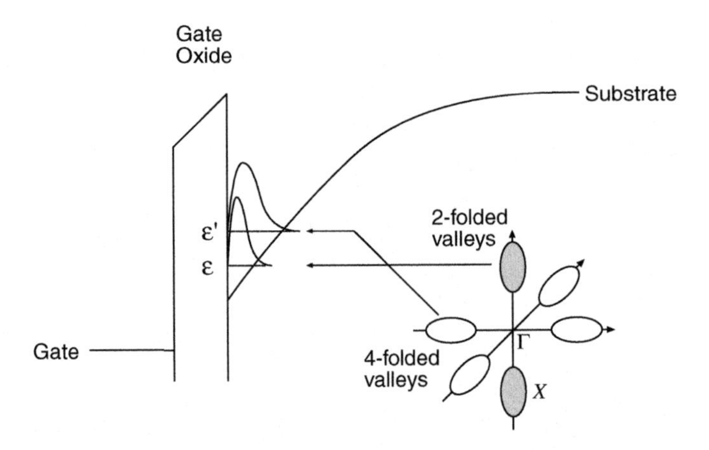

FIGURE 4.4 A schematic of band splitting in the conduction band of a bulk MOS structure.

potential barrier of SiO_2 is relatively high in silicon MOS structure. Some of the models of the surface quantization are based on the triangular well approximation.

The confinement energy depends on the effective mass perpendicular to the Si-SiO_2 interface. Since the six equivalent valleys in the conduction band are anisotropic, the valleys split into two sets of valleys in energy as shown in Figure 4.4. Here, (100) plane is assumed for the Si-SiO_2 interface, which is commonly used in silicon CMOS for VLSI. Now, the valleys are split into twofold degenerate valleys and fourfold degenerate valleys according to the perpendicular masses. Twofold valleys stay at lower energy than fourfold valleys due to higher perpendicular mass. Similarly, energy levels of the light hole band and the heavy hole band in the inversion layer are split.

4.2.3 Carrier Confinement in Thin SOI MOS Structures

Stronger carrier confinement can be achieved in fully depleted (FD) silicon-on-insulator (SOI) structures than in bulk MOS structures. Figure 4.5 shows a schematic of the conduction band in thin-film FD-SOI structure. Electrons are confined in the thin SOI layer that is sandwiched between gate oxide and buried oxide (BOX). The electron energy in the perpendicular direction is quantized and the energy of the ground state rises. As in the case of bulk MOS structure, threshold voltage increases in the ultrathin SOI MOS structure. In this case, threshold voltage depends on both impurity concentration and SOI thickness.

In ultrathin SOI MOS structures, the potential energy can be approximated by a quantum well, because the potential difference between front interface and back interface is small. This is the simplest approximation, but it is often necessary to use perturbation theory for higher accuracy in thicker SOI MOS structure.

4.2.4 Mobility of Confined Carriers

The mobility of the inverted carriers is also affected by the quantum confinement. Carrier mobility is mainly limited by carrier-phonon interactions (phonon scattering)

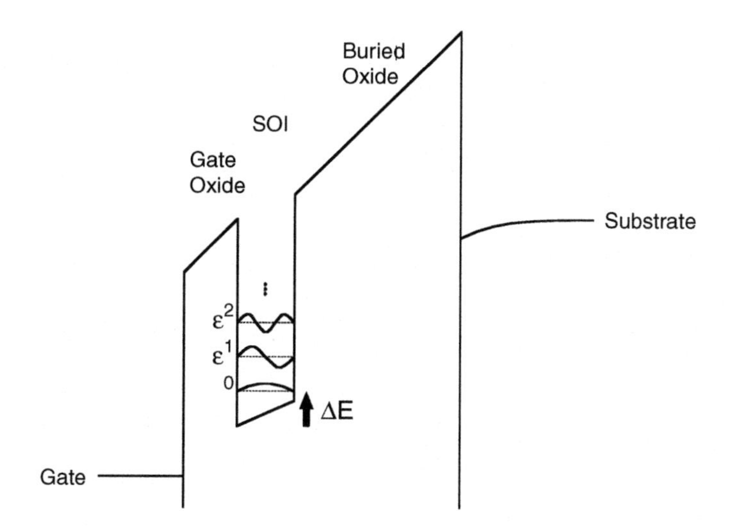

FIGURE 4.5 A schematic of the conduction band in a thin FD-SOI MOS structure.

at room temperature. Carriers have interactions with phonons and are scattered from their initial states to other states. They are scattered to the states within the same valley or to the states in different valleys, which are called the intravalley phonon scattering and the intervalley phonon scattering, respectively. Carrier mobility is determined by all these scattering processes.

4.3 INFLUENCES OF QUANTUM EFFECTS IN MOSFETS

4.3.1 Threshold Voltage Increase in Bulk MOSFETs

In the early 1990s, Ohkura[6] and von Dort et al.[7] investigated threshold voltage in bulk MOSFETs with high impurity concentration. They found that the threshold voltage in MOSFETs with relatively high impurity concentration increases more than the estimation that was calculated by the classical theory based on the Poisson equation. They showed that their experimental data agrees with the calculation which takes quantum confinement effects into account using both the Poisson and the Schrodinger equations, and concluded that an additional threshold voltage increase was due to the quantum confinement in the inversion layer.

Carriers are confined by the gate oxide and steep potential slope at the interface in bulk MOSFETs, as shown in Figure 4.3. Theoretical investigations of the surface quantization have been reported from the simplest triangular potential well approximation to the numerical calculations which solve the Poisson and steady-state Schrodinger equations consistently.[8,9] The results from these calculations indicate that the threshold voltage increase due to the surface quantization will be one of the dominant factors over threshold voltage if the impurity concentration becomes very high. The results have now been included in contemporary device simulators[10,11] because the effect is not negligible any more.

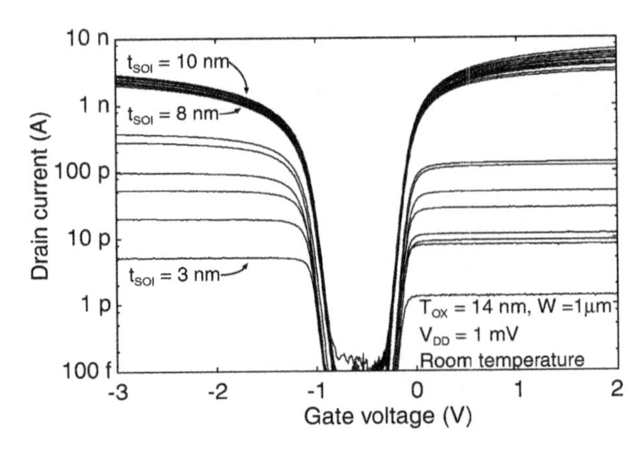

FIGURE 4.6 Gate voltage dependence of drain current in thin FD-SOI MOSFETs with different SOI thicknesses. As SOI becomes thinner, threshold voltage increases. Data of 3-nm-thick devices are largely distributed due to the SOI thickness variations.

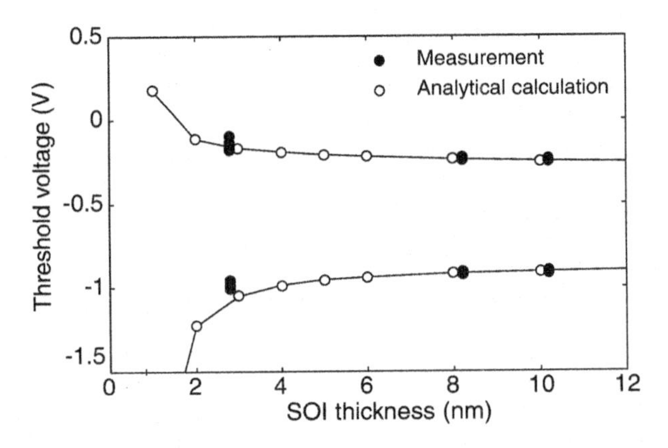

FIGURE 4.7 SOI thickness dependence of threshold voltage in thin FD-SOI MOSFETs. Calculation results are also shown.

4.3.2 Threshold Voltage Increase in FD-SOI MOSFETs

The increase in the threshold voltage by the quantum confinement in ultrathin FD-SOI MOSFETs was first reported by Omura et al.[12] and confirmed by other groups.[13,14] Figure 4.6 shows the drain current in ultrathin FD-SOI MOSFETs as a function of gate voltage.[14] Data of both n-type and p-type MOSFETs are shown. The absolute values of threshold voltage increase in both n-type and p-type MOSFETs as the SOI thickness is thinned. Figure 4.7 shows the relationship between threshold voltage and SOI thickness.[14] The results of analytical calculations that take quantum confinement effects into account are also plotted. The experimental data and calculation results are in agreement, and it is confirmed that the increases in threshold voltage are caused by the quantum confinement.

4.3.3 Mobility in Ultrathin FD-SOI MOSFETs

Takagi et al. and Shoji et al. predicted that the electron mobility enhancement was expected in ultrathin FD-SOI MOSFETs when the SOI thickness is around 3 nm.[15,16] The basic idea was to obtain the mobility enhancement by reducing the intervalley phonon scattering between twofold valleys and fourfold valleys by controlling the subband structure. The intervalley phonon scattering is reduced by separating the energy level of these two kinds of valleys by the carrier confinement in the ultrathin SOI as shown in Figure 4.5. It is well known that the maximum mobility is given by the universal curve in silicon bulk MOSFETs.[17] However, if we take advantage of the quantum mechanical effects, there is a possibility of obtaining higher mobility than the universal mobility. This idea is very attractive because higher drive current can be obtained in MOSFETs with higher mobility even in the very short channel regime.[18]

4.4 Quantum Effects in Ultranarrow Channel MOSFETs

4.4.1 Advantage of Quantum Effects in Ultranarrow Channel MOSFETs

In the previous section, quantum effects in bulk and FD-SOI MOSFETs are described. However, these devices have wide channel width. Therefore, carriers are confined only vertically and the confinement is only one-dimensional. Moreover, the quantum effects depend only on the SOI thickness in the case of FD-SOI MOSFETs. In order to take full advantage of the quantum effects, much stronger quantum effects and more flexibility in device design will be necessary. In this section, quantum effects in ultranarrow channel MOSFETs whose channel width is less than 10 nm are described. In a narrow channel, the carriers are confined not only vertically but also horizontally, and the confinement is two-dimensional, and more confinement effects are expected. The quantum effects depend not only on the SOI thickness but also on channel width and channel direction. Ultranarrow channel MOSFETs are fabricated and their electrical characteristics are measured at room temperature.

4.4.2 Threshold Voltage Increase in n-Type Narrow Channel MOSFETs

Figure 4.8 shows an SEM image of ultranarrow channel.[19] The channel is patterned by using electron-beam (EB) lithography and dry etching. The channel width of this specific device is less than 10 nm. The width is varied from almost 0 to 45 nm. Using this narrow channel, n-type MOSFETs are fabricated. Figure 4.9 shows Ids-Vgs characteristics of all the fabricated devices at room temperature. It is clearly shown that devices with smaller drain current exhibit high threshold voltage.

Figure 4.10 shows the width dependence of threshold voltage in narrow channel MOSFETs. It is very hard to estimate the actual channel width. In this study, the

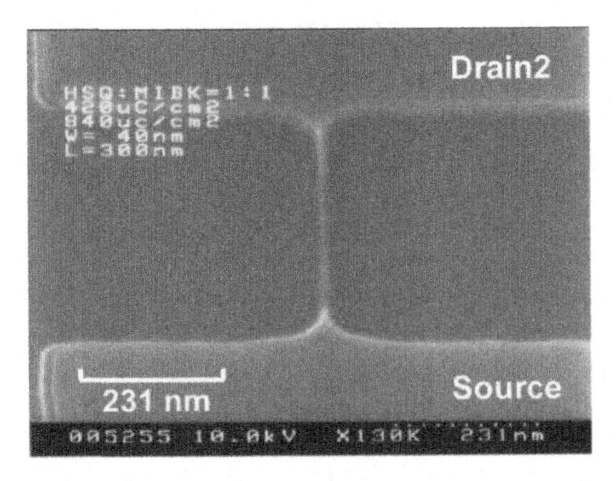

FIGURE 4.8 SEM image of ultra-narrow channel. The width is less than 10 nm in this specific device.

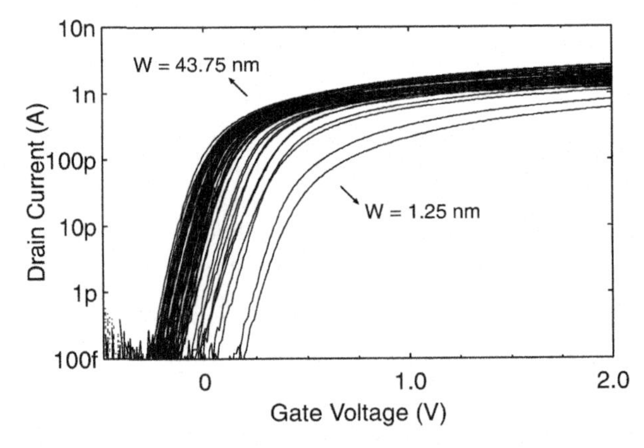

FIGURE 4.9 Ids vs. Vds characteristics of fabricated n-type narrow channel MOSFETs at room temperature. Drain voltage is 1 mV. Devices show good subthreshold swing around 100 mV/dec.

channel width is estimated by the following method.[20] The channel widths of fabricated devices are set by the EB lithography at intervals of 2.5 nm in a wide range. Since some of the channels are too narrow, the channels are cut off during the fabrication process and they show no current. The narrowest device with finite drain current is assumed to have the channel width of 1.25 nm, that is, a half of the interval. Since the standard deviation of the width is less than 2 nm and there is good linearity between EB design line width and resist width, this estimation of the channel width is quite reliable.

In Figure 4.10, threshold voltage increases apparently when the channel width is less than 10 nm. The simulation results that are obtained by solving a two-dimensional Schrodinger equation with the finite element method are also shown.[19] Since the experimental results and simulation are in good agreement, we concluded

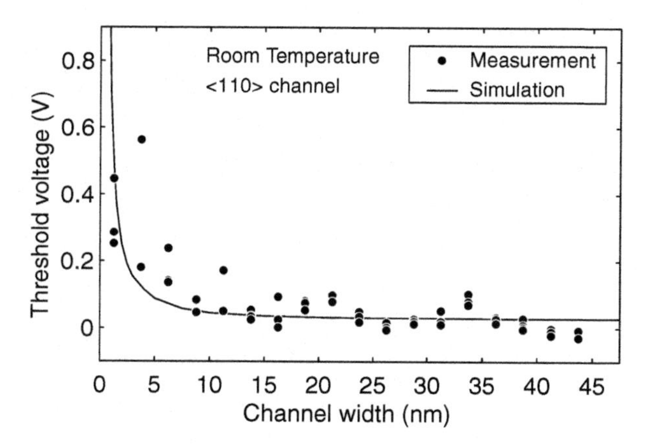

FIGURE 4.10 Channel width dependence of threshold voltage in narrow channel n-type MOSFETs. Simulation result is also shown.

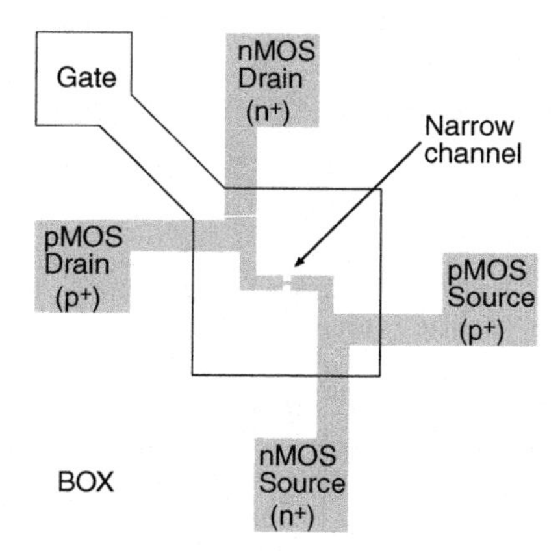

FIGURE 4.11 A schematic of the device that has both n-type source/drain and p-type source/drain. The device acts as both n-type and p-type MOSFET.

that this threshold voltage increase is caused by the quantum confinement effects. We call this phenomenon in ultranarrow channel MOSFETs "quantum mechanical narrow channel effect."

4.4.3 Threshold Voltage Increase in n-Type and p-Type Narrow Channel MOSFETs

In order to investigate the quantum effects in both n-type and p-type nano-scale MOSFETs, a device with a special configuration was fabricated,[14] as illustrated in Figure 4.11. The device has both n-type source/drain and p-type source/drain. When a positive voltage is applied to the gate, electrons are induced in the channel and the device acts as an n-type MOSFET if the current between n-type source and drain is measured. On the other hand, when a negative voltage is applied to the gate, the

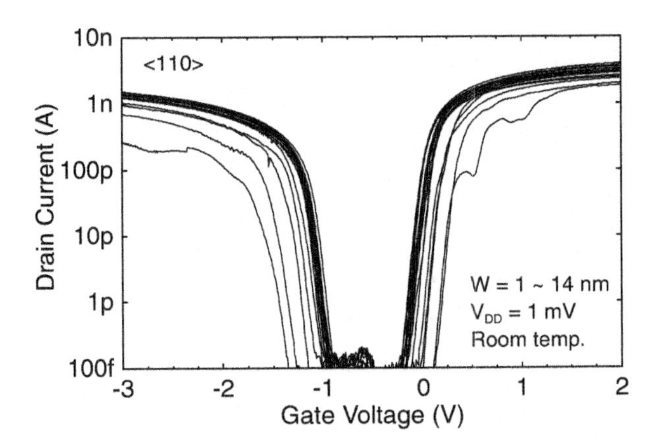

FIGURE 4.12 Ids vs. Vgs characteristics of <110> oriented n- and p-type narrow channel MOSFETs.

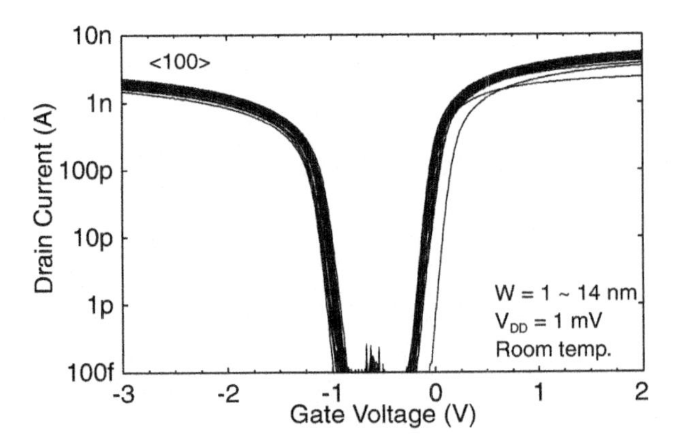

FIGURE 4.13 Ids vs. Vgs characteristics of <100> oriented n- and p-type narrow channel MOSFETs.

device acts as a p-type MOSFET if the current between p-type source and drain is measured. Using this device structure, both electron and hole transport (that is, the potential profiles of both conduction band and valence band) can be examined in the identical device. Ultranarrow channel MOSFETs with two different channel orientations (<110> direction and <100> direction) are fabricated, and channel width is varied from almost 0 to 14 nm.

Figure 4.12 and Figure 4.13 show Ids-Vgs characteristics of <110> oriented and <100> oriented narrow channel MOSFETs, respectively. The n-type operation is observed in the positive gate voltage regime, and p-type operation in the negative gate voltage regime. In <110> oriented devices, some devices with small drain current show high threshold voltage, and this tendency is clearly observed both in n-type and p-type operations. On the other hand, threshold voltage increase is not clearly observed in <100> oriented devices.

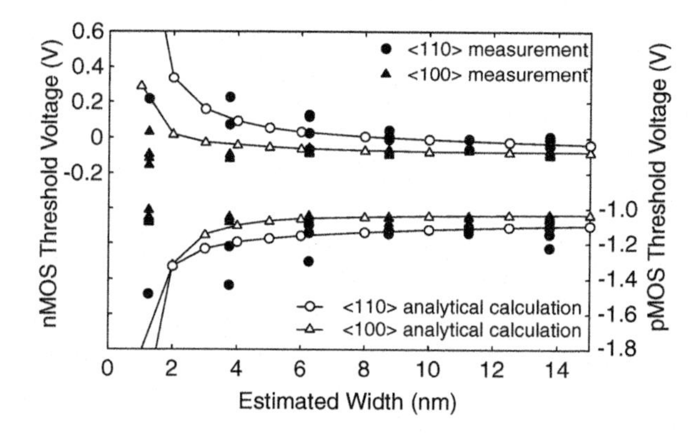

FIGURE 4.14 Channel width dependence of threshold voltage in the device having both n-type and p-type source/drain. The calculation results are also shown.

Figure 4.14 shows the threshold voltage as a function of estimated channel width in <110> oriented and <100> oriented narrow channel MOSFETs.[14] Both n- and p-type device data are shown and calculation results on threshold voltage increase are also shown. The threshold voltage in <110> oriented devices increases as the width becomes narrower due to the quantum confinement in both n-type and p-type ultra-narrow channels. It has been confirmed experimentally that not only electrons but also holes are confined in ultranarrow channels, causing the threshold voltage increase in MOSFETs. The threshold voltage increase in <100> oriented devices was not observed in this experiment, which is not in agreement with calculation.

4.4.4 Threshold Voltage Adjustment Using Quantum Effects

Utilizing the threshold voltage shift due to the quantum confinement effect, we have proposed a new threshold voltage adjustment method in both n-type and p-type MOSFETs.[14] In nano-scale FD-SOI MOSFETs, it is very hard to adjust threshold voltage using impurity concentration, which is a general way to control threshold voltage in conventional MOSFETs, because the volume of the body is very small and the effect of impurity concentration on threshold voltage is too small. A new threshold voltage adjustment method is strongly required.

Figure 4.15 shows calculated threshold voltage shift due to quantum confinement effects in n-type and p-type MOSFETs. The vertical axis is for the p-type and the horizontal axis is for the n-type. It is found that depending on the channel orientation the threshold voltage shifts of n- and p-type narrow channel MOSFETs are different. In a <110> oriented device, n-type has larger threshold voltage shift than p-type. On the other hand, in a <100> oriented device, p-type has the larger threshold voltage shift. When channel width is wide and only the SOI thickness is thinned (that is, normal FD-SOI MOSFETs), threshold voltage shift of a p-type device is larger than that of an n-type. It is also found that in order to obtain the same threshold voltage shift in n-type and p-type devices, a square scaling in which the height and width of the channel are kept constant is effective, as shown in Figure 4.15.

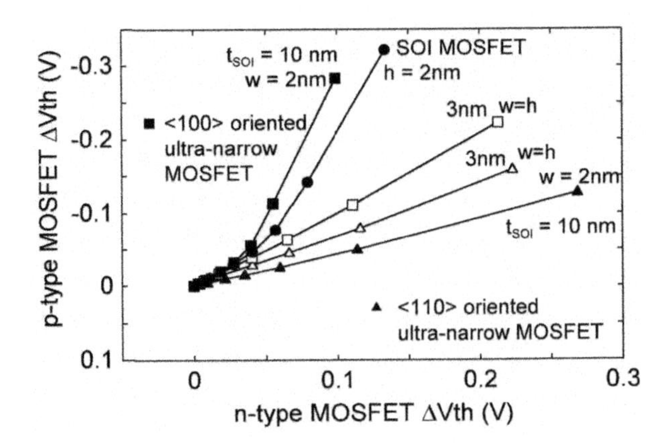

FIGURE 4.15 Threshold voltage shift due to quantum effects in n- and p-type narrow channel MOSFETs.

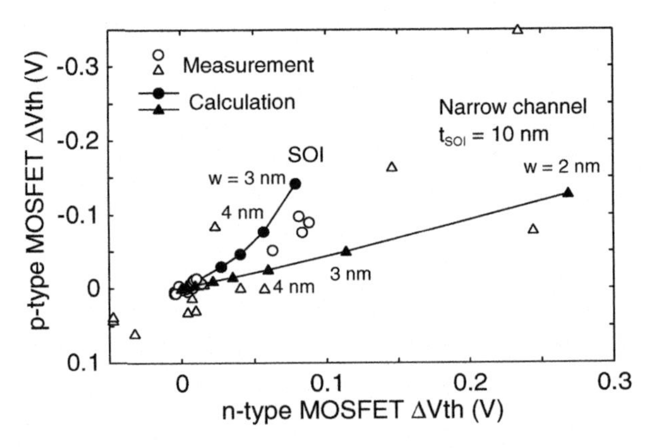

FIGURE 4.16 Comparison of calculations and experiments of threshold voltage shift due to quantum effects in n- and p-type narrow channel MOSFETs.

Figure 4.16 shows the comparison of calculation results and experimental data. Data of <110> oriented devices and wide channel FD-SOI MOSFETs are shown. Although the experimental data are distributed, they are roughly in agreement with calculation results. These results suggest that the quantum confinement effect is a promising way to control threshold voltage of nano-scale MOSFETs if the size of the channel is precisely controlled.

4.4.5 Mobility Enhancement due to Quantum Effects

We have also investigated the electron mobility and hole mobility in ultranarrow channel MOSFETs. Although in wide channel thin FD-SOI MOSFETs the mobility enhancement is expected mainly by the reduction of intervalley phonon scattering, both the intervalley and intravalley scattering play important roles in mobility in

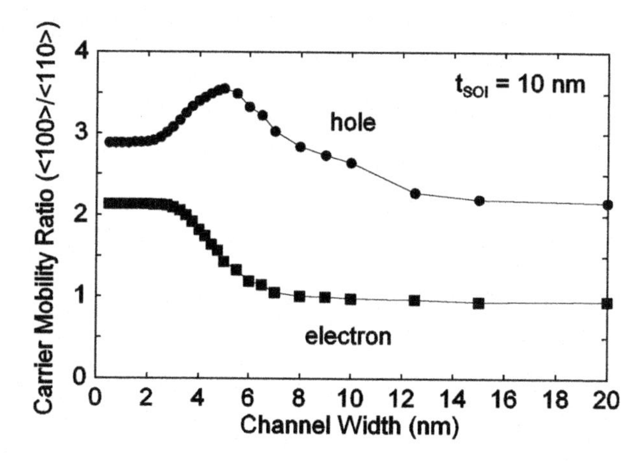

FIGURE 4.17 Calculated channel width dependence on electron and hole mobility in narrow channel MOSFETs. The vertical axis is the ratio of mobility in <100> orientation to <110> orientation.

ultranarrow channel MOSFETs. Thanks to the anisotropy of electron mass and hole mass, the intravalley phonon scattering affects the total mobility. We have shown by simulation that both the electron and hole mobilities in <100> oriented devices are larger than those of <110> oriented devices.[14] Figure 4.17 shows the ratio of mobility of <100> oriented narrow channel MOSFETs to that of <110> oriented devices. Conventional MOSFETs have the channel of <110> direction, and the <100> direction is rotated from the <110> direction by 45 degrees. This larger mobility in <100> oriented devices is due to the anisotropic effective mass of silicon. In an n-type MOSFET, this phenomenon is explained by the electron population of valleys in the conduction band.

Figure 4.18 shows the conduction band structure in <110> oriented and <100> oriented channels. Due to the anisotropy of electron mass, the rise of valley energy is different depending on the direction of the ultranarrow channel. In the <110> oriented ultranarrow channel, six equivalent valleys split into two sets of valleys: twofold valleys and fourfold valleys. When the channel width becomes narrower, the energy of the twofold valleys increase more than the fourfold valleys, and therefore, more electrons are populated in the fourfold valleys. However, the fourfold valley has larger mass and smaller mobility along the <110> direction than the twofold valleys, as shown in Figure 4.19. Accordingly, the total mobility of <110> oriented narrow channel MOSFETs is smaller than the <100> oriented device.

In the <100> oriented narrow channel, on the other hand, six equivalent valleys split into three sets of twofold valleys as shown in Figure 4.18. Most of the electrons are populated in valley (c) because the energy of this valley is smallest, and valley (c) has the smallest mass and the highest mobility, as shown in Figure 4.19. That is why the total mobility of <100> oriented narrow channel MOSFETs is a higher mobility than <110> oriented narrow channel MOSFETs. As a result, when the channel width is scaled into the nano-scale, we should change the channel direction from the conventional <110> direction to <100> direction to take full advantage of the quantum effects.

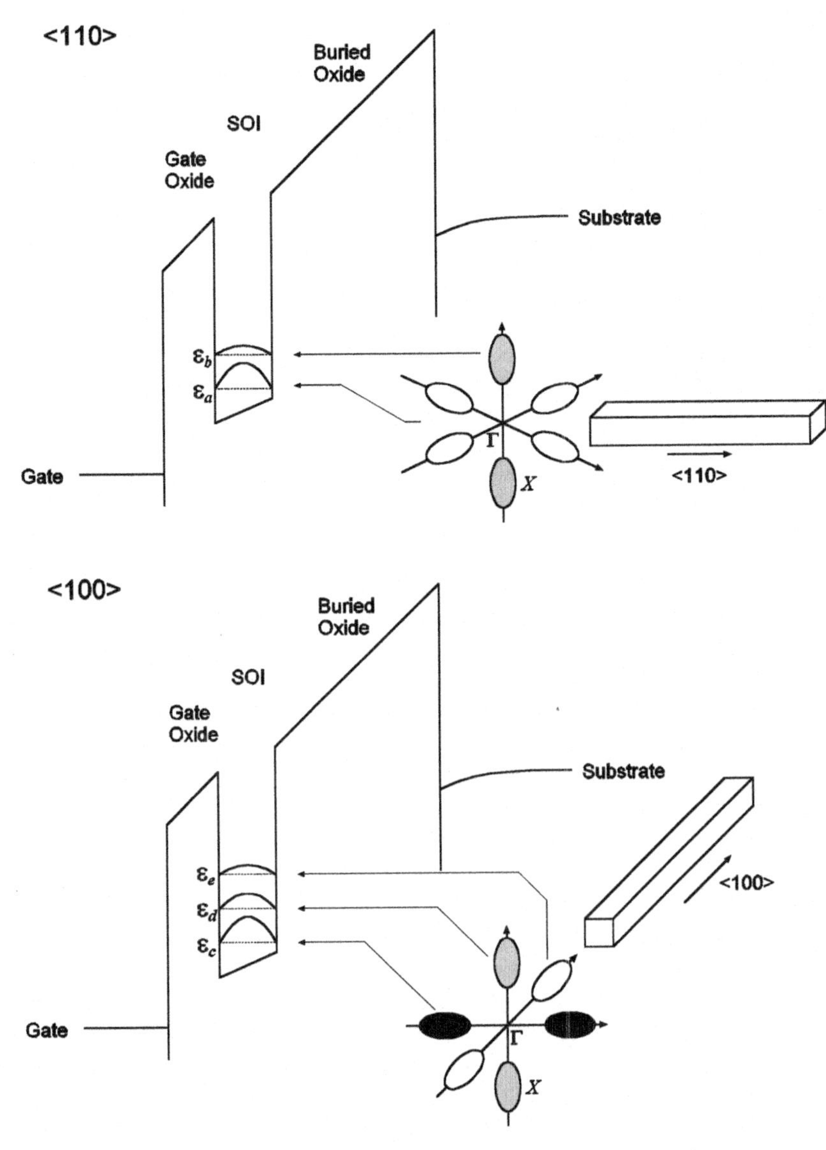

FIGURE 4.18 Schematics of conduction band structures in <110> oriented and <100> oriented narrow channels. The energy separation of the valleys is also shown.

4.5 SUMMARY

Quantum effects in nanoscale MOSFETs are described. The carrier confinement in very thin layers or ultranarrow channels largely affects the device characteristics. It is important to positively utilize the quantum effects that appear in nanoscale MOSFETs. The threshold voltage increase, the threshold voltage adjustment, and the mobility enhancement will be expected in nanoscale MOSFETs, and these phenomena should be incorporated in the design of nanoscale MOSFETs.

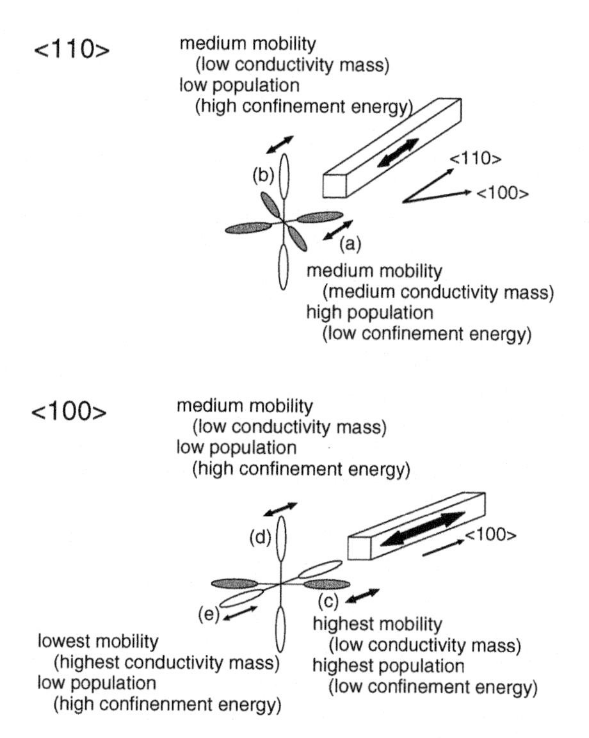

FIGURE 4.19 Schematic of conduction band structures in <110> oriented and <100> oriented narrow channels, showing the energy level, electron population, electron mass, and mobility in each valley.

REFERENCES

1. Doris, B., Ieong, M., Kanarsky, T., Zhang, Y., Roy, R.A., Dokumaci, O., Ren, Z., Jamin, F.-F., Shi, L., Natzle, W., Huang, H.-J., Messapelle, J., Mocuta, A., Womack, S., Gribelyuk, M., Jones, E.C., Miller, R.J., Wong, H.P., and Haensch, W., Extreme Scaling with Ultrathin Si Channel MOSFETs, IEEE Intl Electron Devices Meeting, San Francisco, U.S., December, 2002, p. 267–270.
2. Wakabayashi, H., Yamagami, S., Ikezawa, N., Ogura, A., Narihiro, M., Arai, K.I., Ochiai, Y., Takeuchi, K., Yamamoto, T., and Mogami, T., Sub-10-nm Planar-Bulk-CMOS Devices Using Lateral Junction Control, IEEE Intl Electron Devices Meeting, Washington, D.C., U.S., December, 2003.
3. Stern, F. and Howard, W.E., Properties of semiconductor surface inversion layers in the electric quantum limit, *Physical Review,* 163, 816–835, 1967.
4. Dresselhaus, G., Kip, A.F. and Kittel, C., Cyclotron resonance of electrons and holes in silicon and germanium crystals, *Physical Review*, 98, 368–384, 1955.
5. Luttinger, J.M., Quantum theory of cyclotron resonance in semiconductors: General theory, *Physical Review*, 102, 1030–1041, 1956.
6. Ohkura, Y., Quantum effects in Si n-MOS inversion layer at high substrate concentration, *Solid-State Electronics*, 33, 1581–1585, 1990.

7. van Dort, M.J., Woerlee, P.H., Walker, A.J., Juffermans, C.A.H., and Lifka, H., Quantum Mechanical Threshold Voltage Shifts of MOSFETs Caused by High Level of Channel Doping, IEEE Intl Electron Devices Meeting, Washington, D.C., U.S., December, 1991, pp. 495–498.
8. Stern, F., Self-consistent results for *n*-Type Si inversion layers, *Physical Review B,* 5, 4891–4899, 1972.
9. Trellakis, A., Galick, A.T., Pacelli, A., and Ravaioli, U., Iteration scheme for the solution of the two-dimensional Schrodinger-Poisson equations in quantum structures, *Journal of Applied Physics,* 81, 7880–7884, 1997.
10. van Dort, M.J., Woerlee, P.H., and Walker, A.J., A simple model for quantisation effects in heavily-doped silicon MOSFETs at inversion conditions, *Solid-State Electronics,* 37, 411–414, 1994.
11. Medici, Ver. 4.1, a device simulator, Avant! Corp., 1998.
12. Omura, Y., Kurihara, K., Takahashi, Y., Ishiyama, T., Nakajima, Y., and Izumi, K., 50-nm channel nMOSFET/SIMOX with an ultrathin 2- or 6-nm thick silicon layer and their significant features of operations, *IEEE Electron Device Letters,* 18, 190–193, 1997.
13. Choi, Y.-K., Ha, D., King, T.-J., and Hu, C., Ultrathin Body PMOSFETs with Selectively Deposited Ge Source/Ddrain, Symposium on VLSI Technology, Kyoto, Japan, 2001, PP. 19–20.
14. Majima, H., Saito, Y., and Hiramoto, T., Impact of Quantum Mechanical Effects on Design of Nano-Scale Narrow Channel n- and p-type MOSFETs, IEEE Intl Electron Devices Meeting, Washington, D.C., U.S., December, 2001, pp. 733–736.
15. Takagi, S., Koga, J., and Toriumi, A., Subband structure engineering for performance enhancement of Si MOSFETs, IEEE Intl Electron Devices Meeting, Washington, D.C., U.S., December, 1997, pp. 219–222.
16. Shoji, M. and Horiguchi, S., Phonon-Limited Electron Mobility in Ultrathin SOI MOSFETs, International Conference on Solid State Devices and Materials, Hamamatsu, Japan, September, 1997, pp. 156–157.
17. Takagi, S., Toriumi, A., Iwase, M., and Tango, H., On the universality of inversion layer mobility in Si MOSFET's: Part I-Effects of substrate impurity concentration, *IEEE Transactions on Electron Devices*, 41, 2357–2362, 1994.
18. Antoniadis, A., MOSFETs Scalability Limits and "New Frontier" Devices, Symposium on VLSI Technology, Honolulu, U.S., 2002, pp. 2–5.
19. Majima, H., Ishikuro, H., and Hiramoto, T., Experimental evidence for quantum mechanical narrow channel effect in ultranarrow MOSFETs, *IEEE Electron Devices Letters,* 21, 396–398, 2000.
20. Majima, H., Ishikuro, H., and Hiramoto, T., Threshold Voltage Increase by Quantum Mechanical Narrow Channel Effect in Ultra-Narrow MOSFETs, IEEE International Electron Devices Meeting, Washington, D.C., U.S., December, 1999, pp. 379–382.

5 Ballistic Transport in Silicon Nanostructures

Hiroshi Mizuta, Katsuhiko Nishiguchi and Shunri Oda

5.1 INTRODUCTION

For the past two decades novel transport phenomena in nanometer scale semiconductor structures have been explored along with tremendous progress of nanofabrication and material growth technologies. Both wave and particle natures of electrons have come out in various ways and have brought very unique transport properties to us, which had not been shown by electrons in the micro scale structures. The electric characteristics we usually obtain for the present ultra-large scale integration (ULSI) devices are the statistical average of a large number of electrons (to the order of Avogadro's number) in motion, which suffer from frequent scattering events caused by various mechanisms. Under these circumstances the electric characteristics are described by using macroscopic quantities such as the mobility which is determined mainly by elastic impurity scattering and inelastic phonon scattering. Electrons hardly maintain the memory of the phase of their wave functions as the energy of electrons changes due to the inelastic phonon scattering with the phonon emission/absorption. The motion of electrons is therefore described well as classical particles under the electric fields.

When a device structure, however, is made smaller than the phase coherence length of the electron wave function, the wave nature of electrons should appear, and the effects of the phase of the electron wave function can no longer be neglected. Such a regime is called as a *mesoscopic* system. The quantum confinement effect in the semiconductor superlattice is one of the examples that results from the electron wave nature. As the elastic scattering does not change the energy of electrons, the phase of electron wave function is maintained, and the associated quantum-mechanical phenomena of electrons are still observed even if the *elastic* mean free path is shorter than the device dimensions. Reducing the device dimensions further down below the elastic mean free path for impurity scattering, the motion of electrons across the device becomes *ballistic*. Under these circumstances electrons behave in a manner similar to light propagating in a loss-free waveguide or electrons traveling in vacuum. In a perfect ballistic system the current conduction should be determined purely by the channel geometry (i.e., the boundary of the channel) without any statistical fluctuation caused by randomly distributed dopant atoms. As the disappearance of the conductivity fluctuations has already been observed,[1] ballistic transport is expected to be increasingly important along with further downscaling devices and improving the crystallinity.

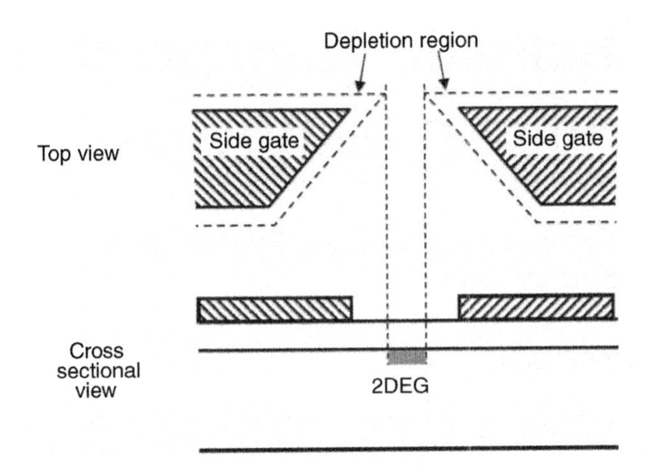

FIGURE 5.1 A lateral quantum point contact (QPC) structure formed in two-dimensional electron gas by using two in-plane Au gates.

In this short review we investigate physics of ballistic transport in Si nanostructures by focusing on the phenomenon of conductance quantization observed for an extremely short and narrow constriction structure called a quantum point contact (QPC). Due to lack of space and time we do not cover ballistic (and quasi-ballistic) transport phenomena expected to occur in MOSFETs by decreasing their dimensions into a few tens of nanometers, which is also an important subject apparently. Section II provides a brief history of research on the conductance quantization in lateral semiconductor QPC structures with a simple theoretical footing for explaining the phenomenon. In Section III we introduce a vertical nano FET structure, which is also a QPC formed in a vertical direction. We discuss device fabrication, observation of the conductance quantization, and the effects of a magnetic field on the conductance characteristics.

5.2 BALLISTIC TRANSPORT IN QUANTUM POINT CONTACTS

Early studies on electron transport in mesoscopic systems have often adopted metallic structures, but it was difficult to observe ballistic transport since the electron mean free path l_e in metal is usually less than 10 nm. On the other hand, recent rapid progress of semiconductor nanotechnologies has achieved ballistic transport in the semiconductor nanostructures at low temperatures. For example, two-dimensional electron gas (2-DEG) with mobility as high as 106 cm/Vs can now easily be formed at 4.2 K in AlGaAs/GaAs modulation doped structures. In addition, development of nanofabrication techniques such as electron beam lithography, focused ion beam technique and reactive ion etching facilitate making the 2-DEG structures smaller than 1 μm while hardly degrading the high electron mobility. It is indeed possible to fabricate structures smaller than le and also the ballistic electron systems by combining these modern fabrication technologies.

Conductance quantization was first discovered by van Wees et al.[2] and Wharam et al.[3] independently in 1988 by using the device with a split gate formed on the

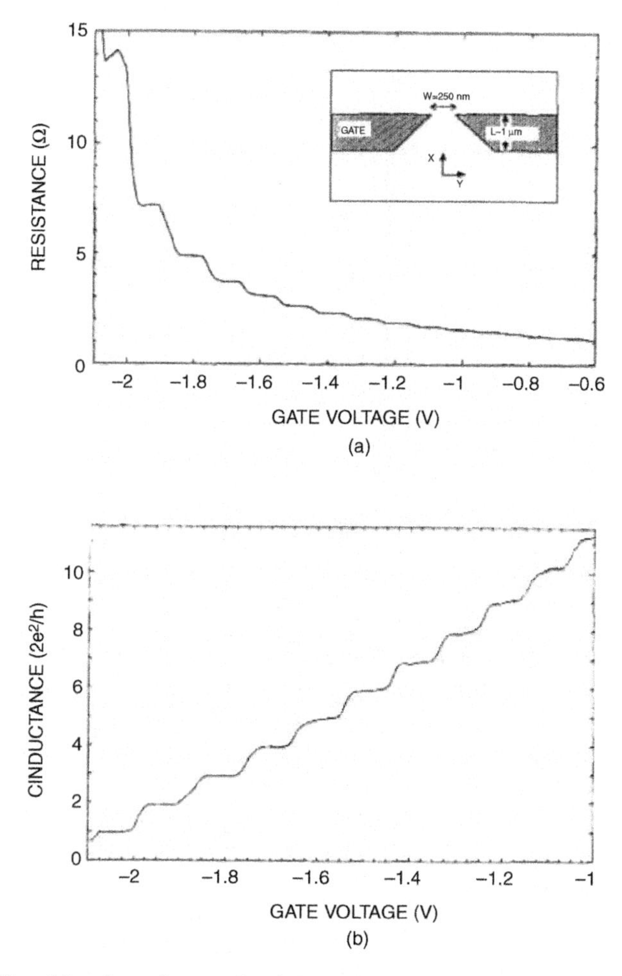

FIGURE 5.2 Gate bias dependence of resistance (a) and conductance (b) observed for the QPC in Figure 1. After van Wees et al.[2] with permission.

AlGaAs/GaAs heterostructures (Figure 5.1). This device structure is called *quantum point contacts* (QPCs) which features an extremely short and narrow conducting channel formed between the split gate. By applying a negative bias to the split gate, the 2-DEG under the gates was depleted and a short 1-D (one-dimensional) channel was formed; the effective channel length was estimated to be less than 0.3 μm including the depletion length which is much shorter than an le of 8 μm for the 2-DEG. The width of the channel and, hence, the number of the subbands under the Fermi energy were controlled via the applied gate bias. They observed a series of steps in the conductance in a beautiful manner as the width of the QPC channel was varied by means of the gate bias (Figure 5.2). The steps were found to be nearly multiples of $2e^2/h$ after correction for parasitic series resistance in large-area source and drain regions. Similar conductance quantization characteristics were observed by modulating the electron density (i.e., the Fermi energy) and keeping the channel

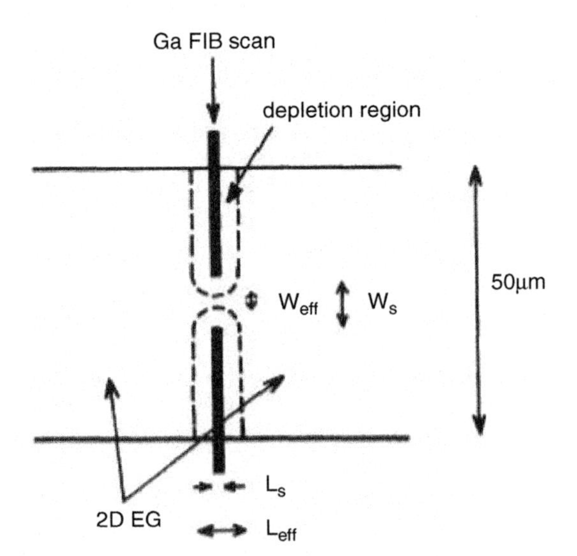

FIGURE 5.3 A QPC structure formed by using focused ion beam technique. After Hirayama et al.[4] with permission.

width constant. Hirayama et al.[4,5] have used the focused ion beam (FIB) of Ga ions and formed the lateral pn-junction gates as the implanted Ga ions convert n-type GaAs regions into highly resistive p-type (Figure 5.3). The beam diameter was less than 100 nm, and an extremely short QPC channel was formed at an interval in a single beam scan. The QPCs with a wide range of the gap of The Schottky gate were then formed above the QPC structure. Figure 5.4 shows the gate bias dependence of the conductance observed for the QPC at various temperatures. The stepwise increase in the conductance was clearly observed each time the Fermi energy passed across the subband energies while increasing the gate bias.

Let us first take a look at the theoretical basis of conductance quantization in the QPC. Suppose a perfect 1-D semiconductor wire without any scattering events. Lateral confinement forms the 1-D electronic subbands in the wire, and the density of states (DOSs) for the subbands, $N_n(E)$ (n= 1, 2, 3,.....) are given by

$$N_n(E) = \frac{g_s g_v}{h}\left[\frac{m^*}{2(E-E_n)}\right]^{1/2} \tag{5.1}$$

where g_s and g_v are the spin and valley degeneracy factors; $g_s = 2$, $g_v = 1$ for the direct bandgap semiconductors such as GaAs, and $g_s = 2$, $g_v = 2$ for the indirect bandgap semiconductors such as Si. Although the subband energies En depend on the shape of the confinement potential, Equation (5.1) works for any 1-D structure; the schematic DOSs are shown in Figure 5.5(a). Suppose an infinitesimal voltage of V is applied between both ends of the wire. If the electron transmission probability through the ith subband is unity, the current Ii at zero temperature is expressed as

$$I_i = N_i(E_F)ev(eV) = g_x g_v e^2 V / h \tag{5.2}$$

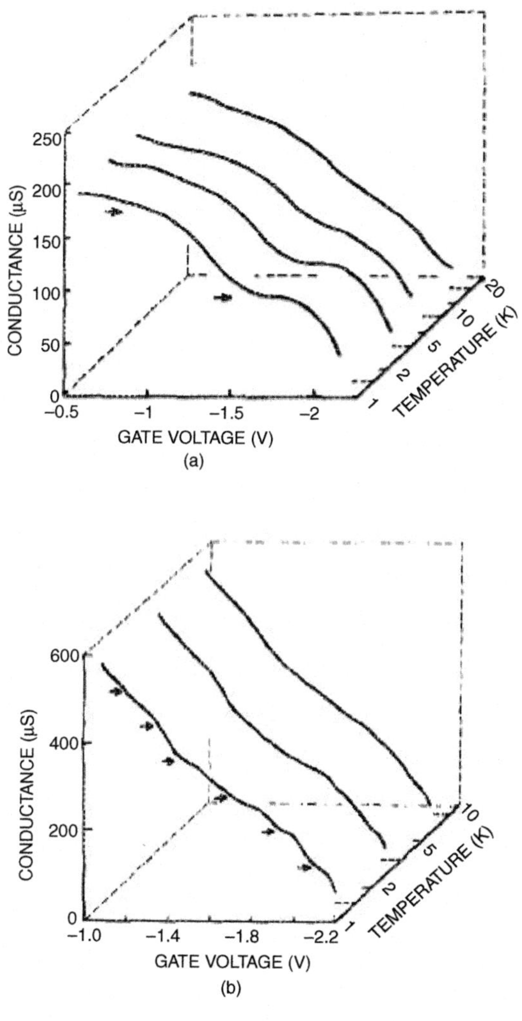

FIGURE 5.4 Gate bias dependence of conductance observed at various temperatures for the QPC structures formed by using the FIB technique: (a) with a narrow QPC channel and (b) with a wide QPC channel. After Hirayama et al.[5] with permission.

where v is the velocity of electrons at the Fermi energy in the ith subband, $v = \sqrt{2(E_F - E_i)/m^*}$. If there is no scattering in the wire, there is no subband mixing, and, therefore, the total current through the wire and associated conductance are given as

$$I = \sum_{i=1}^{n} I_i = g_s g_v e^2 V n / h \tag{5.3}$$

and

$$G = I / V = g_s g_v e^2 n / h \tag{5.4}$$

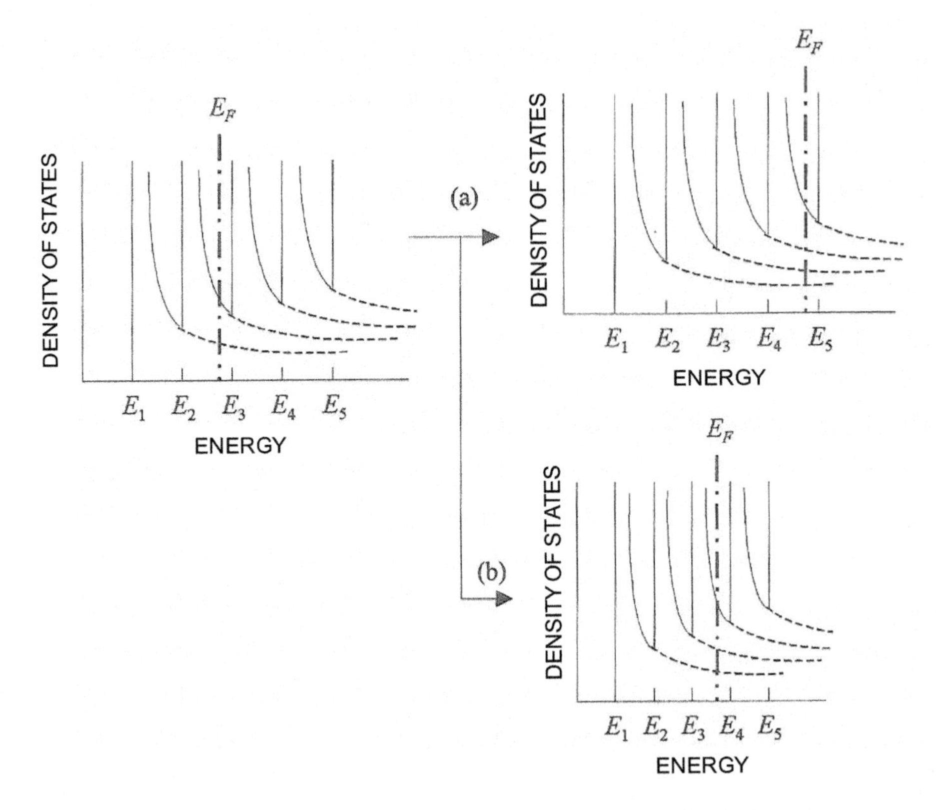

FIGURE 5.5 Schematic density of States (DOSs) for a one-dimensional channel: an increase in the number of subbands (a) with increasing the Fermi energy E_F and (b) with varying the DOSs.

where n represents the number of subbands located below the Fermi energy that are fully occupied by electrons. Equation (5.4) shows that the conductance though the scattering-free 1-D wire is quantized, and the quantization unit is e^2/h except for the degeneracy factors.

A more generalized expression of the conductance is derived from the Landauer formula for the subband channels with the transmission probability as

$$G = \frac{g_s g_v e^2}{h} \sum_{i=1}^{n} T_i \tag{5.5}$$

For the QPC structures, the transmission probability can be approximated by the following expression[6]

$$T_i(E_i) = \frac{1}{1 + \exp\left[-\beta_i \left(E_i - E_F\right)\right]} \tag{5.6}$$

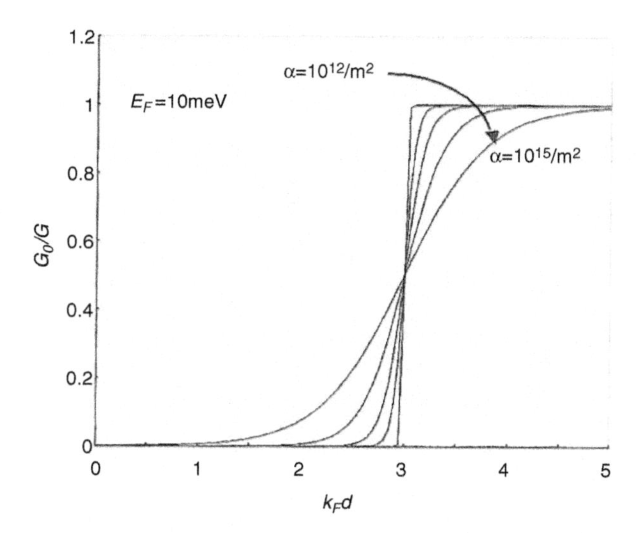

FIGURE 5.6 Conductance of the QPC as a function of the channel width d with various values of the parameter α in Equation (5.7).

where

$$\beta_i = \sqrt{\frac{2m^*}{\alpha\hbar^2 E_i}} \tag{5.7}$$

assuming a quadratic saddle-point potential around the point contact channel region. The constant α in Equation (5.7) depends upon the QPC geometry: α decreases as the shape of the QPC becomes blunter and, therefore, the effective channel length longer. The effect of α on the conductance is shown in Figure 5.6. It should be noted that the conductance step becomes more gradual with increasing α. The conductance through the QPC at a finite temperature is then given as

$$G = \frac{g_s g_v e^2}{h} \sum_{i=1}^{n} \frac{1}{1 + \exp\left[\dfrac{-\beta_i (E_i - E_F)}{k_B T}\right]} \tag{5.8}$$

Equation (5.8) indicates that a higher temperature also results in a less steep rise at the conductance steps.

In order to observe clear ballistic transport in QPCs, the condition $l_e > l_{ch}$, w_{ch} should be met, where l_{ch} and w_{ch} are the channel length and width, respectively. Practically, l_{ch} and w_{ch} values of about 20 nm are more or less the smallest achievable dimensions based on the most recent nanofabrication techniques. On the other hand, l_e is determined by any backscattering in the QPC channel, especially by impurity

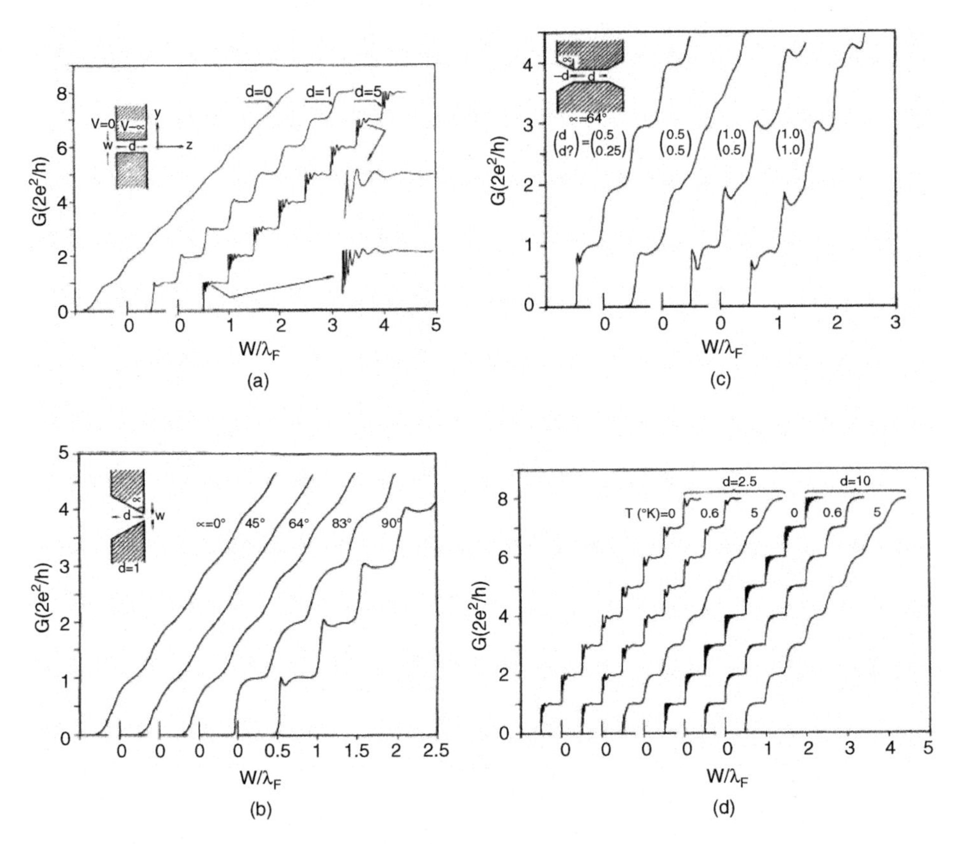

FIGURE 5.7 Conductance quantization characteristics calculated for the QPCs with various channel geometries (a) – (c) and temperatures (d). After Takman et al.[8] with permission.

scattering and boundary scattering at channel wall irregularities. In the case of QPCs fabricated on the 2-DEG formed in the modulation doped structures, impurities play as scattering centers only in a remote manner, and the remote impurity scattering may be reduced significantly by introducing a thick buffer layer between the 2-DEG and the electron supply layer. This is not the case when a simply doped channel is used to form QPCs, and more careful considerations are needed for optimizing the doping concentration. Lower doping concentrations are obviously favorable to reduce the impurity scattering, but they cause a high resistance and, more seriously, a considerable freeze-out of electrons at low temperatures. The reduced screening of the impurity potential should also be taken into consideration for lower doping concentrations. If the parasitic resistance is not small enough compared to the QPC channel resistance, the overall conductance is reduced greatly, and extraction of intrinsic conductance becomes difficult.[7] These requirements to the channel dimensions might make us believe that an ideal geometry to observe perfect conductance quantization is a slit-like QPC. It has, however, been shown that the conductance steps are distorted if the junctions between the QPC channel and reservoirs are too abrupt (see Figure 5.7). This results from the quantum-mechanical electron reflection at the junctions, and the conductance decreases as if there were scattering events in the channel. It also causes the conductance overshoot in a similar manner to the tunneling current through resonant tunneling diodes. Theoretical study of those effects has been reported by Tekman et al.,[8] which indicates the QPC geometry should be designed carefully.

In contrast to a number of studies done on ballistic transport in the GaAs/AlGaAs structures, there have been a limited number of reports[9–13] on that in Si nanostructures as l_e in Si is in general shorter than that in GaAs. First observation was conducted by Warren et al.[11] by using a MOS (Metal-Oxide-Semiconductor) FET (Field-Effect Transistor) structure with double gates. They adopted grating gate strips embedded in a gate oxide layer under a normal upper gate (Figure 5.8). By controlling the bias voltages applied to the double gates, the electron channel may be formed either beneath the strip gates or between the adjacent two strip gates. The quantized conductance was observed at 1.4 K only for the second channel configuration since the channel width was wider for the first channel configuration than that for the second one (Figure 5.9). A very small Si-based QPC (Figure 5.10) has also been fabricated recently in a more direct manner[13] by using the e-beam writing & etching technique. The conductance quantization observed for these Si structures has not been as clear as those for GaAs-based structures because the channel dimensions were not small enough compared with l_e, which is determined mainly by elastic impurity scattering and boundary scattering; also, the parasitic resistance of the source and drain regions connected to the QPC was relatively high.

5.3 BALLISTIC TRANSPORT IN ULTRA-SHORT CHANNEL VERTICAL SILICON TRANSISTORS

5.3.1 Fabrication of Nanoscale Vertical FETs

Study of conductance quantization by using *lateral* QPC structures in Section II indicated that further size reduction as well as more precise structural control is needed for clarifying ballistic transport in Si. As this is a fairly hard requirement for the present nanofabrication techniques, and a *vertical* Si FET structure with a wrap-around gate has been examined[14,15] as an alternative to the conventional lateral QPCs. In the following sections the details of fabrication and characterization of the vertical Si FETs are discussed.

The nanoscale vertical FET was fabricated by using electron beam lithography and chemical vapor deposition (CVD), in which both the width and length of the channel are designed to be smaller than the elastic electron mean free path in Si. Figure 5.11 shows the fabrication process flow for the vertical nano FETs. The devices were fabricated in an SiO_2(20 nm)/polySi(20 nm)/SiO_2(20 nm)/(100) Si substrate (Figure 5.11(a)) where the lower SiO_2 was formed using thermal oxidation. The polySi layer is phosphorous doped by 10^{18} /cm^3 and acts as a gate electrode. For suppressing random telegraph signals, that is, fluctuations of the conduction current, the density of electron trap states should be reduced both at the amorphous Si (a-Si) grain boundaries (GBs) and inside the Si grains. A solid phase crystallization (SPC) method was therefore adopted to make the grain size larger and consequently

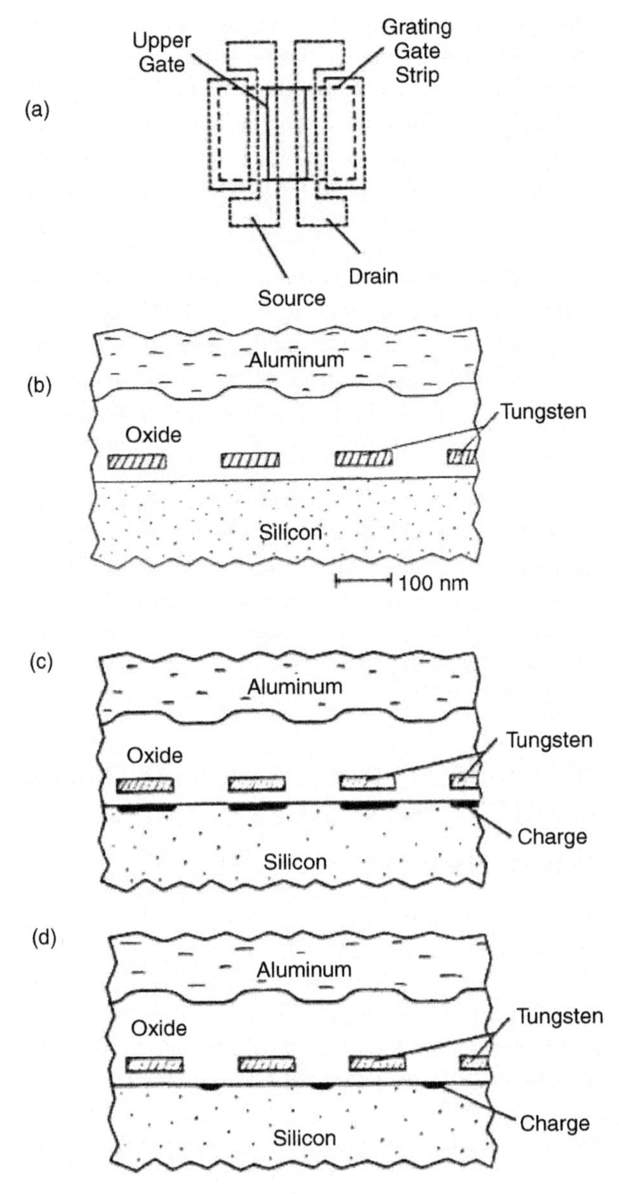

FIGURE 5.8 A lateral silicon ballistic transistor with the 1st Tungsten grating gate strip and the 2nd Aluminum top gate: (a) a schematic plan view, (b) a cross-sectional view, an operation mode with a channel formed (c) under the 1st gate strips and (d) between them. After Warren et al.[11] with permission.

the number of the GBs in the device area becomes smaller. An a-Si film was first deposited by using an Hg-sensitized photo-induced CVD method since a high density of hydrogen-terminated Si-bonds tend to make the grains larger. The sample was then annealed at 700°C for 4 hours for conducting the SPC process,[16,17] followed

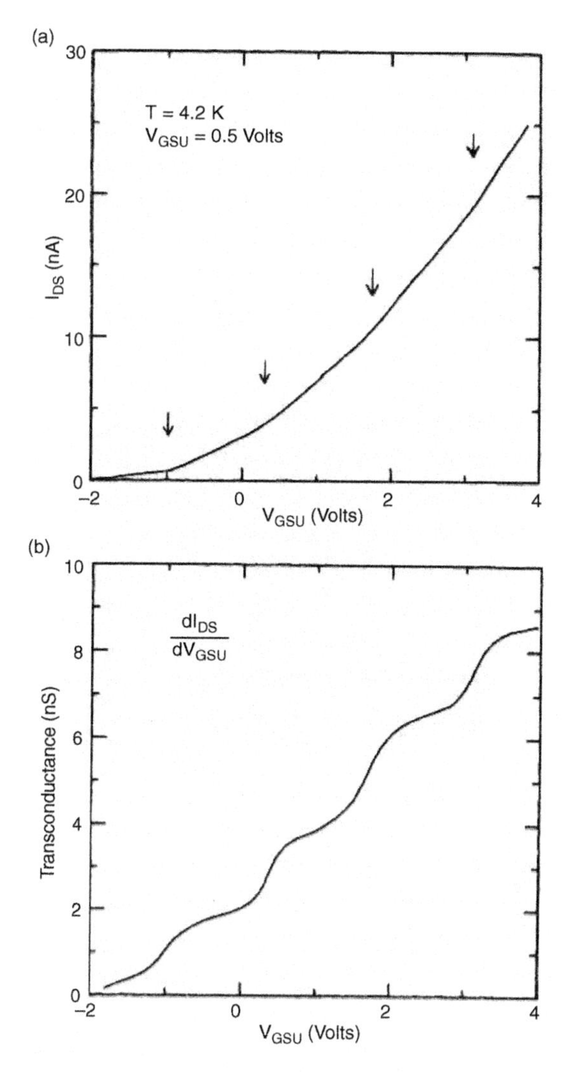

FIGURE 5.9 Drain current (a) and transconductance (b) as a function of gate bias observed for the Si ballistic transistor in Figure 8. After Warren et al.[11] with permission.

by annealing at 900°C for activating the dopant atoms. A polySi film with the grain sizes over 200 nm is formed by using this method. The upper SiO_2 layer was finally formed with tetraethoxysilane (TEOS) by using the plasma-enhanced CVD followed by annealing at 1100°C for 2 hours.

For the vertical channel, a square hole of 60 × 60 nm in area was first formed by using electron-beam (EB) lithography. An upper SiO_2 layer was etched by using anisotropical electron-cyclotron-resonance reactive-ion-etching (ECR-RIE) with CH_4 gas. A wrap-around gate was then formed by etching the polySi layer by using isotropical plasma etching with a $CH_4 + O_2(10\%)$ gas mixture. Side etching of the polySi makes the hole size bigger than that defined with the EB lithography, and

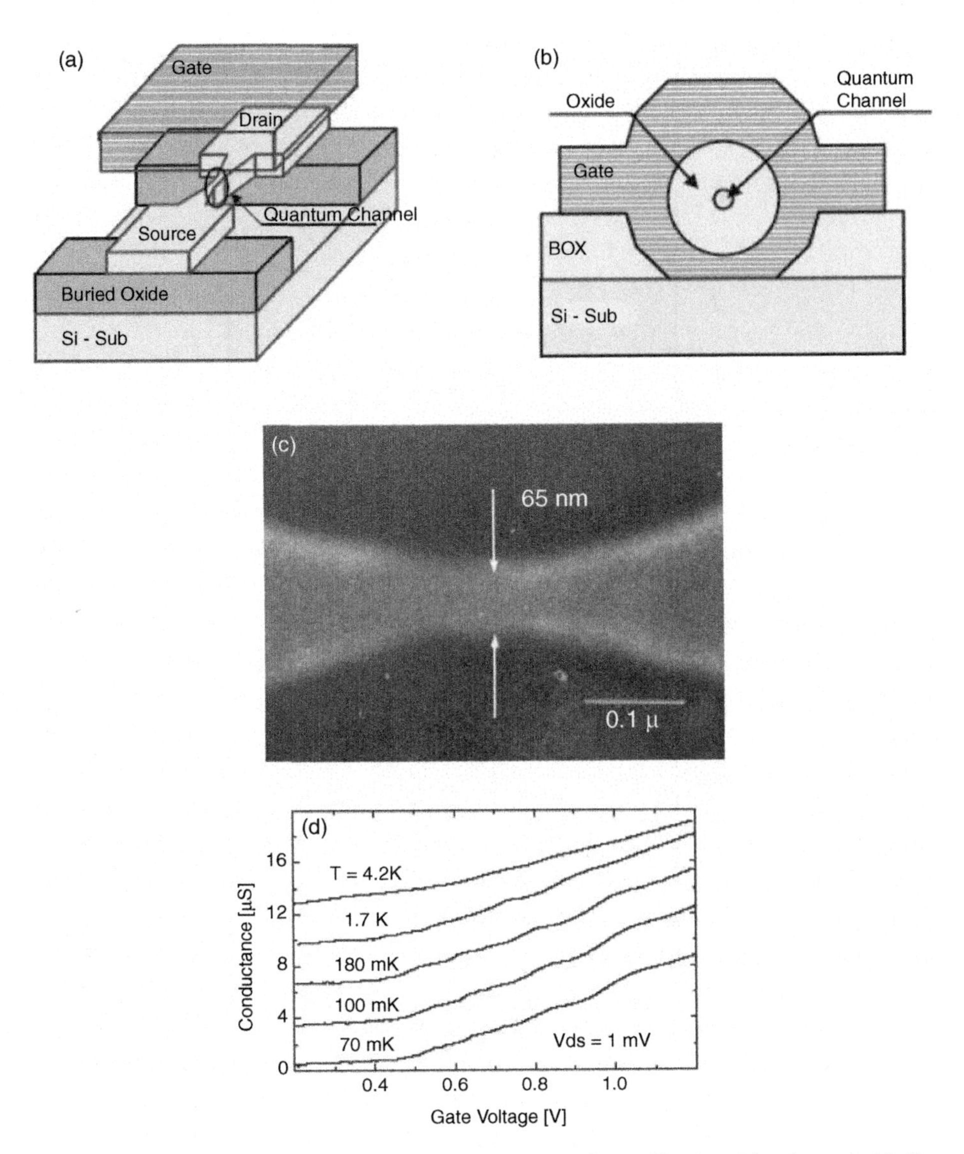

FIGURE 5.10 A lateral silicon ballistic transistor by using a Si wire: (a) schematic bird's-eye view (b) and cross-sectional view, (c) a SEM micrograph of the Si wire, and (d) the gate bias dependence of the conductance. (After Je et al.[13] with permission.)

the side etching length determines roughly the gate oxide thickness formed at a later step. The lower SiO_2 was then anisotropically etched by ECR-RIE as shown in Figure 5.11(b).

Next, a 20-nm-thick TEOS SiO_2 layer was deposited as the gate oxide, followed by annealing. The TEOS SiO_2 layer was etched by ECR-RIE as shown in Figure 5.11(c). The thickness of the gate oxide is, therefore, not given simply by thickness of the deposited TEOS SiO_2 layer of 20 nm but by a series of the process steps

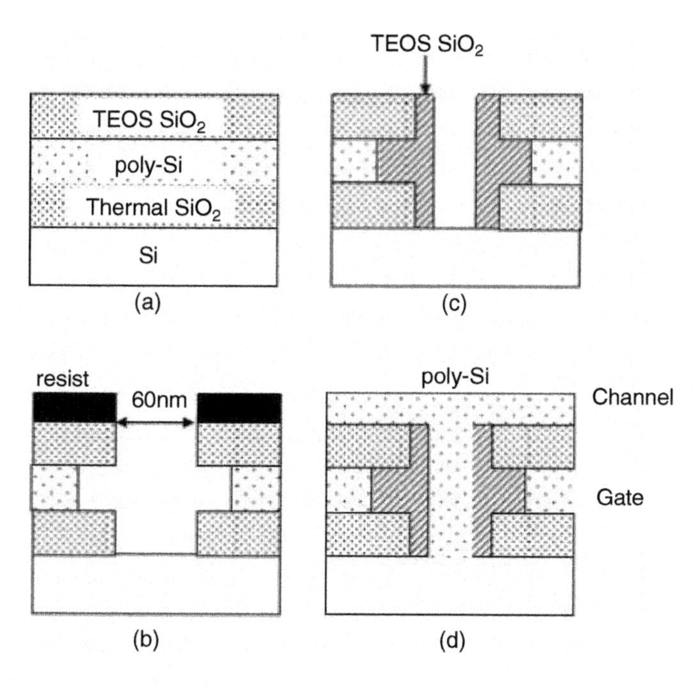

FIGURE 5.11 Fabrication processes. (a) The initial structure before lithography process. (b) The structure after the formation of the channel hole by EB lithography and etching processes. (c) The formation of the gate oxide. (d) The final structure of the vertical structure transistor.

described above; the actual thickness is estimated from observed electrical characteristics. The polySi channel was finally formed with SPC of the CVD a-Si deposited (Figure 5.11(d)) under the conditions similar to those used for the formation of the gate polySi. Because the size of grains in the polySi channel is larger than 200 nm, most of the FETs fabricated have a channel free from scattering due to GBs, which would affect the electron mean free path strongly. Although the Si grains are crystallized from a-Si with random crystal orientations, the SPC process provides more chance to have a GB-free polySi channel due to the following reason: Before depositing a-Si, the surface of the substrate was covered with little or no SiO_2. Thus, a-Si was expected to crystallize epitaxially to the Si (100) surface exposed at the bottom of the holes, leading to the absence of GBs in the channel. In fact, the electric characteristics observed for most samples were found to be consistent with theoretical ones obtained assuming no GBs in the channel. Finally, gold films were deposited both on the top polySi and the bottom Si substrate after an isolation process for the devices for obtaining the ohmic contacts.

5.3.2 Conductance Quantization in Nanoscale Vertical FETs

For the electrical measurement, the Si substrate was electrically grounded and a source-drain bias voltage was applied to the top polySi layer. By applying the gate bias the channel is depleted laterally, and the vertical current is modulated. Figure 5.12 shows the channel conductance measured at a temperature of 5 K and with a

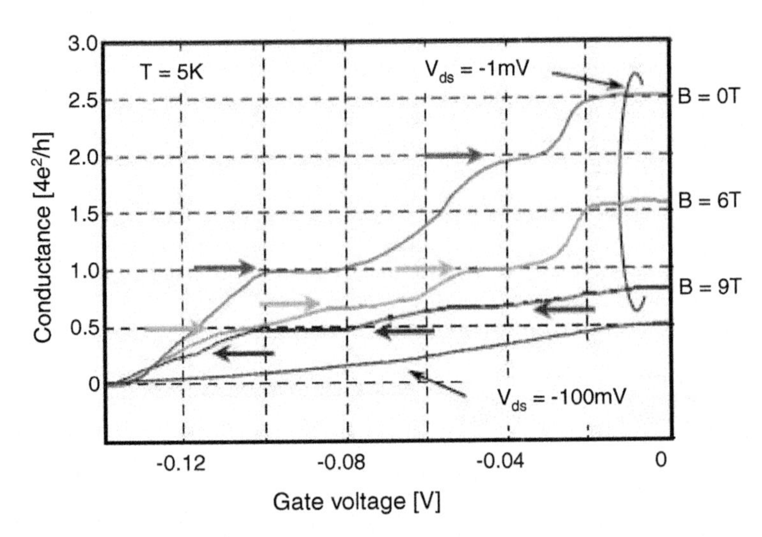

FIGURE 5.12 The conductance characteristics as a function of the gate voltage at various magnetic fields. The measurement is performed at 5 K with a source-drain bias voltage of –1 mV and –100 mV.

source-drain bias voltage of 1 mV. The conductance exhibits staircase-like characteristics with a step height of $4e^2/h$. It also shows the characteristics observed under a magnetic field of 6 and 9 T, which was applied perpendicular to the channel. The channel length of the vertical nano FET is defined by the thickness of the CVD film, and the channel width is controlled by the gate voltage. It is a great advantage over the conventional QPCs to realize a very short and narrow channel easily and with superior control. The electron mean free path was measured to be 45 nm by Hall measurement at 5 K for the large area (>100 μm) polySi film which should contain a number of GBs. In the vertical nano FETs, however, the effective electron mean free path could be longer than the length evaluated for the large area film since the device dimensions are now smaller than that. In addition, the scattering probability at the oxide interface, caused by surface roughness and interface states, is supposed to decrease significantly[2,3] since the depletion layer separated the channel away from the gate oxide interface.

The heavily doped channel enhances the confinement effect in the channel. The depletion layer thickness, which is narrow in the heavily doped polySi channel, can be controlled by the gate voltage. The low electric field in the depletion layer, which depends on the density of the dopants in the channel, forms the parabolic-like lateral confinement potential profile. In that case, energy intervals between subband levels near the depletion layer are reduced, leading to scattering between subbands. The strong electric field in the more heavily doped channel makes the potential profile steep in the channel with the same depletion length and prevents subbands near the depletion layer from being smeared. Thus, the scattering between energy subbands can be reduced. These features of the vertical nano FETs facilitate observing ballistic transport effects.

All curves in Figure 5.12 saturated above a gate voltage of –20 mV. This saturation effect is presumably attributed not to ballistic transport but to the shielding effect caused by ionized impurities in the gate oxide, because the saturation was observed even for a source-drain bias voltage V_{ds} of –100 mV, where no ballistic transport was observed. The impurities may be included in the gate oxide during the etching process of the gate oxide. With a source-drain bias of 100 mV, the device worked like a conventional normally-on field effect transistor (FET). By comparing the characteristics for V_{ds} of –1 mV and –100 mV, a higher current drivability is apparent in the ballistic transport regime. From the conventional FET characteristics observed for V_{ds} of –100 mV, the device dimensions were evaluated. The quantized subband energies were then calculated from the device dimensions based on the effective mass approximation and used for analyzing the characteristics in ballistic transport regime. First, let us discuss the effect of the channel shape. The original channel shape patterned by EB lithography was square. However, it is literally difficult to reproduce the patterned shape in the actual etching process since the overall dimensions are very small. If the channel shape were symmetric, the subband energies in the channel would be degenerate with energies given by the one-dimensional confinement model. In that case, the second conductance step height in the ballistic transport regime should be twice as large as the unit value of $4e^2/h$. However, the observed steps showed the same height, and this fact indicates that the channel shape was asymmetric.

Another deviation between the patterned and the actual structure of great consequence to ballistic transport is an angular misorientation of the channel against the Si substrate.[18] The direction of the current path is expected to be along the Si [100] axis because the polySi film is deposited on the Si [100] substrate. However, the planar view orientation of the channel depends not on the Si substrate but on the mechanical setting on the EB instrument. In the conduction band of Si, there are six ellipsoidal valleys in k-space. We, hereafter, define axes in xyz space as follows: x-axis is [100], y-axis [010], z-axis [001]. In Si, the valley in k-space is not at the Γ point but near the X point. Thus, effective mass equations in each axis are given as follows:

$$\left[\frac{1}{2m_l}\left(-i\hbar\frac{\partial}{\partial x}\pm\hbar k_{Si}\right)^2-\frac{\hbar^2}{2m_t}\frac{\partial^2}{\partial y^2}-\frac{\hbar^2}{2m_t}\frac{\partial^2}{\partial z^2}+V(y,z)\right]\phi=E\phi \quad \text{(for } x\text{-axis)} \tag{5.9}$$

$$\left[-\frac{\hbar^2}{2m_t}\frac{\partial^2}{\partial x^2}+\frac{1}{2m_l}\left(-i\hbar\frac{\partial}{\partial y}\pm\hbar k_{Si}\right)^2-\frac{\hbar^2}{2m_t}\frac{\partial^2}{\partial z^2}+V(y,z)\right]\phi=E\phi \quad \text{(for } y\text{-axis)} \tag{5.10}$$

$$\left[-\frac{\hbar^2}{2m_t}\frac{\partial^2}{\partial x^2}-\frac{\hbar^2}{2m_t}\frac{\partial^2}{\partial y^2}+\frac{1}{2m_l}\left(-i\hbar\frac{\partial}{\partial z}\pm\hbar k_{Si}\right)^2+V(y,z)\right]\phi=E\phi \quad \text{(for } z\text{-axis)} \tag{5.11}$$

where k_{Si} is the distance of the center of the ellipsoids from the Γ point in k-space, $m_t = 0.19\ m_0$ the transverse mass, $m_l = 0.98\ m_0$ the longitudinal mass, $\hbar$ the Planck's constant divided by 2π, E the energy eigenvalue, ϕ the wave function. V(y,z) represents the cross-sectional potential in the channel. The channel is surrounded by the depletion layer with a high electric field and SiO_2. Thus, we assume that the channel is a one-dimensional confined system surrounded by an infinite potential barrier. By solving these equations, the subband energies in the narrow channel are evaluated for various shapes of the channel. In this case, the most critical dimension is that of the narrowest constriction along the channel. Figure 5.13(a) plots the subband energies as a function of a ratio of L_y to L_z calculated for the constant cross-sectional square ($L_y \times L_z = 10 \times 10$ nm^2) where L_y and L_z are the widths of the channel along the y- and z-axes, respectively. Figure 5.13(b) shows the subband energies as a function of the angle θ as depicted in the inset. At an angle of 45°, y- and z-valleys are degenerate because of symmetry in the *yz* plane. This theoretical analysis indicates that the channel shape and angle are very important parameters, and fabrication technologies with high precision are needed for controlling the device characteristics.

The device structural parameters were estimated using the characteristics observed in the nonballistic transport regime, where the device works like a normally-on FET, and those in the ballistic transport without a magnetic field, where the conductance steps of $4e^2/h$ are observed. When the lateral channel geometries are assumed as depicted in Figure 5.14(a), the calculations reproduced the experimental characteristics in the nonballistic and ballistic regimes as shown in Figure 5.14(b) and Figure 5.14(d), respectively. As shown in Figure 5.14(a), the lateral geometries of the channel are modeled to be asymmetric, and the planar view orientation is best fitted with a tilt of 12° against the flat Si substrate orientation. Because samples were prepared after cutting an Si wafer into pieces and set into the EB lithography system, it is plausible to assume such deviations in the shape and angle from the designed pattern. Figure 5.14(b) also shows the dependence of the channel widths on the gate voltage. The subband energies are then evaluated in the effective mass approximation as shown in Figure 5.14(c). The Fermi energy shown in Figure 5.14(c) was estimated from the step position of the observed characteristic at B = 0 T. The doping level and the resistivity used as fitting parameters are 4×10^{17} /cm^3 and 0.0045 Ωcm, respectively. As mentioned earlier, the shielding effect in the gate oxide is expected to shift the conductance characteristics by a gate voltage of –20 mV, and this gate voltage shift was included in the simulation. In Figure 5.14(b), the calculated curve agrees quite well with the experimental results. However, the calculated curve in the ballistic transport regime agrees only for the height of the conductance plateau and the approximate position of voltage step as shown in Figure 5.14(d). The difference may be attributed to the effects of thermal fluctuation of electrons as explained in Section II and those of the finite source-drain bias voltage.[19] At finite temperatures, electrons travel through discrete subband levels in the point contact with a spread of energy due both to the finite temperature and the source-drain bias voltage. The characteristics can thus be broadened by the energy spread of the incoming electrons, leading smeared experimental curves. Figure 5.15(b) shows a dramatic smearing of the quantized conductance

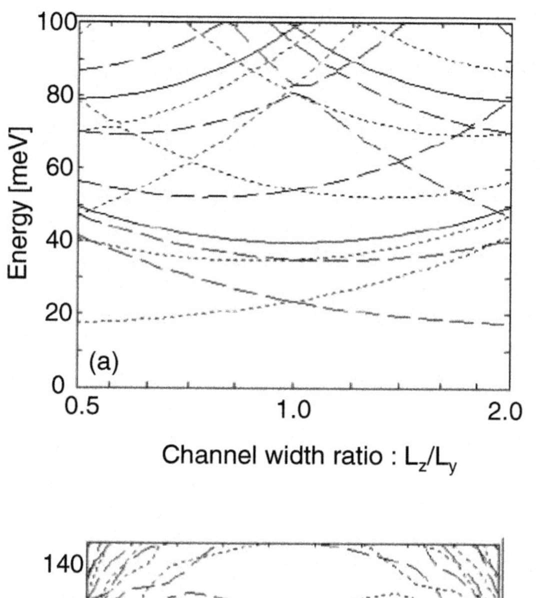

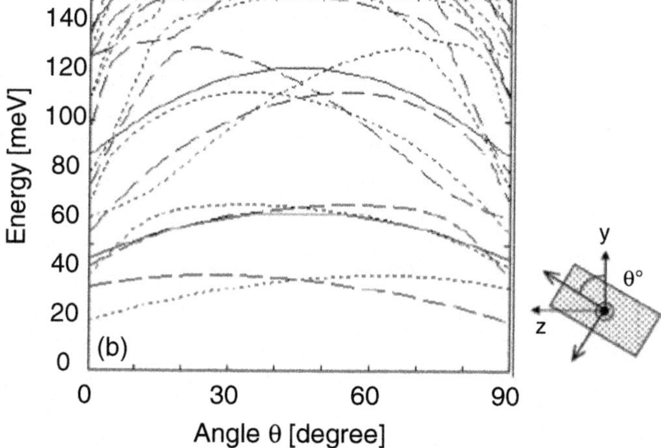

FIGURE 5.13 (a) Calculated subbands as a function of L_z/L_y at $L_y \times L_z$=100 nm^2. The solid, dashed, and dotted curves indicate the subbands resulted from the x-, y-, and z-axis. (b) Calculated subbands as a function of θ at L_y=12 nm and L_z=8 nm. θ is an angle between y-axis and L_y as shown in the inset.

characteristics observed for another sample. Taking into account the present experimental circumstances of the thermal energy $4k_BT$ of 1.7 meV and voltage-gained energy of 1 meV, it is likely that the thermal fluctuation of electrons dominates the difference between the experimental and theoretical results.

5.3.3 Characteristics under a Magnetic Field

With an increasing magnetic field, the conductance decreased in such a way that values of the quantized conductance became half or quarter, as shown in Figure

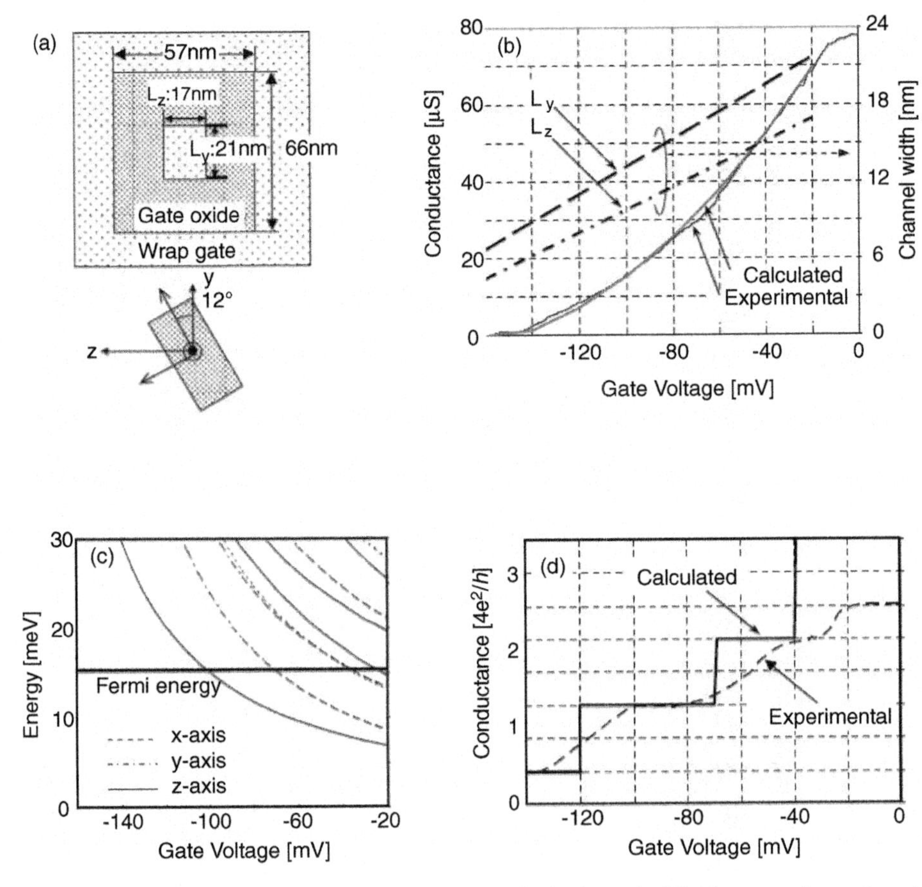

FIGURE 5.14 (a) Plan view structure of the vertical channel. (b) The experimental and calculated conductance characteristics as a function of the gate voltage at the source-drain bias voltage of 100 mV. The channel size is also shown at the source-drain bias voltage of 1 mV. (c) The calculated subbands in the point contact as a function of the gate voltage. Based on the consideration of the shielding effect from ionized impurities in the gate oxide, the curves are shifted by –20 mV in x-axis. (d) The experimental and calculated conductance characteristics as a function of the gate voltage at the source-drain bias voltage of –1 mV in ballistic transport regime. The bold and the dashed line indicate the calculated and experimental characteristic, respectively.

5.12. These characteristics likely resulted from spin and valley splitting in a strong magnetic field.[20–22] Spin splitting follows the Zeeman effect; $\mu_B g^* B$, where μ_B is the Bohr magnetron and g^* the effective Landé g factor. The thermal energy at 5 K (0.43 meV) is smaller than the difference of spin splitting of 0.7 meV at a magnetic field of 6 T. Thus, it is possible indeed to observe spin splitting. On the other hand, the condition for observation of the Landau level ($B\mu > 1$ where μ is electron mobility) is satisfied. In Si, there are three pairs of valleys. So there are two kinds of subband shift on each pair of valleys, positive and negative. There are many models for valley

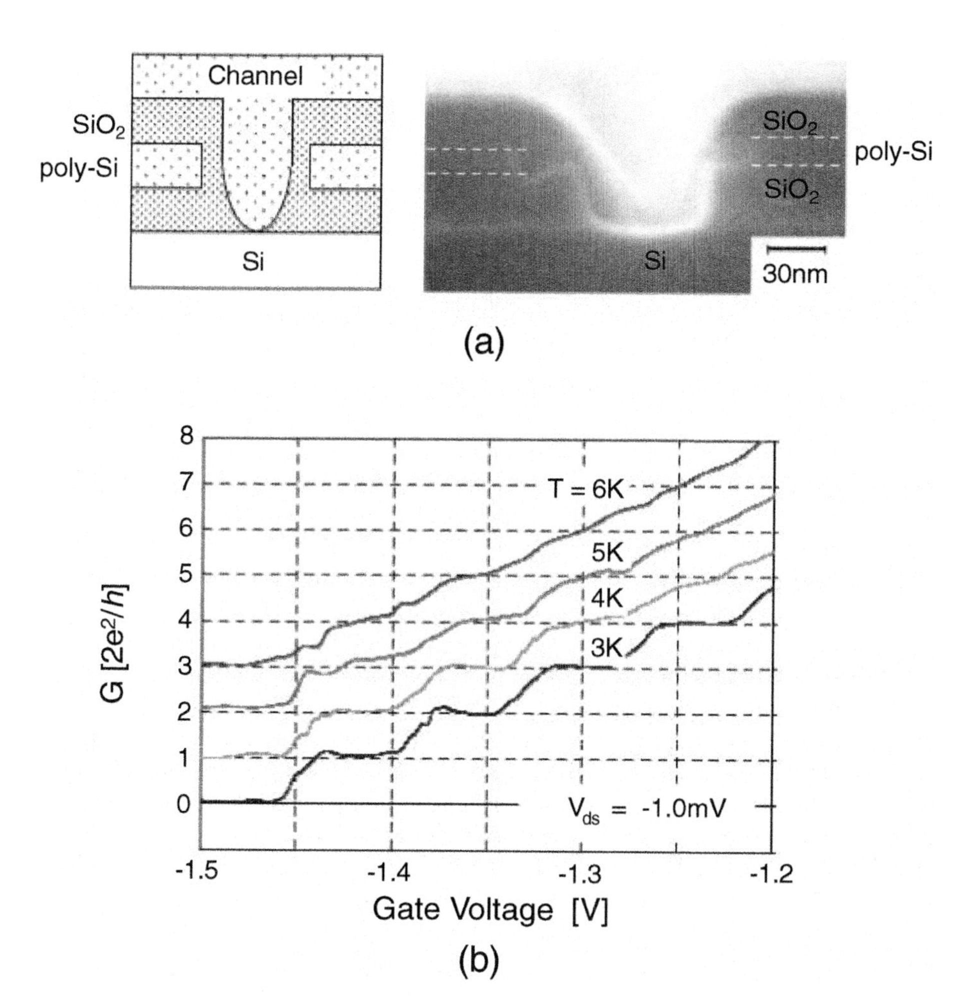

FIGURE 5.15 The calculated subbands as a function of the gate voltage at a magnetic field of 6 T and 9 T. The calculation is based on the device structure shown in Figure 4 (a). The Fermi energy is located at 15 meV.

splitting in strong inversion layers in metal-oxide-semiconductor (MOS) structures, that is, electron tunneling effect and surface scattering.[22] For a simple description of valley splitting, the vector potential ***A*** is added to Equations (5.9), (5.10), and (5.11). Since the magnetic field was applied perpendicular to the channel, the vector potential is defined as $A = (zB_y - yB_z, 0, 0)$, where B_y and B_z are the magnetic field along the *y*- and *z*-axes, respectively, and they satisfy the conduction $B^2 = B_y^2 + B_z^2$. In this case, effective mass equations are

$$\left[\frac{1}{2m_l}\left\{-i\hbar\frac{\partial}{\partial x} \pm \hbar k_{Si} + e(zB_y - yB_z)\right\}^2 - \frac{\hbar^2}{2m_t}\frac{\partial^2}{\partial y^2} - \frac{1}{2m_t}\hbar^2\frac{\partial^2}{\partial z^2} + V(y,z)\right]\phi = E\phi$$

(for *x*-axis) (5.12)

$$\left[-\frac{1}{2m_t}\left\{-i\hbar\frac{\partial}{\partial x}+e(zB_y-yB_z)\right\}^2+\frac{1}{2m_l}\left(-i\hbar\frac{\partial}{\partial y}\pm\hbar k_{Si}\right)^2-\frac{\hbar^2}{2m_t}\frac{\partial^2}{\partial z^2}+V(y,z)\right]\phi=E\phi$$

(for y-axis) (5.13)

$$\left[-\frac{1}{2m_t}\left\{-i\hbar\frac{\partial}{\partial x}+e(zB_y-yB_z)\right\}^2-\frac{\hbar^2}{2m_t}\frac{\partial^2}{\partial y^2}+\frac{1}{2m_l}\left(-i\hbar\frac{\partial}{\partial z}\pm\hbar k_{Si}\right)^2+V(y,z)\right]\phi=E\phi$$

(for z-axis) (5.14)

Figure 5.16(a) and Figure 5.16(b) show subband levels as a function of the gate electrode at $B_z = 6$ T and 9 T, respectively. The Fermi energy, derived from the conductance characteristics without magnetic field in Section III.2, is also shown. These curves indicate that there are more than eight subbands at the gate voltage of –20 mV, and therefore, the conductance should be more than $8 \times e^2/h$. However, observed curves indicated that conductance at 6 T and 9 T was less than $8 \times e^2/h$ and $4 \times e^2/h$, respectively. These deviations are quite large and worth further discussion. A plausible explanation is the increase of the series resistance connecting to the small point, which was contact caused by magnetic resistance or electron scattering resulting from cyclotron movement, as shown in Figure 5.17(a). Usually, cyclotron movement can reduce the electron scattering from the impurities. However, when the radius of the cyclotron movement is comparable to the size of leads connecting to the point contact, electron-scattering at channel edges and reflection at the point contact increase.[23] In that case, the measured conductance G_m is related to the conductance of the point contact G_p and the series resistance G_s^{-1} as

$$G_m = \frac{1}{G_p^{-1}+G_s^{-1}} \tag{5.15}$$

G_p can be obtained from Equation (5.15). Since experimental values of G_m were not available, G_m was treated as a fitting parameter. Figure 5.17(b) shows the corrected conductance of the point contact under various magnetic fields, with $G_0 = 4e^2/h$. Calculated curves are roughly the same as the observed curves. The deviation results from the smearing by finite temperature. Thus, it suggests that the observed characteristics were influenced by both the ballistic transport effect from the point contact and the series resistance from the Si film. Further discussion, however, is needed for the assumed values of G_s because the series resistance may be changed by not only magnetic field but also the channel size controlled by the gate electrode. The model of the calculation also should be discussed because there is, as yet, no completely satisfactory model of valley splitting even though many models are proposed, for example, the one-dimensional zero effective mass theory can account for the coupling of two valleys.[21]

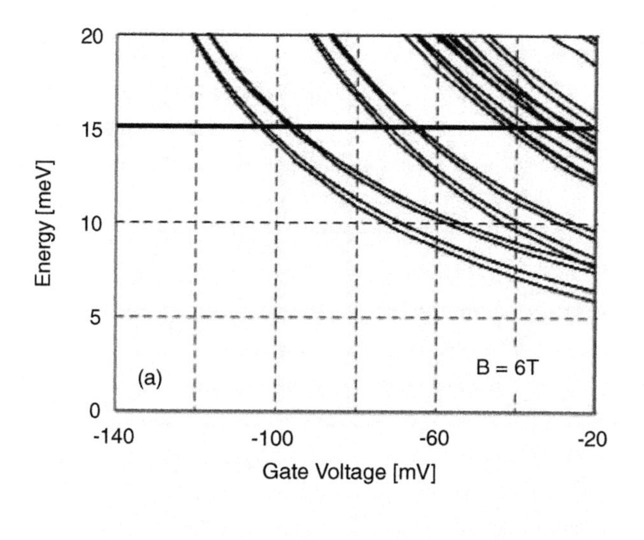

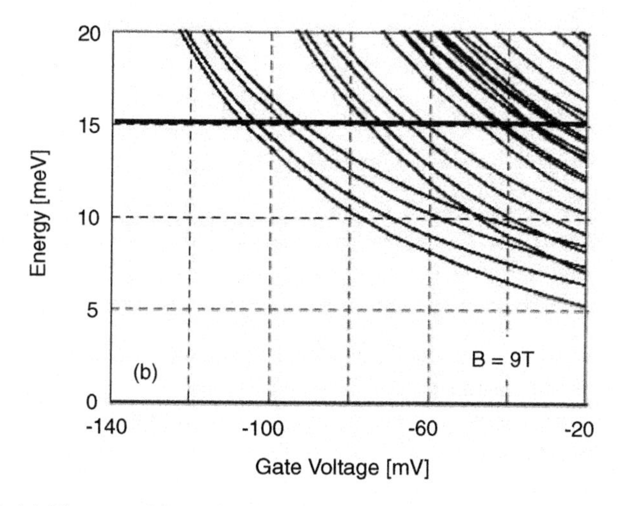

FIGURE 5.16 (a) The parasitic resistance elements caused under magnetic field. (b) The experimental and calculated conductance characteristics at a magnetic field of 6 T and 9 T. For clarity, curves at 9 T are shifted by $2 \times 4e^2/h$. The bold and the dotted lines indicate the calculated from effective mass equations and the observed characteristic, respectively. The dashed lines indicated the curves subtracted the series resistance, which is increased by applying a magnetic field. G_0 indicates $4e^2/h$.

5.3.4 Effects of Cross-Sectional Channel Geometries

In this section, we investigate the effect of the cross-sectional channel shape perpendicular to the substrate. The channel is formed after the etching process. In some conditions, the etched shape becomes tapered like Figure 5.15(a). The sample with

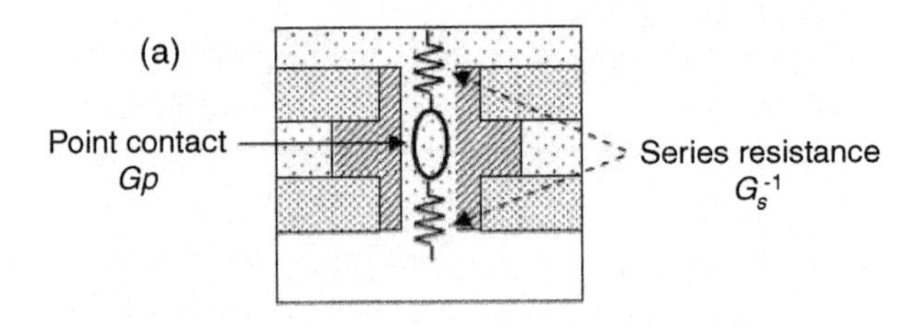

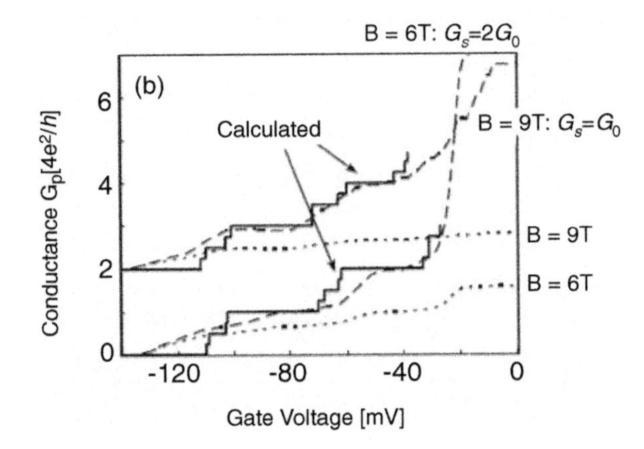

FIGURE 5.17 (a) The cross-sectional structure of a vertical transistor with a tapered channel and its SEM image. (b) The conductance characteristics from the device with a tapered channel at various temperatures. The source-drain bias voltage is 1 mV. For clarity, all curves are shifted by $0.5 \times 4e^2/h$.

tapered channel shows quantized conductance characteristics resulting from ballistic transport effect as shown in Figure 5.15(b). However, the step height of quantized conductance became one-half of the height for a point contact. The reason for this phenomenon is not clear. One speculation is the formation of double point contact, leading to half conductance. Figure 5.18 shows the change of depletion layer by the gate electrode. In proper device structures, two point contacts are formed near the top and the bottom of the polySi gate electrode. When the distance between two point contacts is comparable to the electron mean free path, the conductance becomes half.[24] However, the probability of double point contact formation is expected to be small. Here we discuss another possible reason for the observed step height. Figure 5.19 depicts the energy profile in the device. E_n and E_{n+1} indicate subband energy with index n and n+1, respectively, in the point contact. E_{FL} and E_{FR} indicate the Fermi energy of left and right lead connecting to the point contact, respectively. At the small source-drain bias voltage V_b like Figure 5.19(a), the conductance is $n \cdot G_0$, where G_0 is the unit conductance of ballistic transport. When V_b increases and one subband E_{n+1} is sandwiched between the Fermi energies of two leads connecting to the point contact as shown in Figure 5.19(b), quantized conductance has a fractional form, $(n+1/2) \cdot G_0$. Further increase of V_b returns G to $n \cdot G_0$. This phenomenon strongly

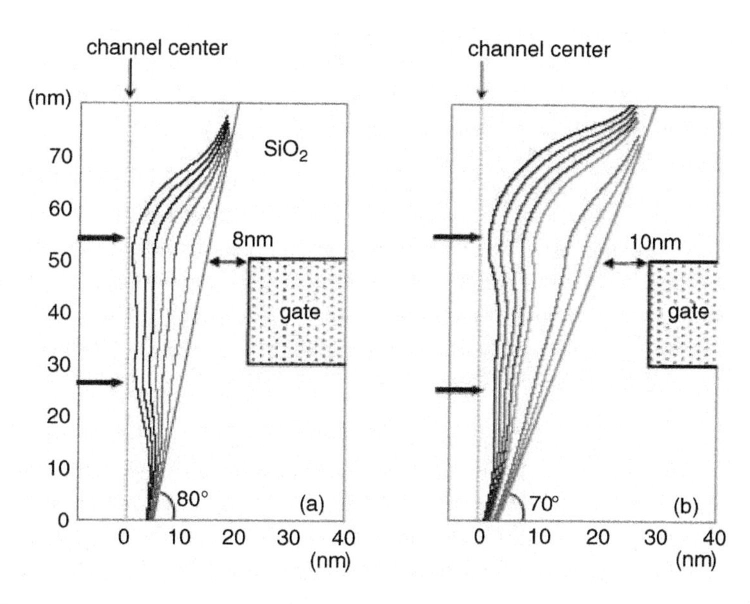

FIGURE 5.18 The change of the depletion layer edge for two structures with taper angle of the channel of 70° and 80° when the gate voltage is changed from −1.45 V to 0 V. The dashed lines indicate the center of the channel. The arrows indicate a point contact formed by the depletion layer.

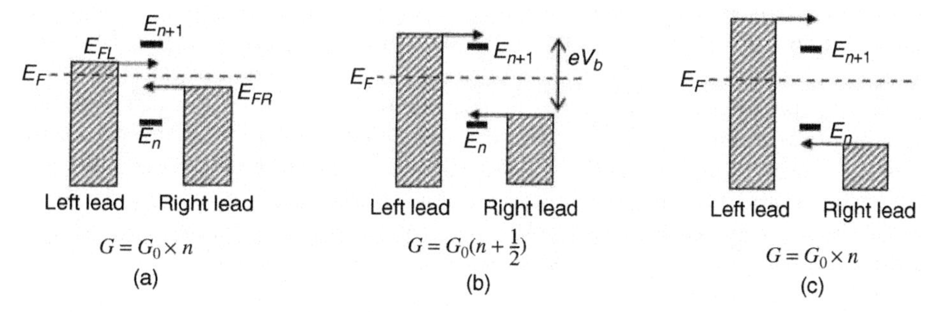

FIGURE 5.19 The energy profile in the device. E_n and E_{n+1} indicate subband energy with index n and n+1, respectively, in the point contact. E_{FL} and E_{FR} indicate the Fermi energy of left and right lead connecting to the point contact, respectively. When the source-drain bias voltage V_b is small as shown in (a), the conductance G is $n \times G_0$, where G_0 is unit conductance of ballistic transport $4e^2/h$. The increase of V_b makes $G=(n+1) \times G_0$ like (b), and more increase makes $G=n \times G_0$ like (c).

depends on the relation between subband levels and two Fermi energies as shown in Figure 5.19. Subband levels in the channel depend on the channel geometry as mentioned in Section III.2. Thus, only a sample with a tapered channel displayed the ballistic transport effect with a step size of one-half quantized conductance.

5.4 SUMMARY AND FUTURE SUBJECTS

This chapter has been devoted to the experimental and theoretical study of ballistic transport in Si. Because the electron mean free path in Si is generally shorter than those obtained for compound semiconductors, early studies based on the lateral QPC structure have not clarified the phenomenon fully. The vertical nano FETs have then been introduced, which enabled us to control the structural parameters of the QPC channel in the decananometer regime in a more precise manner. Clear conductance quantization has been observed at temperature of 5 K. An increase in the current drivability has been demonstrated in the ballistic transport regime compared with the conventional FET characteristics in the nonballistic transport regime. In addition, the splitting of the valley and spin degeneracy has been observed under an applied magnetic field.

For any future applications of the ballistic conduction in Si, device operation at higher temperatures and with a higher source-drain voltage is necessary, which leads to higher device drivability. It is vital to make a device even smaller and the mean free path longer. In the vertical nano FET structures the channel length may easily be shrunk by decreasing the thickness of gate electrode films. For achieving a long mean free path, a lightly doped channel is desired. However, this leads to high parasitic resistance. Ideally, only the intrinsic part of the channel surrounded by the gate oxide should be lightly doped. The nano FETs fabricated so far have adopted the heavily doped polySi channel, which provided the scattering centers, resulting in smeared ballistic transport effect. The present investigation indicates that it is important to control the dopant distribution precisely as well as the structural parameters.

Before closing this chapter we briefly introduce a recent study of ballistic electron transport in Si nanostructures which are formed using nanocrystalline silicon (nc-Si) dots. We have so far looked at the Si nanostructures fabricated by using modern nanolithography techniques. In addition to this *top-down* approach, alternative *bottom-up* technologies recently attract lots of attention for fabricating Si based nanostructures. Using Si nanodots (SiNDs) and nanowires (SiNWs)[25–28] as a building block for forming Si nanodevices, precisely controlled nanostructures may be formed by assembling the SiNDs and SiNWs. The SiNDs may be fabricated by using a VHF plasma enhanced deposition of silane with a hydrogen gas pulse sequence.[29] This technique facilitates separating the nucleation and crystal growth process and helps to fabricate nc-Si particles with diameters less than 10 nm and dispersions of 1 nm.[30] Particle diameter down to 3 nm has also been reported recently.[31] The interparticle tunnel barriers can be formed by in-situ oxidation or nitridation in a controlled manner. Another popular approach for obtaining the SiNDs is to adopt porous Si[32] formed by using photoanodization of the Si substrate. The surface of the nc-Si islands formed in the substrate can be oxidized selectively by electrochemical oxidation. Formation of a linear chain of the nc-Si dots with a diameter as small as 5 nm has been observed.[33] Electron transport properties of these SiNDs interconnected with thin oxide layers have aroused growing interest due to the experimental observation of ballistic electron emission.[34] High-energy electron emission from an SiND diode have been observed and have been investigated intensively as a potential candidate for cold cathodes in future flat panel displays due to its low turn-on voltage and high current stability. The energy distribution measured for the emitted electrons implies that the electrons travel through the SiNDs without suffering from the frequent energy loss processes and are emitted into vacuum in a quasi-ballistic manner. A microscopic mechanism of the quasi-ballistic transport and emission phenomenon should be different from that we studied for the QPC structures. A possibility of the reduced electron-phonon interactions in the SiND clusters has been suggested theoretically as a physical origin.[35] The electronic and phononic states calculated for a one-dimensional array of SiNDs interconnected with thin oxide layers predicted that the system has a wide variety of interesting features such as phonon bandgap, phonon confinement, reduction of electron-phonon scattering potential, and strong energy loss suppression.[35,36] These unique physical properties of the assembled SiND systems may open a different path to observe and control ballistic transport in Si nanostructures, which is unlikely to be realized as long as we play with the conventional *top-down* approaches for device miniaturization.

REFERENCES

1. Higurashi, H., Iwabuchi, S., and Nagaoka, Y., Conductance fluctuation in mesoscopic quantum wires near the ballistic regime, *Surf. Sci.*, 263, 382, 1992.
2. van Wees, B.J., van Houten, H., Beenaker, C.W.J., Williamson, J.G. Kouwenhoven, L.P., van der Marel, F. and Foxon, C.T., Quantized conductance of point contacts in a two-dimensional electron gas, *Phys. Rev. Lett.*, 60, 848, 1988.
3. Wharam, D.A., Thornton, T.J., Newbury, R., Pepper, M., Ahmed, H., Frost, J.E.F., Hasko, D.G., Peacock, D.C., Ritchie, D.A., and Jones, G.A.C., One-dimensional transport and the quantisation of the ballistic resistance, *J. Phys.*, C21, L209, 1988.
4. Hirayama, Y., Saku, T. and Horikoshi, Y., Electronic transport through very short and narrow channels constricted in GaAs by highly resistive Ga-implanted regions, *Phys. Rev.*, B39, 5535, 1989.
5. Hirayama, Y. and Saku, T., Conductance characteristics of ballistic one-dimensional channels controlled by a gate electrode, *Appl. Phys. Lett.*, 54, 2556, 1989.
6. Fertig, H.A. and Halperin, B.I., Transmission coefficient of an electron through a saddle-point potential in a magnetic field, *Phys. Rev.*, 36, 7969, 1987.
7. Nakajima, Y., Takahashi, Y., Horiguchi, S., Iwadate, K., Namatsu, H., Kurihara, K., and Tabe, M., Quantized conductance of a silicon wire fabricated by separation-by-implanted- oxidation technology, *Jpn. J. Appl. Phys.*, 34, 1309, 1995.
8. Tekman, E. and Ciraci, S., Novel features of quantum conduction in a constriction, *Phys. Rev.*, B39, 8772, 1989.
9. Takeuchi, K. and Newbury, R., Periodic conductance fluctuations in quasi-one-dimensional metal-oxide-semiconductor field-effect transistors with shallow-trench isolations, *Phys. Rev.*, B43, 7324, 1991.
10. Wang, S.L., van Son, P.C., van Wees, B.J., and Klapwijk, T.M, Quantum conductance of point contacts in Si inversion layers, *Phys. Rev.*, B46, 12873, 1992.
11. Warren, A.C., Antoniadis, D.A., and Smith, H.I., Quasi one-dimensional conduction in multiple, parallel inversion line, *Phys. Rev. Lett.*, 56, 1858, 1986.
12. Wieser, U., Kunze, U., Ismail, K. And Chu, J.O., Quantum-ballistic transport in an etch-defined Si/SiGe quantum point contact, *Appl. Phys. Lett.* 81, 1726, 2002.

13. Je, M., Han, S., Kim, I., and Shin, H., A silicon quantum wire transistor with one-dimensional subband effects, *Solid State Electron.*, 44, 2207, 2000.
14. Nishiguchi, K., and Oda, S., Conductance quantization in nanoscale vertical structure silicon field-effect transistors with a wrap gate, *Appl. Phys. Lett.*, 76, 2922, 2000.
15. Nishiguchi, K. and Oda, S., Ballistic transport in silicon vertical transistors, *J. Appl. Phys.*, 92, 1399, 2002.
16. Geis, M.W., Flanders, D.C., and Smith, H.I., Crystallographic orientation of silicon on an amorphous substrate using an artificial surface-relief gating and laser crystallization, *Appl. Phys. Lett.*, 35, 71, 1979.
17. Geis, M.W., Tsaur, B-Y., and Franders, D.C., Graphoepitaxy of germanium on gratings with square-wave and sawtooth profiles, *Appl. Phys. Lett.*, 41, 526, 1982.
18. Horiguchi, S., Naksjima, Y., Takahashi, Y., and Tabe, M., Energy eigenvalues and quantized conductance values of electrons in Si quantu, wires on {100} plane, *Jpn. J. Appl. Phys.*, 34, 5489, 1995.
19. Koester, S.J., Brar, B., Bolognesi, C.R., Caine, E.J., Patlach, A., Hu,E.L., and Kroemer, H., Length dependence of quantized conductance in ballistic constrictions fabricated on InAs/AlSb quantum wells, *Phys. Rev.*, B 53, 13063, 1996.
20. Ando, T., Electronic properties of two-dimensional systems, *Rev. Mod. Phys.,* 54, 437, 1982.
21. Ohkawa, F.J. and Uemura, Y., Valley splitting in an n-channel (100) inversion layer on p-type substrate, *Surf. Sci.*, 58, 254, 1976.
22. Sham, L.J. and Nakayama, M. Effect of interface on the effective mass approximation, *Surf. Sci.*, 73, 272, 1978.
23. de Picciotto, R., Stormer, H.L., Pfeiffer, L.N., Baldwin, K.W., and West, K.W., Four-terminal resistance of a ballistic quantum wire, *Nature*, 411,51, 2001.
24. Hirayama, Y., and Saku, T., Transport characteristics of series ballistic point contacts, *Phys. Rev.*, B 41, 2927, 1990.
25. Cui Y. and Lieber., C.M., Functional nanoscale electronic devices assembled using silicon nanowire building blocks, *Science*, 291, 851, 2001.
26. Huang, Y., Duan, X., Wei, Q., and Lieber, C.M., Directed assembly of one-dimensional nanostructures into functional networks, *Science*, 291, 630,, 2001.
27. Duan, X., Niu, C., Sahi, V., Chen, J., Parce, J.W., Empedocles, S., and Goldman, J.L., High-performance thin-film transistors using semiconductor nanowires and nanoribbons, *Nature*, 425, 274, 2003.
28. Piscanec, S., Cantoro, M., Ferrari, A.C., Zapien, J.A., Lifshitz, Y., Lee,S.T., Hofmann, S., and Robertson, J., Raman spectroscopy of silicon nanowires, *Phys. Rev.*, B68, 241312, 2003.
29. Oda, S. and Otabe, M., Preparation of nanocrystalline silicon by pulsed plasma processing, *Mater. Res. Soc. Proc.*, 358, 721, 1995.
30. Ifuku, T., Otabe, M., Itoh, A., and Oda, S., Fabrication of nanocrystalline silicon with small spread of particle size by pulsed gas plasma, *Jpn. J. Appl. Phys.*, 36, 4031, 1997.
31. De Blauwe, J., et al., A novel aerosol-nanocrystal floating gate device for non-volatile memory applications, *IEDM Tech. Dig.*, 683, 2000.
32. Koshida, N. and Koyama, H., Visible electroluminescence from porous silicon, *Appl. Phys. Lett.*, 60, 347, 1992.
33. Cullis, A.G., Canham, L.T., and Calcott, P.D.J., The structural and luminescence properties of porous silicon, *J. Appl. Phys.*, 82, 909, 1997.
34. Koshida, N., Sheng, X., and Komoda, T., Quasiballistic electron emission from porous silicon diodes, *Appl. Surf. Sci.*, 146, 371, 1999.

35. Uno, S., Nakazato, K., Yamaguchi, S., Kojima, A., Koshida, N., and Mizuta, H., New insights in high-energy electron emission and underlying transport physics of nanocrystalline Si, *IEEE Trans. Nanotechnol.*, 2, 301, 2003.
36. Uno, S., Mori, N., Nakazato, K., Koshida, N. and Mizuta, H., Reduction of acoustic phonon deformation potential in one-dimensional array of Si quantum dot interconnected with tunnel oxide, to be published in *J. Appl. Phys.*, 2005.

6 Resonant Tunneling in Si Nanodevices

Michiharu Tabe, Hiroya Ikeda, and Yasuhiko Ishikawa

6.1 INTRODUCTION

6.1.1 OUTLINE OF RESONANT TUNNELING

6.1.1.1 Early Work on Resonant Tunneling

The framework of resonant tunneling formalism in the realistic three-dimensional system was proposed by Tsu and Esaki[1] in 1973 in the general treatment of a finite superlattice with N periods. The voltage dependence of the tunneling current was calculated by assuming the conservation of the total electron energy and of its momentum transverse to the tunneling direction. Resonance is shown to give rise to current maxima at such applied voltages that the Fermi energy of the electrode coincides with the quasi-bound electronic states in the potential well. In 1974, Chang, Esaki and Tsu reported,[2] for the first time, resonant tunneling of electrons in double-barrier structures of AlGaAs/GaAs/AlGaAs grown by molecular beam epitaxy. The resonance manifests itself as peaks or humps in the tunneling current at voltages near the quasibound states of the potential well, although these features were observed only at low temperatures.[2] Thereafter, characteristics of double-barrier resonant tunneling diodes (RTD) had been significantly improved, leading to room temperature observation of negative differential conductance (NDC),[3] and possible applications of RTDs were proposed based on the high-speed nature of RTD such as detection and generation of high-frequency electromagnetic waves[4] and a hot electron transistor.[5]

In RTDs, the diode current is usually considered to be the sum of two competing components:

$$J = J_{RT} + J_{EX}$$

The term J_{RT} is the resonant tunneling current through the quasi-bound state in the potential well and gives rise to a peaked structure in a current-voltage curve. It has been demonstrated by Tsuchiya and Sakaki[3] that J_{RT} depends on such parameters as barrier widths L_B, barrier heights V_B and well width L_W. All other nonresonant current components are referred to as the excess current J_{EX}, including inelastic tunneling current, nonresonant tunneling current through higher-lying quasicontinuum levels, and thermionic current over the barriers.

6.1.1.2 Resonant Tunneling in Si-Based Materials — Si/SiGe and Si/SiO_2

Thus, resonant tunneling through semiconductor double-barrier structures has been extensively studied because of the physical and technological interests, and, as mentioned above, NDC has been reported mostly in III-V compound semiconductor. Among Si-based materials, Si/SiGe heterostructures have been most intensively studied. SiGe RTD structures were first designed for hole tunneling, since one can readily obtain large valence-band offsets.[6] However, since the hole effective mass is large, the observed current peak smeared out at high temperatures. Ismail et al. reported electron tunneling with a current peak, observed even at room temperature.[7] Thereafter, Suda and Koyama[8] reported a triple barrier structure for electron tunneling, and they obtained a high peak-to-valley ratio of more than seven, indicating that the double-well (or triple-barrier) coresonance is useful in device application.

In contrast, there has been little work on resonant tunneling in Si/SiO_2 systems, even though there is a wider variety of potential applications involving integrated circuits. In Si/SiO_2 systems, NDC had not been observed until the recent work by the authors,[9,10] and only nonlinear current-voltage curves with weak inflections, ascribed to resonant tunneling, were observed.[11,12] The absence of NDC is primarily because of difficulties in formation of a high-quality, single-crystalline Si layer sandwiched by amorphous barriers, in contrast to heteroepitaxial systems.

In this chapter, the recent progress of resonant tunneling studies in $SiO_2/Si/SiO_2$ double-barrier structures is presented. When capacitances of the tunnel barriers are extremely small, as in the order of aF (10^{-18} Farad) or the lateral size of RTD enters nanometer-scale, resonant tunneling and Coulomb blockade may coexist and both characteristics will appear in *I-V* curves. Such experiments of zero-dimensional tunneling are also described.

6.1.2 Quantum Confinement Effect in a Thin Si Layer[13]

Quantum confinement effect in a thin two-dimensional Si layer should be the basis of resonant tunneling, because electrons tunnel through quasibound states in the Si layer. There are a number of reports on classical and quantum confinement effects in ultrasmall Si structures by studying photoluminescence, photoemission spectroscopy or electrical characteristics. Most of the studies deal with Si dot or dot-like systems, such as porous Si and deposited Si nuclei, and there have been only a few reports on quantum confinement effects in 2D systems. In this section, we briefly describe our previous work of x-ray photoelectron spectroscopy (XPS) for two-dimensional single-crystalline Si.[13] There are also other relevant references by Lu et al.[14] on the quantum confinement effect of SiO_2/amorphous-Si superlattices and by Miyazaki et al.[15] on SiN/ amorphous-Si.

We studied valence-band spectra of a thin (10 to 1 nm-thick) top Si layer on a buried SiO_2 layer in a silicon-on-insulator (SOI) wafer. We particularly focused on the energy shift of the valence-band maximum (VBM) over the course of repetitive steps of Si thinning by oxidation and oxide removal. The SOI wafer was fabricated by a wafer bonding technique in our laboratory, and the details are described in the following section. (See Section 6.2.1 and Reference 16). As the first step in thinning,

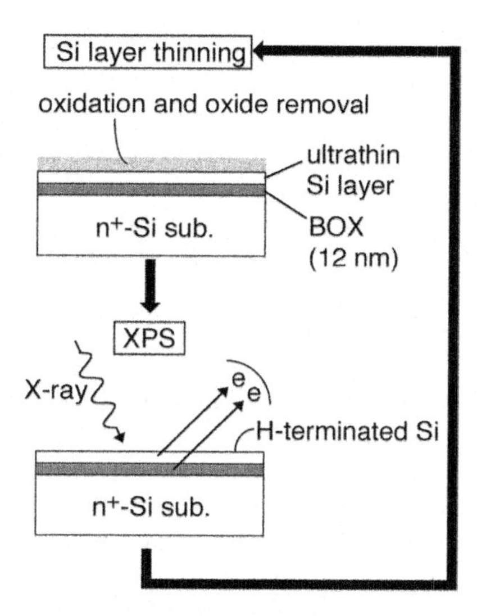

FIGURE 6.1 The cycle of Si layer thinning and XPS measurement steps. (From Tabe et al.[13])

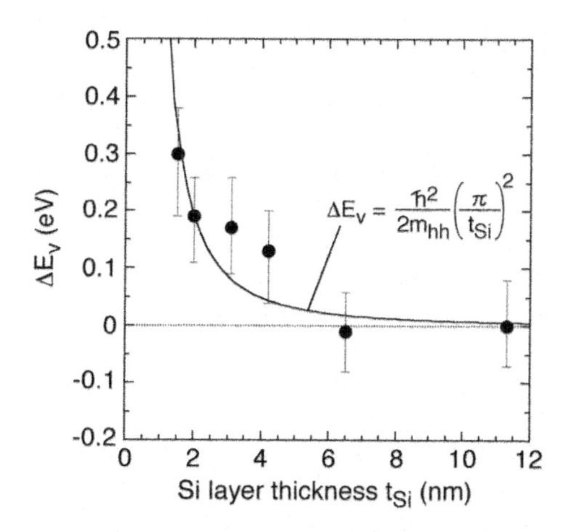

FIGURE 6.2 The VBM shift as a function of Si thickness. (From Tabe et al.[13])

the top Si layer was thinned to 10 nm by thermal oxidation and subsequent removal of the grown oxide. Next, the sample was loaded into an ultrahigh vacuum (UHV) chamber and XPS measurements of the valence band were performed using an Mg-K line. The cycle of Si layer thinning and XPS measurement was repeated, as shown in Figure 6.1.

As a result, we found that the VBM obviously shifts toward higher binding energy with decreasing Si layer thickness, measured from 11.3 to 1.5 nm. Figure 6.2 shows the VBM shift, E_V, as a function of Si thickness, t_{Si}. Here, the error bars arise from the arbitrariness in the determination of VBM by drawing extrapolation lines in the valence-band spectra. The overall thickness dependence is almost in accordance with the calculated curve (solid curve), which represents the ground-level energy for a heavy hole (effective mass $m_{hh} = 0.49\ m_0$; m_0 is the electron rest mass) obtained by solving the Shrödinger equation for Si quantum wells with infinite potential barriers. We believe that this is the first observation of the quantum confinement effect in a 2D Si crystal. If the thickness of the Si layer is extremely uniform, valence-band spectra near the VBM should have a shape of multistep density-of-states, which is one of general 2D features. However, in our sample structure, the thickness uniformity was good enough for VBM shift detection but was insufficient for observing the multistep shape.

6.1.3 Double-Barrier Structures of SiO_2/Si/SiO_2 Formed by Anisotropic Etching[11,12]

Since interference of electron wave plays a substantial role in resonant tunneling, any origins, which disturb electron coherence, should be avoided. In that sense, the atomic structure of the Si well is favorably single-crystalline without structural defects such as twin boundaries, dislocations and other point defects.

To achieve this condition, Yuki et al.[11] fabricated a thin Si single-crystal plate as a quantum well by adopting anisotropic wet chemical etching and thermal oxidation in a fabrication process, as shown in Figure 6.3. Because of the minimized etching rate for the (111) plane of ethylenediamine, V-shaped trenches with (111) side walls were formed and finally a thin Si plate was formed between trenches. The width of the Si plate was controlled by the etching time. Thermal oxidation readily formed thin tunnel barriers on both sides of the Si plate. Heavily doped polySi served as electrodes. The resultant *I-V* curve measured at $T = 3.3$ K is redrawn from Reference 11 in Figure 6.4, where the well was as thick as 43 nm. Although the *I-V* curve shows a weak nonlinearity, the resonant tunneling effect is not clear in the absence of NDC probably because of narrow energy spacing in the potential well due to the thick well, and also because of diffused energy levels due to thickness fluctuation.

Thereafter, Namatsu et al.[12] reported resonant tunneling *I-V* curves for a thinner (111) Si plate of 5 nm by using a (110) Si wafer, as shown in Figure 6.5. By using such a special orientation wafer, electron-beam lithography and anisotropic etching of an aqueous KOH solution, they insisted that a highly accurate Si plate with smooth (111) side-planes is formed standing in the vertical direction and that pattern-size fluctuations can be minimized. The *I-V* curve is shown in Figure 6.6, which was measured at 45 K. Some inflections are observed in the *I-V* curve, which become more prominent in the differential curve. They calculated resonant peak voltages of the SiO_2/Si/SiO_2 diode with 1.6-nm-thick SiO_2 films on the basis of the band diagram in Figure 6.5, as briefly described below.

An *i*th resonant peak occurs when the bottom of the conduction band E_c in the emitter-side electrode coincides with the *i*th energy level E_i in the Si potential well. The voltage for the *i*th resonant current peak is provided by the equation

$$V = E_i + V_{ox},$$

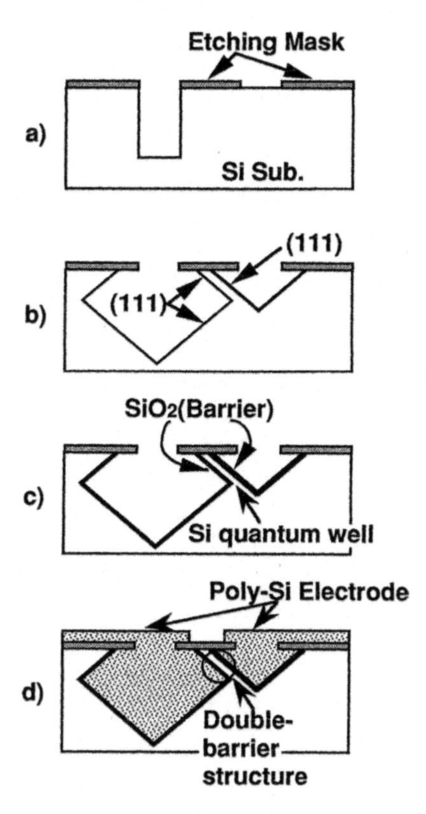

FIGURE 6.3 Fabrication steps of an RTD by anisotropic etching. (From Yuki et al.[11])

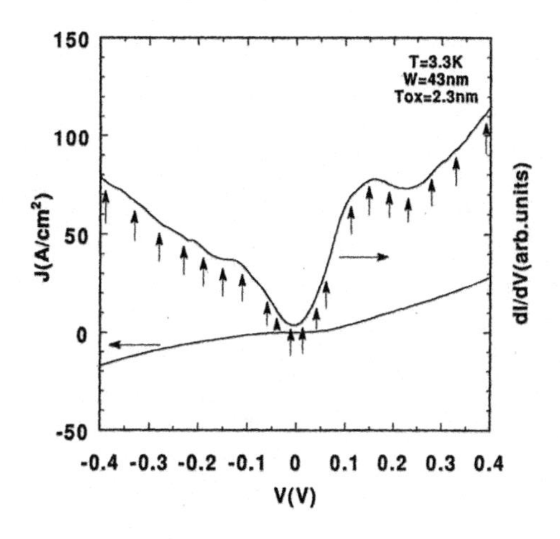

FIGURE 6.4 The *I-V* curve of the diode in Figure 6. 3. (From Yuki et al.[11])

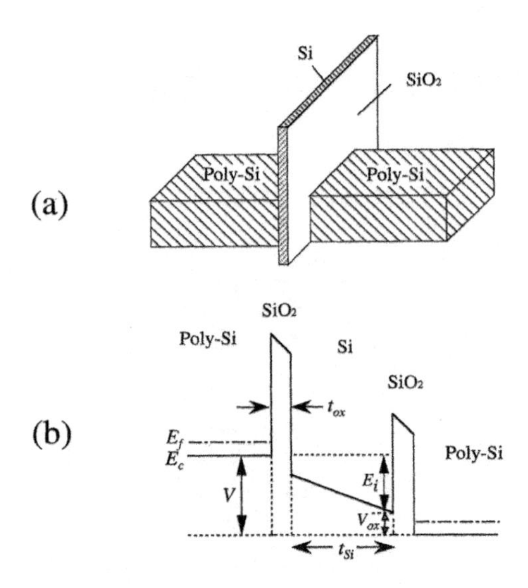

FIGURE 6.5 An RTD structure by using a (110) Si wafer and the corresponding potential diagram. (From Namatsu et al.[12])

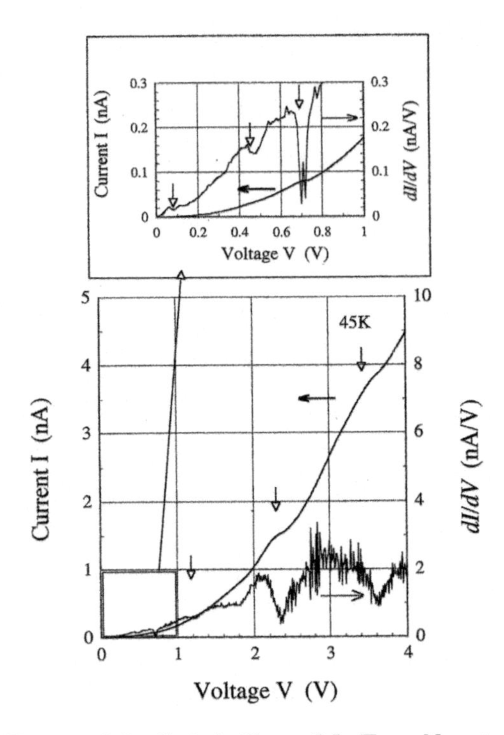

FIGURE 6.6 The *I-V* curve of the diode in Figure 6.5. (From Namatsu et al.[12])

where V_{ox} is the voltage across the SiO_2 film and is given by

$$V_{ox} = Vt_{ox}\varepsilon/(t_{Si} + 2t_{ox}\varepsilon)$$

Here, ε is defined to be $\varepsilon_{Si}/\varepsilon_{ox}$. From these equations, the relationship between E_i and V is given by

$$E_i = V(t_{Si} + t_{ox}\varepsilon)/(t_{Si} + 2t_{ox}\varepsilon)$$

In order to compare theoretical voltages providing current peaks and experimental values, it is necessary to solve the Shrödinger equation. Namatsu et al.[12] solved numerically with the potential barrier height of 3.1 eV, the effective mass in SiO_2 of $0.5m_0$ and the normal effective mass in the (111) plane of $0.26m_0$. Agreement between theory and the experiment was quite satisfactory, although NDC was not observed and only small inflections were observed.

It should be noted that, in these two articles, although the Si wells are single crystalline, there are strong restrictions in the experiment for observing resonant tunneling: The Si potential plate has only (111) side walls, since anisotropic chemical etching has a nature of the slowest etching rate in the <111> direction. Furthermore, since the emitter and the collector materials are poly-Si, giving rise to a mixture of different $\boldsymbol{k}$-vectors, it is questionable whether the conservation of transverse momentum is actually satisfied through tunneling from the emitter to the collector. In addition to such a complex situation in physics, a technical problem may also be present: anisotropic etching perhaps hardly provides satisfactorily smooth plane within a fluctuation of a few atomic layers. Therefore, it is necessary to open up a door to a new technique for significant improvement of the Si RTD.

6.2 RESONANT TUNNELING IN SIO_2/SI/SIO_2[9,10]

In order to clearly observe resonant tunneling, or NDC, in the Si/SiO_2 system, we fabricated SiO_2/single-crystalline-Si/SiO_2 double-barrier diodes by means of a wafer bonding technique. That is, a bonded silicon-on-insulator (SOI) wafer having a thin (~2 nm) buried SiO_2 layer as well as a thin Si layer was prepared for the RTD experiment. As a result, we observed, for the first time, NDC due to resonant tunneling.

6.2.1 Fabrication of an RTD

Figure 6.7 summarizes the fabrication procedure of the sample.[16] Wafer bonding was used for sample preparation, because this seems to be the natural way of fabricating the extremely thin buried oxide layer sandwiched between Si crystals. First, a commercially available SOI (001) wafer was bonded to an n^+ Czochralski (CZ) Si (001) base wafer at room temperature, and then the sample was annealed at 1000°C for 2 hours in N_2, in order to stabilize the bonding. Prior to the bonding,

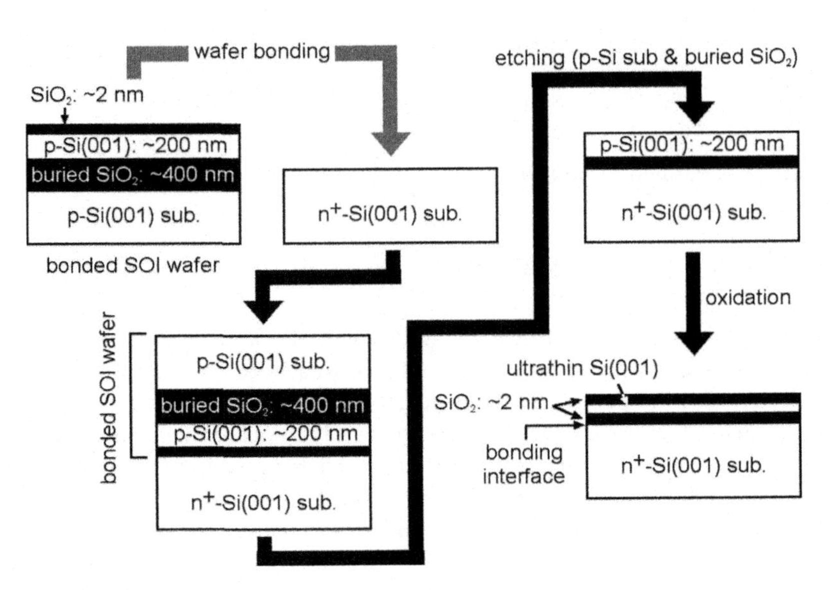

FIGURE 6.7 Fabrication procedure of the thin Si sample. (From Ishikawa et al.[16])

a 2-nm-thick SiO_2 layer was grown on top of the SOI wafer by dry oxidation at 700°C, while the native oxide on the (001) wafer was removed by immersing in a diluted HF solution. In the final structure, the 2-nm-thick SiO_2 layer grown on the SOI surface serves as the lower barrier, which permits electron tunneling between the top Si layer and the n^+ base substrate. The bonded interface is located between the buried oxide and the base wafer.

Next, the (001) Si substrate of the commercial SOI wafer was back-etched in a KOH solution (10%, 70°C), while the bottom of the base (001) wafer was covered with an SiO_2 layer. Note that the etching rate is 660 nm/min for the (001) Si and only 1.4 nm/min for thermal SiO_2. Therefore, the thick buried oxide of the SOI wafer acts as an etch stopper during this etching process. By etching the thick buried SiO_2 layer in an HF solution, a new SOI structure with a thin buried SiO_2 layer was obtained (see the upper right of Figure 6.7).

The double barrier structure was obtained by thinning of the top Si layer to nanometer range and by oxidation to form an upper barrier layer (~2-nm-thick SiO_2). In fact, the transmission electron microscope (TEM) image in Figure 6.8 reveals the ultrathin (~2 nm ± 0.3 nm-rms) single-crystalline-Si layer sandwiched between SiO_2 layers. For the *I-V* measurements, local oxidation was made to isolate the diodes from neighboring ones. Top Al electrodes were formed by the conventional vacuum evaporation, and the sample was annealed in an H_2/N_2 mixture to reduce gap states.

In this bonding procedure, the in-plane rotational angle, or the twist angle, of the top Si layer with reference to that of the base substrate was carefully aligned within the accuracy of ±1 degree. When the twist angle between the emitter Si and the well Si becomes large, constant energy surfaces in ***k***-space are also twisted as shown in Figure 6.9 and resonant tunneling may be partly prohibited by the conservation law of transverse momentum.

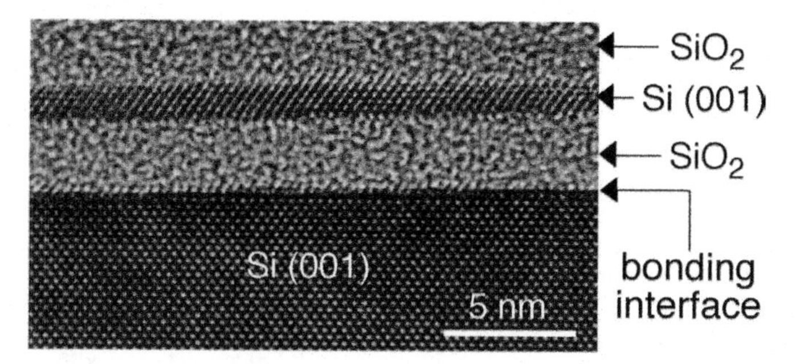

FIGURE 6.8 The cross-sectional TEM image of the sample in Figure 6. 7. (From Ishikawa et al.[16])

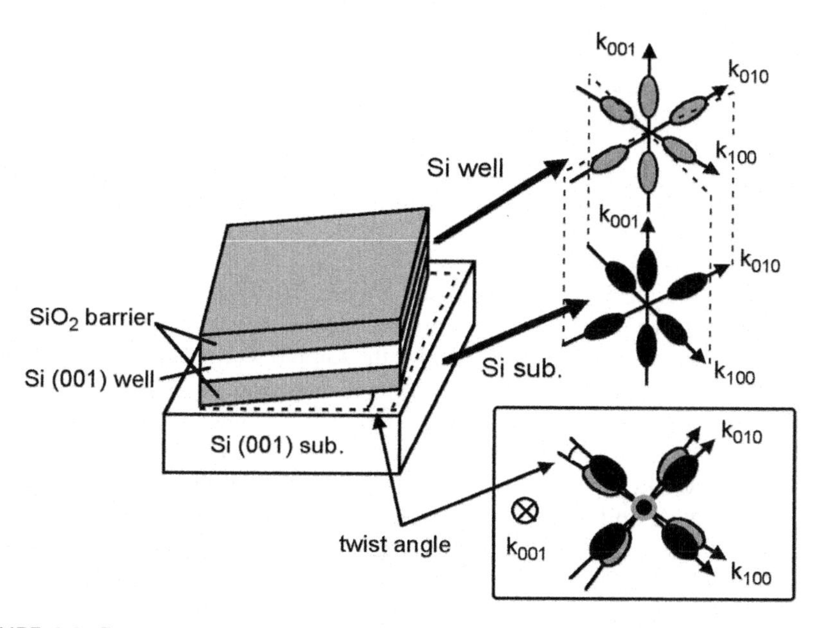

FIGURE 6.9 Constant energy surfaces in ***k***-space for the bonded sample with a twisted interface.

6.2.2 Resonant Tunneling in the Low Voltage Region[9]

Current-voltage measurements were performed for diodes (area ~10^{-5} cm^2) with the Si well thicknesses of ~5 and ~2 nm, respectively. As shown in Figure 6.10(a), positive voltage was applied to the Al electrode, corresponding to electron transport from the n^+-Si substrate. Application of negative voltage did not lead to NDC, since, for the transport from the Al electrode, resonance condition is satisfied at any voltages due to the continuous energy distribution of electrons below the Fermi level.

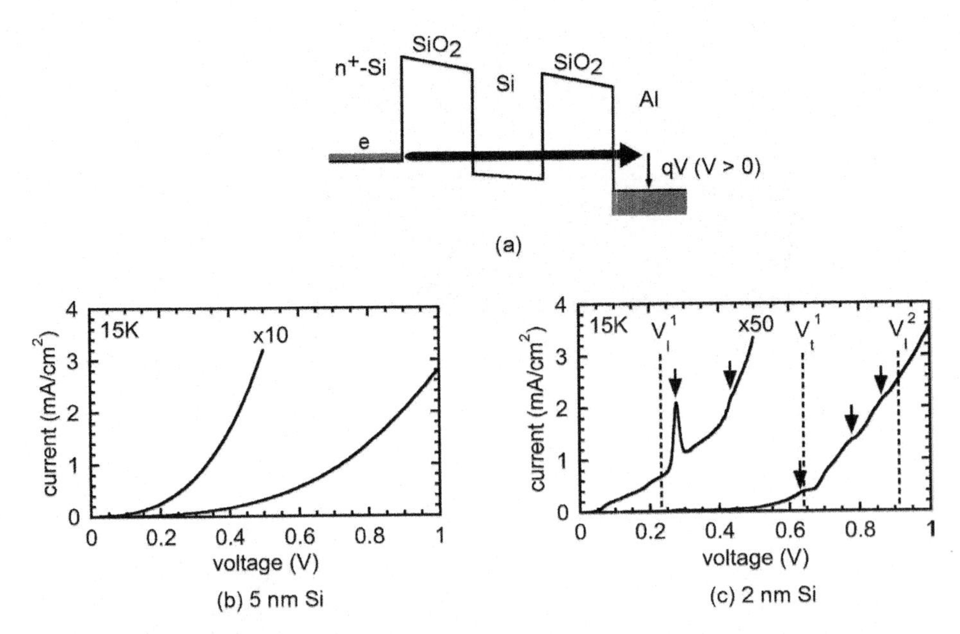

FIGURE 6.10 (a) The potential diagram of the RTD, fabricated by wafer bonding, (b) the *I-V* curve for the 5-nm-thick Si well, and (c) the *I-V* curve for the 2-nm-thick Si well. (From Ishikawa et al.[9])

Figure 6.10(b) and Figure 6.10(c) show typical *I-V* curves, measured at 15 K. As in Figure 6.10(b), the sample with ~5-nm-thick Si showed a featureless curve. On the other hand, as in Figure 6.10(c), for the sample with ~2-nm-thick Si, a clear NDC (peak-to-valley ratio (PVR) of ~1.8) was reproducibly observed at 0.28 V together with inflections indicated by the arrows. We believe that this is the first observation of the NDC in Si/SiO_2 systems. This NDC feature was observed up to ~100 K, as shown in Figure 6.11. Figure 6.11 also reveals broadening of the peak width and negative shift of the peak position with increasing temperature. Such features are adequate as the effect of the thermal energy on the resonant tunneling current, indicating that the observed NDC is ascribed to the resonant tunneling.

Resonance voltages were estimated by a conventional manner,[12] using the band diagram shown in Figure 6.10(a). Here, the anisotropy of electron mass for the Si (001) plane was taken into account (m_l = 0.98 m_0 and m_t = 0.19 m_0 [m_0: electron rest mass]). The effective mass in SiO_2 and the band offset at the Si/SiO_2 interface were assumed to be 0.50 m_0 and 2.9 eV.[17] The thicknesses of the lower and upper barriers were assumed to be 3.0 and 2.5 nm, respectively, from the TEM observation.

Dashed lines in Figure 6.10(c) represent the calculated resonance voltages for the Si thickness of 1.6 nm. (V_l^i is for m_l, while V_t^i is for m_t [i is level index]). These values almost agree with the experiment. However, further study is necessary to explain overall features in the *I-V* curve.

Thus, we fabricated Si/SiO_2 double barrier structures from a bonded SOI wafer with an ultrathin buried oxide layer and we observed, for the first time, NDC (PVR of ~1.8 at 15 K) in the *I-V* characteristics due to the resonant tunneling effect.

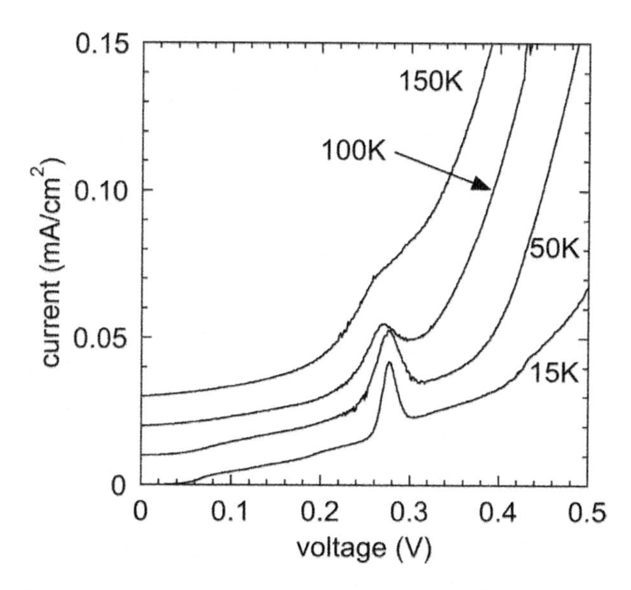

FIGURE 6.11 Temperature dependence of the current peak in Figure 6.10(c). (From Ishikawa et al.[9])

6.2.3 Hot-Electron Storage in the High-Voltage Region[10]

In the preceding section, we have studied *I-V* characteristics of RTDs in a range of relatively low applied voltages, and observed resonant tunneling characteristics. It is not known, however, what happens in the high-voltage region like Fowler-Nordheim tunneling injection. In this section, we study *I-V* characteristics of RTDs as a function of kinetic energy of incident electrons from the emitter to the Si well. For this purpose, we fabricated RTDs with varied thicknesses of the lower oxide and measured *I-V* characteristics in a wide range of applied voltage. As a result, we have found that the diode characteristics strongly depend on the kinetic energies of injected electrons in the Si well; in low voltages, resonant tunneling is observed as described in the preceding section, whereas in high voltages a large current peak is observed but it is ascribed not to resonant tunneling but to electron storage in the Si well.

We fabricated RTDs with a structure of Al/upper SiO_2 barrier/p^--Si well/lower SiO_2 barrier/n^+-Si substrate in the same way as in the preceding section. In the final RTD structure, the lower tunnel barrier oxide has a thickness range of 3 to 7 nm, while thicknesses of the top Si layer and of the upper tunnel barrier oxide were both about 2 nm. The Al-electrode area was about 150 μm^2 and the measurement temperature was 15 K. In the *I-V* measurement, again, positive voltages were applied to the top Al electrode so that electrons flow from the n^+-Si substrate to the top Al electrode.

A typical *I-V* curve (Figure 6.12) observed in an RTD with a buried oxide thickness of 3 nm was essentially the same as that in Figure 6.10. In both figures, the applied voltage range corresponds to relatively low injected-electron kinetic energies in the Si well of 2.7 eV or less. The current peak and inflections are in

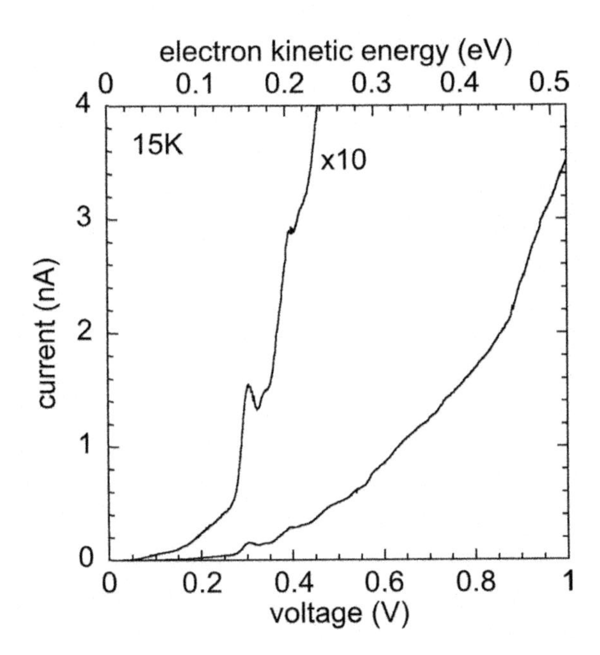

FIGURE 6.12 A typical *I-V* curve observed in an RTD with a 3-nm-thick buried oxide. This is another example of NDC for Figure 6.10(c). (From Ikeda et al.[10])

good agreement with the theoretical resonance voltages. Hence, electrical characteristics of RTDs with a buried oxide layer thickness below 4 nm are dominated by the ordinary resonant tunneling mechanism, where the kinetic energy of injected electrons in the Si well is relatively low (< 2.7 eV) in comparison with the Si/SiO_2 barrier height (2.9 eV).

Diode properties for RTDs with a BOX layer of 5 and 7 nm in thickness are shown in Figure 6.13(a) and Figure 6.13(b). The measurement duration time in these figures is defined by the sum of holding and sampling time of a current at each voltage step ($V = 10$ mV). Some specific characteristics, which are greatly different from Figure 6.10, are observed: (i) a large current peak in upward voltage ramping with the duration-time-dependent peak height, (ii) large hysteresis and (iii) an opposite (negative) current in downward ramping. As for hysteresis in RTDs, a parasitic series resistance causes a hysteresis, as previously reported in AlAs/GaAs RTDs,[3] but this kind of hysteresis should be independent of ramping rate (or duration time). Another origin of a hysteresis is reported in CaF_2/Si RTDs,[18] that is, electron storage effect in the Si well is responsible for the hysteresis. The latter may be partly effective in explanation of our work. However, both reports deal with the direct tunneling mode, while it should be noted that our results in Figure 6.13 were obtained in the Fowler-Nordheim tunneling mode involving high-kinetic energies of electrons in the Si well.

The specific characteristics for thick-buried-oxide diodes shown in Figure 6.13 can be explained by the electron storage due to the hot electron injection in the Si well. Under the Fowler-Nordheim tunneling condition, the injected electrons in the

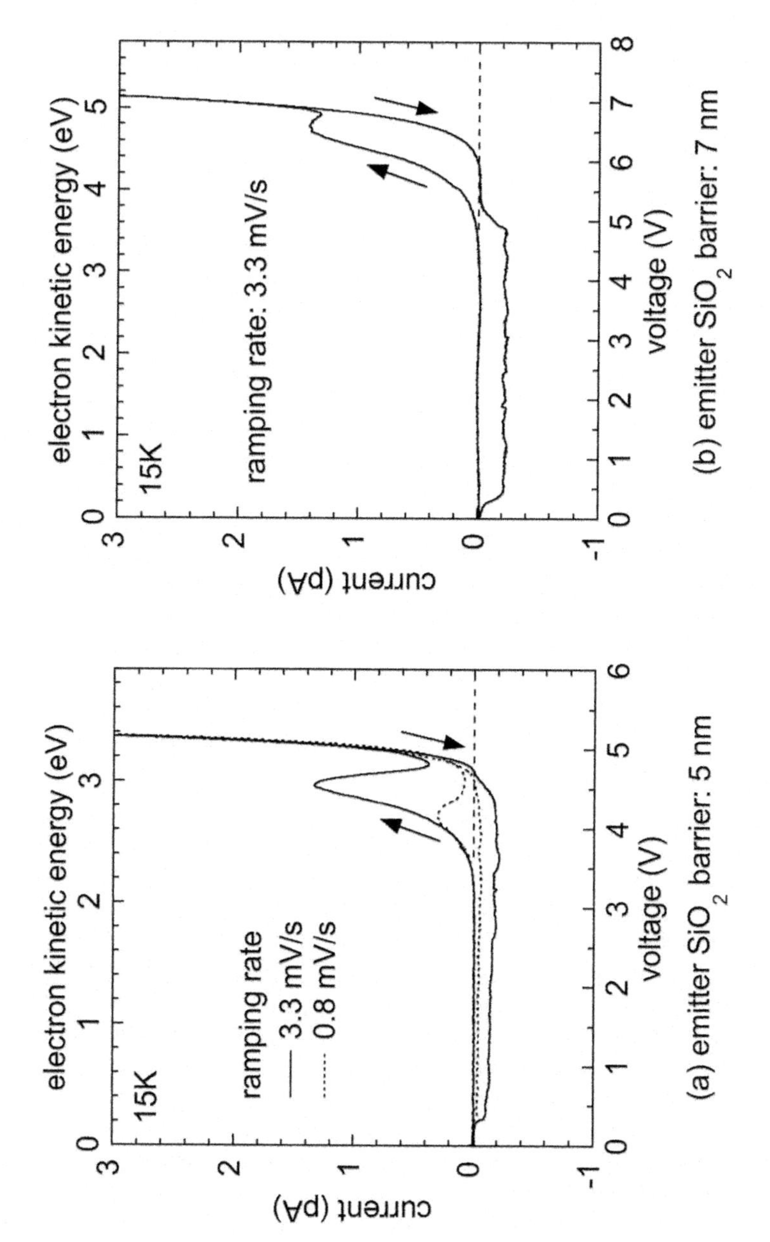

FIGURE 6.13 (a) and (b) show *I-V* curves for RTDs with emitter-side barrier thicknesses of 5 nm and 7nm, respectively. (From Ikeda et al.[10])

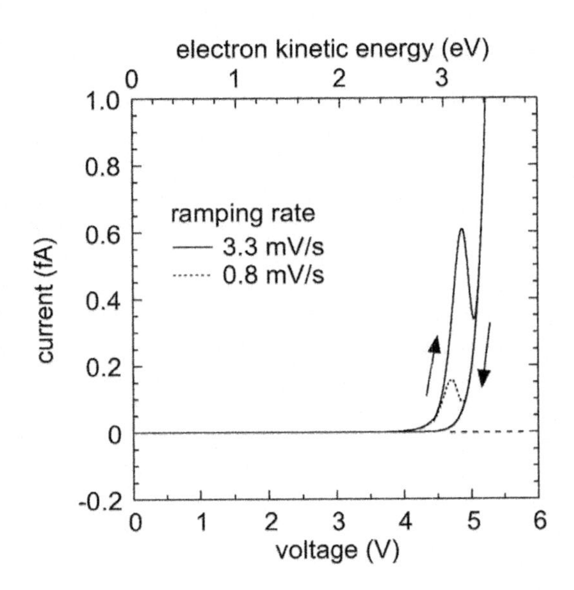

FIGURE 6.14 Calculated I-V curves of a 5-nm-buried oxide for ramping rates of 3 and 12 sec. (From Ikeda et al.[10])

Si well have relatively high kinetic energies above 2.9 eV. The injected electrons with high kinetic energies easily experience impact ionization scattering and lose their energies, as described later. Once the injected electrons are rapidly stored during upward ramping, the Si well potential will also be rapidly raised and the voltage drop across the buried oxide layer (emitter-side barrier) will be reduced. This leads to the reduction in the tunneling current in upward voltage ramping. Figure 6.14 shows calculated I-V curves of a 5-nm buried oxide diode for ramping rates (duration time) of 3 and 12 sec, taking into account the electron storage effect reflecting impact ionization scattering, under the assumption that the stored electrons are fixed in the well (not released). The hysteresis in downward voltage ramping is also reproduced in Figure 6.14. From this agreement between Figure 6.13 and Figure 6.14, it is convincing that the result in Figure 6.13 is due to electron storage. However, there is a slight difference between these figures: that is, the opposite (negative) current in downward voltage ramping cannot be reproduced in the calculated curve. Since the opposite current is inversely proportional to the duration time in the whole voltage range as shown in Figure 6.13(a), we believe that the opposite current is primarily a transient current which should have been influenced by parasitic capacitance and resistance, but further study is necessary.

In the scattering process, electron-electron, electron-phonon and electron-impurity collisions may be possible, but impact ionization scattering[19–21] should be the most relevant scattering process, as is usually considered in hot electron issues in MOSFETs, because the injected electrons must efficiently lose their kinetic energy, as large as ~3 eV, during the short transit time (subpicoseconds) in the Si well. The electron storage rate caused by the impact ionization scattering can be proportional to the product of the injected electron flux ($cm^{-2}s^{-1}$), the ionization rate (sec^{-1}) and

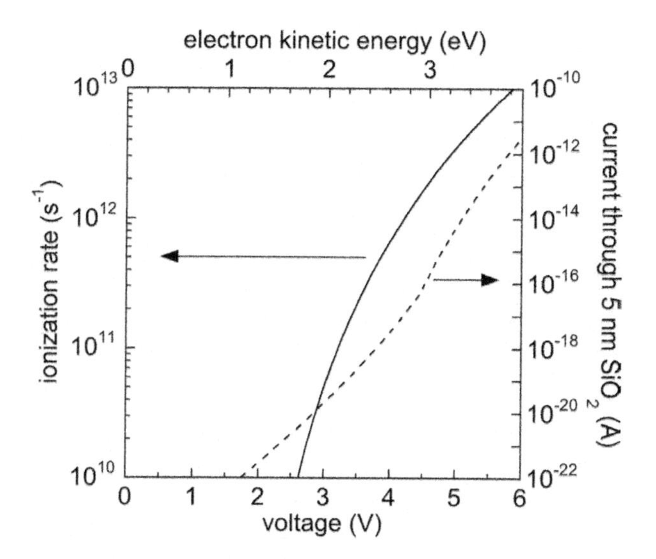

FIGURE 6.15 The ionization rate of an electron in Si (after Reference 21) and the calculated tunneling current through a single barrier of the 5-nm-thick oxide, as a function of applied voltage. (From Ikeda et al.[10])

the transit time in the Si well t_{well}. Figure 6.15 shows the ionization rate of an electron in Si after Reference 21 and the calculated tunneling current through a single barrier of the 5-nm-thick SiO_2 film, as a function of applied voltage to an RTD. The tunneling current is calculated using the direct and Fowler-Nordheim tunneling equations, assuming the effective mass $m^* = 0.5\ m_0$ and the band offset 2.9 eV. Both ionization rate and tunneling current increase exponentially with increasing applied voltage,[19–21] and t_{well} is inversely proportional to the square root of the voltage across the buried oxide layer. Thus the electron storage rate should exponentially increase with increasing voltage, and hence the electron storage in the well becomes prominent only at high voltages.

6.2.4 Switching of Tunnel-Modes: Comparison with a Single Barrier

It is interesting to compare these results with that of the Si/SiO_2 single barrier system. Figure 6.16 shows a tunnel-mode diagram of single-tunnel-barrier MOS diodes reported by Itsumi et al.[22] Heavily doped poly-Si was used as a gate electrode. In this figure, the horizontal axis is the oxide thickness and the vertical axis is electric field across the tunnel SiO_2 barrier. The tunneling modes are divided by boundaries determined by experimental data, in which UD denotes current-undetectable (smaller than 0.1 pA), DT is direct tunneling, FN is Fowler-Nordheighm, and DB is dielectric breakdown. The boundaries between UD and DT and between UD and FN will shift depending primarily on the diode area.

Our experimental conditions of the double-barrier structure described in Section 6.2.2 (in low voltages) and Section 6.2.3 (in high voltages) approximately correspond

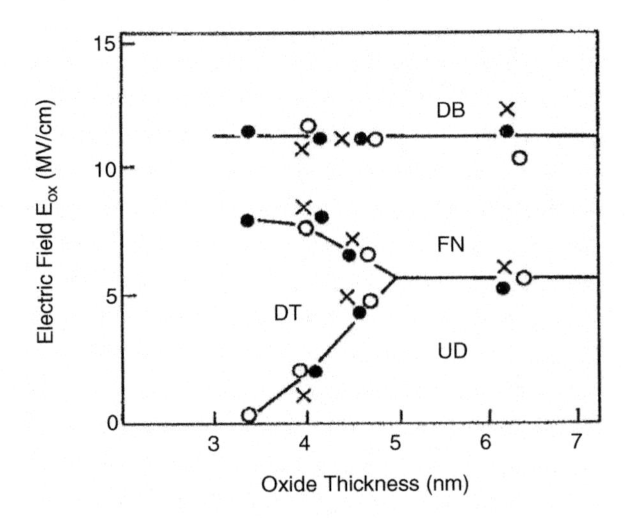

FIGURE 6.16 A tunneling-mode diagram of single-tunnel-barrier MOS diodes. (From Itsumi et al.[22])

to DT and FN for the emitter-side SiO_2 barrier. As a result, we have reached a conclusion that the DT mode for the emitter barrier results in resonant tunneling, and the FN mode for the emitter barrier gives rise to electron storage. Such comparison between single-barrier tunneling and double-barrier tunneling may be interesting. It should be noted this correspondence is based on a difference in the kinetic energy of incident electrons into the potential well and their energy dissipation due to impact ionization as argued in Section 6.2.3.

6.3 ZERO-DIMENSIONAL RESONANT TUNNELING

6.3.1 Coexistence of Coulomb Blockade and Resonant Tunneling

The potential well structure in RTD is usually 2D-shaped, as described previously. However, once the diode area becomes extremely small, like a dot structure, we need to treat I-V characteristics by means of different analytical expressions taking into account three-dimensional (3-D) confinement in the potential well.[23] Furthermore, when the collector-side (exit) tunnel barrier is relatively thick, the Coulomb blockade effect, which is based on significantly large charging energy of the dot due to the small total capacitance of the surrounding junctions, dominates tunneling events occurring in a small bias region.[24] When the bias voltage becomes large, the exit barrier resistance is efficiently reduced. This may give rise to a change of tunnel modes from sequential tunneling (Coulomb blockade) to resonant tunneling.

Now, we briefly describe an experimental result of coexistence of Coulomb blockade and resonant tunneling in a single *I-V* curve. Cain et al. reported on hole transport through a double quantum dot structure formed by trench isolation from an SiGe/Si heterostructure.[25] Figure 6.17 shows the source-drain *I-V* characteristics at 4.2 K for various gate voltages. A clear zero current region can be seen for $|V_{ds}|$

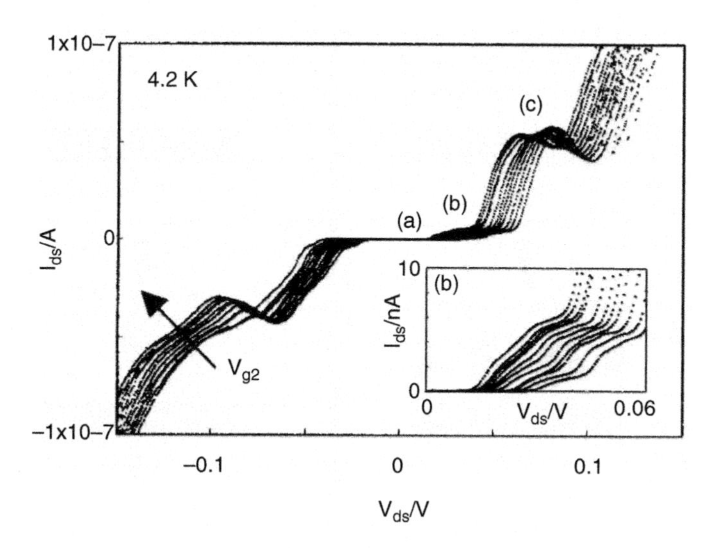

FIGURE 6.17 The source-drain I-V characteristics at 4.2 K of a double quantum dot structure for various gate voltages. (From Cain et al.[25])

< 30 mV. A Coulomb staircase can also be seen just outside the Coulomb gap. The zero current region and the staircase are recognized as typical features of Coulomb blockade effects. At higher V_{ds}, a region of NDC is present for both positive and negative biases. The authors attributed this NDC to resonant tunneling between quasi-bound states in the dots. This is a clear example of coexistence of Coulomb blockade and resonant tunneling in the horizontal zero-dimensional (0-D) dot system.

6.3.2 Fabrication of a SiO_2/Si-Dots/SiO_2 Structure[26]

In order to investigate the Coulomb blockade effect in the SiO_2/Si/SiO_2 system, we purposely changed the 2D Si plate to high-density Si dots and measured *I-V* characteristics through the 0-D dots in the vertical direction.[26] Since crystalline imperfection must also be avoided for the Si dot system to suppress defect-induced scattering, we fabricated high-density Si dots by nanometer-scale local oxidation (nano-LOCOS) from a thin single-crystal Si layer of an SOI wafer. Regarding deposition-type 0-D structures, there are controversial arguments on interpretation of nonlinear *I-V* characteristics, and, according to Chou et al.,[27] local electrical breakdowns dominate *I-V* characteristics with few exceptional devices, which might be exceptionally attributed to resonant tunneling. In addition, Fukuda et al.[28] reported *I-V* curves of deposited Si crystallites, measured by a conductive AFM probe, and they attributed observed current peaks to resonant tunneling. However, considering uncertain factors like thermal drift during *I-V* measurement or charge-induced Coulomb force between the tip and the sample, it seems to be difficult to ensure reliability of the measurements. It is even more important that, in such devices, it is difficult to remove defect-related phenomena from experiments. Therefore, we believe that it is important to prepare single-crystalline Si dots for the present purpose.

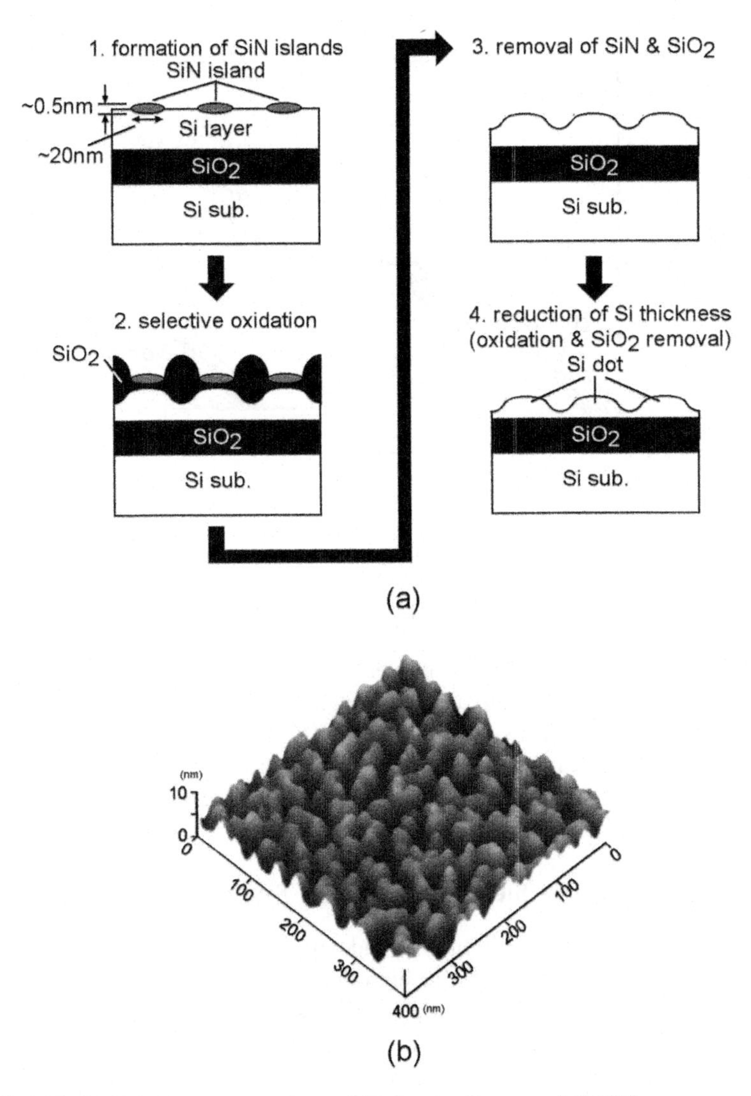

FIGURE 6.18 (a) Fabrication procedure of Si dots by the nano-LOCOS process, and (b) an AFM image of the sample surface. (From Tabe et al.[32])

As the starting substrate, we fabricated an SOI wafer composed of a 150-nm-thick lightly doped p-type top Si layer, a 3-nm-thick buried SiO_2 layer and an n^+-type Si substrate. In the first step of fabrication, the thickness of the top Si layer of the SOI wafer was reduced to about 18 nm by repetitive cycles of thermal oxidation and subsequent chemical etching. Such top Si thickness, 18 nm, is thick enough to prevent the agglomeration of the Si layer[29,30] during high temperature cleaning in the nano-LOCOS process. (Although agglomeration-induced Si dot formation in References 29 and 30 may also be attractive in some cases because of its automatic alignment of Si dots, we shall omit this topic here.)

Figure 6.18(a) shows the fabrication procedure of Si dots by the nano-LOCOS process which consists of four steps: (1) formation of ultrasmall SiN islands, (2) selective oxidation with SiN masks, (3) removal of SiN islands and SiO_2, and (4) thickness adjustment of the interdot Si region. Details of the process are as follows: Prior to loading into an ultrahigh vacuum (UHV) chamber, the SOI substrate is chemically cleaned by H_2SO_4/H_2O_2 and then rinsed in deionized water. After high temperature cleaning (about 930°C) in UHV for removing the surface oxide, ultra-small SiN disks were naturally formed on the Si surface by heating at 750°C for 100 s under an N_2 partial pressure of 1×10^{-5} Torr. Here, nitrogen species automatically excited by a nude ionization gauge are responsible for the SiN formation. According to our previous studies,[31,32] the SiN islands with a lateral size of 10 to 20 nm, a thickness of only 0.5 nm and a density of about 3×10^{11} cm^{-2}, are formed by this treatment.

Next, the surface was selectively oxidized with the SiN masks by conventional furnace oxidation. It is noted that the SiN islands work as oxidation masks despite their thickness of only 0.5 nm.[31,32] The oxidation was carried out at 800°C (1 atm dry O_2), immediately after the nitrided sample was removed from the UHV chamber. Since this oxidation process produced a 12-nm-thick SiO_2 layer on the top Si surface, resultant Si coupled dots had a height of about 5 nm, and the thickness of the interdot coupling region ($t_{interdot}$) was about 13 nm. Then, the sample was dipped in an H_2SO_4/H_2O_2 mixture and subsequently in a diluted HF solution. By this treatment, the SiN masks as well as SiO_2 were removed. The $t_{interdot}$ can be adjusted by further dry oxidation and subsequent chemical etching (HF solution). In multidot single-electron tunneling devices, which work laterally, precise control of $t_{interdot}$ is important because the interdot region serves as tunnel barrier.[33,34] For the present purpose of vertical tunneling, however, it is favorable that the Si dots are mutually isolated with $t_{interdot} = 0$.

Figure 6.18(b) shows a typical atomic force microscopy (AFM) image of fabricated dots. From the AFM image, the lateral size, height and density of Si dots are evaluated to be about 20 nm, 4 nm and 2×10^{11} cm^{-2}, respectively. The values of the lateral size and density of the dots are almost in agreement with that of SiN disks, indicating that the SiN disks certainly work as oxidation masks.

6.3.3 *I-V* Characteristics of an SiO_2/Si-Dots/SiO_2 Tunnel Diode

Thus, we fabricated a double-barrier tunnel diode with a multidot in the potential well layer by the nano-LOCOS technique, and measured *I-V* characteristics. Figure 6.19 shows a resultant *I-V* curve, in which a staircase can be observed. We believe that this staircase is due to Coulomb blockade, and, thus, electron tunneling via Si dots in the potential well layer is governed by Coulomb blockade. When the applied voltage becomes so large that tunneling resistances of the both-side-barriers are significantly reduced, resonant tunneling characteristics are expected to be observed. Unfortunately, however, in our device, dielectric breakdown severely occurred for higher voltages, and, therefore, we have not yet observed coexistence of Coulomb blockade and resonant tunneling in the vertical 0-D RTD structure. It should be noted that dielectric breakdown, in general, readily occurs in dot systems, because

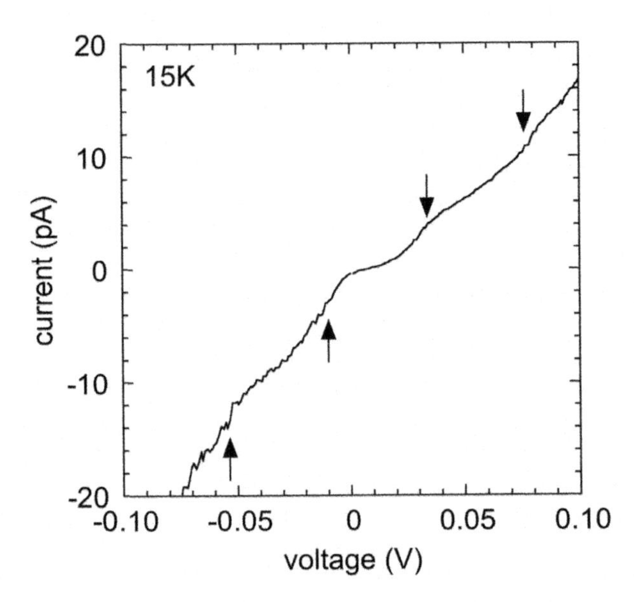

FIGURE 6.19 An *I-V* curve of an SiO_2/Si-dots/SiO_2 structure, showing a Coulomb staircase (From Ishikawa et al.[26]).

local electric field is concentrated and enhanced due to the nonplanar surface morphology.[35,36] Local mechanical stress within the lower and the upper barrier oxides may also be responsible for such low-field breakdown of the oxide.

ACKNOWLEDGMENT

The authors are grateful to T. Mizuno (Shizuoka University) for his continuous assistance to the experiment, and to Y. Takahashi, S. Horiguchi, Y. Ono (NTT) and T. Tsuchiya (Shimane University) for their fruitful discussion and support. This work was partly supported by Core Research Evolutional Science and Technology from Japan Science and Technology Agency and by a Grant-in-Aid for Scientific Research from the Japan Society Promotion Science.

REFERENCES

1. Tsu, R. and Esaki, L., Tunneling in a finite superlattice, *Appl. Phys. Lett.*, 22, 562–564, 1973.
2. Chang, L.L., Esaki, L., and Tsu, R., Resonant tunneling in semiconductor double barriers, *Appl. Phys. Lett.*, 24, 593–595, 1974.
3. Tsuchiya, M. and Sakaki, H., Dependence of resonant tunneling current on well widths in AlAs/GaAs/AlAs double barrier diode structures, *Appl. Phys. Lett.*, 49, 88–90, 1986.
4. Sollner, T.C.L.G., Goodhue, W.D., Tannenwald, P.E., Parker, C.D., and Peck, D.D., Resonant tunneling through quantum wells at frequencies up to 2.5 THz, *Appl. Phys. Lett.*, 43, 588–590, 1983.

5. Yokoyama, N., Imamura, K., Muto, S., Hiyamizu, S., and Nishi, H., A new functional, resonant-tunneling hot electron transistor (RHET), *Jpn. J. Appl. Phys.*, 24, L853–L854, 1985; Capasso, F. and Kiehl, R.A., Resonant tunneling transistor with quantum well base and high-energy injection: A new negative differential resistance device, *J. Appl. Phys.*, 58, 1366–1368, 1985.
6. Liu, H.C., Landheer, D., Buchanan, M., and Houghton, D.C., Resonant tunneling in $Si/Si_{1-x}Ge_x$ double-barrier structures, *Appl. Phys. Lett.*, 52, 1809–1811, 1988; Rhee, S.S., Park, J.S., Karunasiri, R.P.G., Ye, Q., and Wang, K.L., Resonant tunneling through a $Si/Ge_xSi_{1-x}/Si$ heterostructure on a GeSi buffer layer, *Appl. Phys. Lett.*, 53, 204–206, 1988.
7. Ismail, K., Meyerson, B.S., and Wang, P.J., Electron resonant tunneling in Si/SiGe double barrier diodes, *Appl. Phys. Lett.*, 59, 973–975, 1991.
8. Suda, Y. and Koyama, H., Electron resonant tunneling with a high peak-to-valley ratio at room temperature in Si1-xGex/Si triple diodes, *Appl. Phys. Lett.*, 79, 2273–2275, 2001.
9. Ishikawa, Y., Ishihara, T., Iwasaki, M., and Tabe, M., Negative differential conductance due to resonant tunneling through SiO_2/single-crystalline-Si double barrier structure, *Electronics Lett.*, 37, 1200–1201, 2001.
10. Ikeda, H., Iwasaki, M., Ishikawa, Y., and Tabe, M., Resonant tunneling characteristics in SiO_2/Si double-barrier structures in a wide range of applied voltage, *Appl. Phys. Lett.*, 83, 1456–1458, 2003.
11. Yuki, K., Hirai, Y., Morimoto, K., Inoue, K., Niwa, M., and Yasui, J., Fabrication of novel Si double-barrier structures and their characteristics, *Jpn. J. Appl. Phys.*, 34, 860–863, 199).
12. Namatsu, H., Horiguchi, S., Takahashi, Y., Nagase, M., and Kurihara, K., Fabrication of $SiO_2/Si/SiO_2$ double barrier diodes using two-dimensional Si structures, *Jpn. J. Appl. Phys.*, 36, 3669–3674, 1997.
13. Tabe, M., Kumezawa, M., and Ishikawa, Y., Quantum-confinement effect in ultrathin Si layer of silicon-on-insulator substrate, *Jpn. J. Appl. Phys.*, 40, L131–L133, 2001.
14. Lu, Z.H., Lockwood, D.J., and Baribeau, J.-M., Quantum confinement and light emission in SiO_2/Si superlattices, *Nature*, 378, 258–260, 1995.
15. Miyazaki, S., Ihara, Y., and Hirose, M., Resonant tunneling through amorphous silicon-silicon nitride double-barrier structures, *Phys. Rev. Lett.*, 59, 125–127, 1987.
16. Ishikawa, Y., Makita, S., Zhang, J., Tsuchiya, T., and Tabe, M., Capacitance-voltage study of silicon-on-insulator structure with an ultrathin buried SiO2 layer fabricated by wafer bonding, *Jpn. J. Appl. Phys.*, 38, L789–L791, 1999.
17. Weinberg, Z.A., On tunneling in metal-oxide-silicon structure, *J. Appl. Phys.*, 53, 5052–5056, 1982.
18. Watanabe, M., Iketani, Y., and Asada, M., Epitaxial growth and electrical characteristics of $CaF_2/Si/CaF_2$ resonant tunneling diode structures grown on Si(111)1°-off substrate, *Jpn. J. Appl. Phys.*, 39, L964–L967, 2000.
19. Chang, C., Hu, C., and Brodersen, R.W., Quantum yield of electron impact ionization in silicon, *J. Appl. Phys.*, 57, 302–309, 1985; Cartier, E., Fischetti, M.V., Eklund, E.A., and McFeely, F.R., Impact ionization in silicon, *Appl. Phys. Lett.*, 62, 3339–3341, 1993.
20. Takagi, S., Yasuda, N., and Toriumi, A., Experimental evidence of inelastic tunneling in stress-induced leakage current, *IEEE Trans. Electron Devices*, 46, 335–341, 1999.
21. Kamakura, Y., Mizuno, H., Yamaji, M., Morifuji, M.,Taniguchi, K., Hamaguchi, C., Kunikiyo, T., and Takenaka, M., Impact ionization model for full band Monte Carlo simulation, *J. Appl. Phys.*, 75, 3500–3506, 1994.

22. Itsumi, M., Shiono, N., and Shimaya, M., Influence of polysilicon gate formation conditions on thin gate oxide (4-6 nm) dielectric and charging properties, *J. Appl. Phys.,* 73, 7515–7519, 1993.
23. Chou, S.Y., Wolak, E., and Harris, J.S., Jr., Resonant tunneling of electrons of one or two degrees of freedom, *Appl. Phys. Lett.*, 52, 657–659, 1988; Liu, H.C. and Aers, G.C., Resonant tunneling through one-, two-, and three-dimensionally confined quantum wells, *J. Appl. Phys.*, 65, 4908–4914, 1989; Bryant, G.W., Resonant tunneling in zero-dimensional nanostructures, *Phys. Rev. B*, 39, 3145–3152, 1989.
24. Su, B., Goldman, V.J., and Cunningham, J.E., Single-electron tunneling in nanometer-scale double-barrier heterostructure devices, *Phys. Rev. B.*, 46, 7644–7655, 1992.
25. Cain, P.A., Ahmed, H., Williams, D.A., and Bonar, J.M., Hole transport through single and double SiGe quantum dots, *Appl. Phys. Lett.*, 77, 3415–3417, 2000.
26. Ishikawa, Y., Ikeda, H., and Tabe, M., Potential-well-roughness-induced transition from resonant tunneling to single-electron tunneling in Si/SiO_2 double-barrier structure, Appl. Phys. Lett. 86, 013508-1–013508-3, 2005.
27. Chou, S.Y. and Gordon, A.E., Steps and spikes in current-voltage characteristics of oxide/microcrystallite-Si/oxide diodes, *Appl. Phys. Lett.*, 60, 1827–1829, 1992.
28. Fukuda, M., Nakagawa, K., Miyazaki, S., and Hirose, M., Resonant tunneling through a self-assembled Si quantum dot, *Appl. Phys. Lett.,* 70, 2291–2293, 1997.
29. Ratno-Nuryadi, Ishikawa, Y., Ono, Y., and Tabe, M., Thermal agglomeration of single-crystalline Si layer on buried SiO_2 in ultrahigh vacuum, *J. Vac. Sci. Technol. B*, 20, 167–172, 2002.
30. Ishikawa, Y., Imai, Y., Ikeda, H., and Tabe, M., Pattern-induced alignment of silicon islands on buried oxide layer of silicon-on-insulator structure, *Appl. Phys. Lett.*, 83, 3162–3164, 2003.
31. Tabe, M. and Yamamoto, T., Initial stages of nitridation of Si (111) surfaces: X-ray photoelectron spectroscopy and scanning tunneling microscopy studies, *Surf. Sci.,* 376, 99–112, 1997.
32. Tabe, M. and Yamamoto, T., Nanometer-scale local oxidation of silicon using silicon nitride islands formed in the early stages of nitridation, *Appl. Phys. Lett.*, 69, 2222–2224, 1996; Tabe, M., Kumezawa, M., Yamamoto, T., Makita, S., Yamaguchi, T., and Ishikawa, Y., Formation of high-density silicon dots on a silicon-on-insulator substrate, *Appl. Surf. Sci.*, 142, 553–557, 1999.
33. Ratno-Nuryadi, Ikeda, H., Ishikawa, Y., and Tabe, M., Ambipolar Coulomb blockade characteristics in a two-dimensional Si multi-dot device, *IEEE Trans. Nanotechnology,* 2, 231–235, 2003.
34. Ratno-Nuryadi, Ikeda, H., Ishikawa, Y., and Tabe, M., Current fluctuation in single-hole transport through a two-dimensional Si multidot, *Appl. Phys. Lett.,* 86, 2005, in press.
35. DiMaria, D.J., Dong, D.W., Falcony, C., Theis, T.N., Kirtley, J.R., Tsang, J.C., Young, D.R., Pesavento, F.L., and Brorson, S.D., Charge transport and trapping phenomena in off-stoichiometric silicon dioxide films, *J. Appl. Phys.*, 54, 5801–5827, 1983.
36. Ishikawa, Y., Kosugi, M., Tuchiya, T., and Tabe, M., Concentration of electric field near Si dot/thermally-grown SiO_2 interface, *Jpn. J. Appl. Phys.*, 40, 1866–1869, 2001.

7 Silicon Single-Electron Transistor and Memory

L. Jay Guo

7.1 INTRODUCTION

As the semiconductor device feature size enters the sub-50-nm range, two new effects come into play. One is the quantum effect, which is rooted in the wave nature of the charge carriers, and gives rise to nonclassical transport effects such as resonant tunneling and quantum interference. The other is related to the quantized nature of the electronic charge, often manifested in the so-called single-electron effect: Charging each electron to a small confined region requires a certain amount of energy in order to overcome the Coulomb repulsion; if this charging energy is greater than the thermal energy, k_BT (k_B Boltzman constant, T temperature), a *single* electron added to the region could have a significant effect on other electrons entering the confined region.

Both quantum effect and single-electron effect are unavoidable in ultrasmall devices. Therefore, a new strategy should be adopted to turn these liabilities into assets—one needs to search for new device structures that can utilize the wave-like and particle-like nature of the electrons. Unlike conventional transistors, quantum effect devices and single-electron devices actually work better as they get smaller, a feature that fits naturally into the device downscaling scheme. The single-electron devices have attracted much attention, because their operation relies solely on the Coulomb interaction and the quantized nature of the electron charge. The single-electron device was considered to be a strong contender in the terabit integration age, because any electronic device, regardless of its constituent material and structural detail, will become sensitive to the single-electron charging effect when they are scaled down to a critical dimension.[1] Such a scale is determined by the capacitance of the charge island (C) where the energy for the addition of one electron to the island (e^2/C) is greater than k_BT, provided that the charged region has sufficient isolation from other parts of the device so that the quantum charge fluctuation is suppressed. The latter condition can be satisfied if the junction resistance between the island and the other device regions is greater than the quantum resistance ($h/e^2 \approx 26$ kΩ).[2]

Since the very early demonstration of the single-electron charging effect,[3,4] a number of advances have been made. Low temperature experiments on the single-electron turnstile,[5,6] and single-electron memory[7,8] have established and proven their working principle. Single-electron devices have also been applied to metrology, where a DC current standard and Coulomb blockade thermometry have been proposed.[9,10]

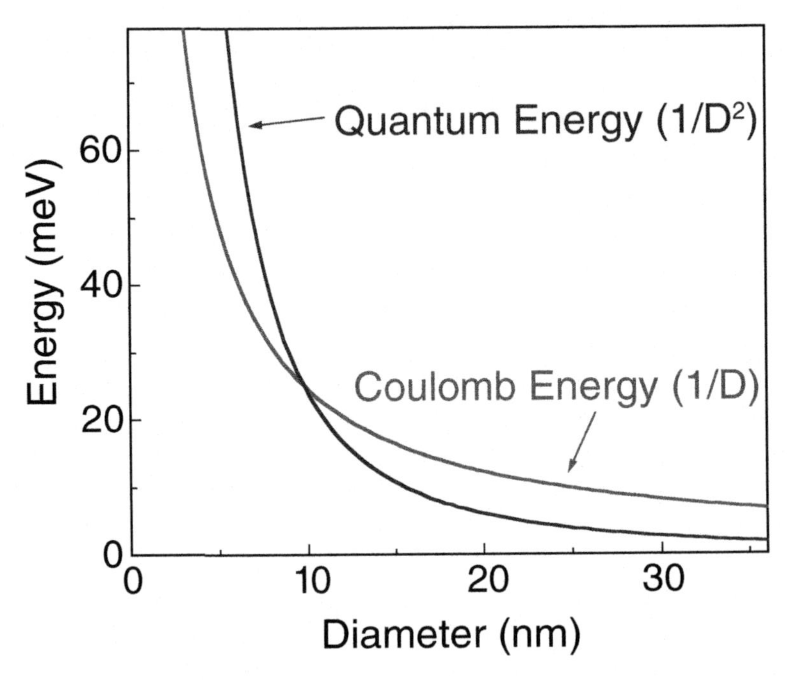

FIGURE 7.1 Comparison of energy due to Coulomb interaction and quantum confinement as a function of the size for a silicon dot.

However, due to the small current drive that can be provided by the single-electron device, new architectures have to be used on a circuit level. In this regard, various schemes of using single-electron devices to implement logic functions have been proposed.[11,12] The single-electron concept has found practical application in a single-electron-transistor scanning electrometer developed at Bell Laboratories.[13]

Single-electron effect is essentially a result of the Coulomb interaction of the electron charges. Since the Coulomb interaction is a classical effect, a single-electron device could work at a scale greater than the phase coherence length, and this could be one of its advantages over the quantum interference devices. For room temperature operation, a capacitance of order 10^{-18} F is required. This corresponds to a confined charge island of 10 nm or less. At such a scale, quantum confinement also enhances the energy level spacing, a fact that is favorable to the device operation. The contributions from both effects are compared in Figure 7.1 as a function of the size of an Si dot island. With ingenuity in device design, room temperature operation has proved achievable.[14–16]

7.1.1 Quantum Dot Transistor

In exploring the limits of semiconductor transistors and searching for innovative devices, various quantum-dot structures have been proposed and demonstrated.[17–19,] The basic idea is to replace the straight channel in the conventional field-effect transistor by a quantum-dot channel as shown in Figure 7.2. The dot is connected to the source

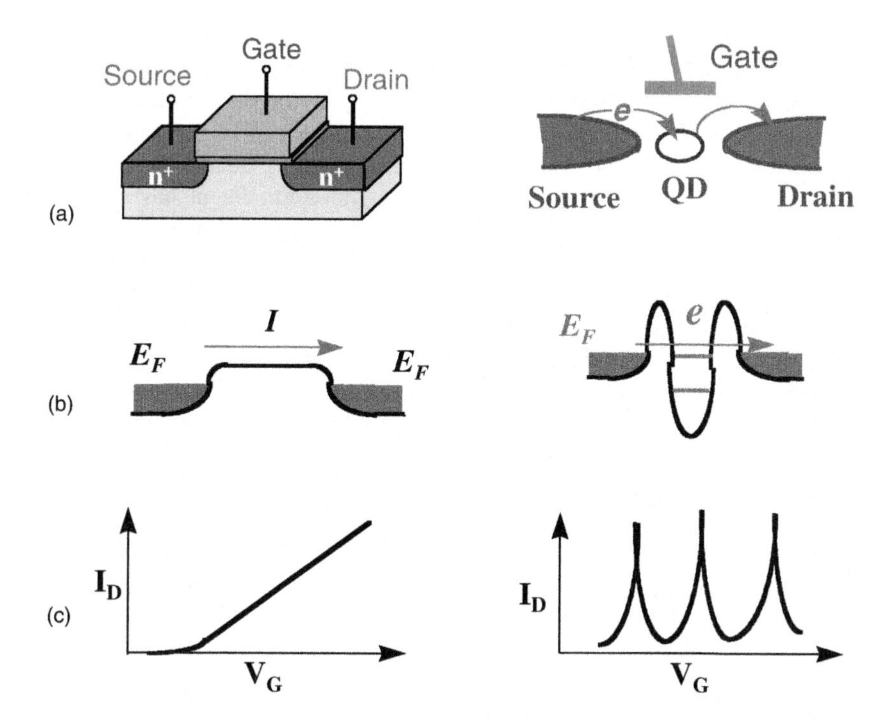

FIGURE 7.2 Comparison of the conventional MOSFET (left column) and the quantum dot transistor (right column) in (a) structure, (b) band diagram, and (c) $I_D - V_G$ characteristics.

and the drain of the device through two tunnel barriers, and is capacitively coupled to a gate. The significance of such a structure is that discrete energy levels are formed inside the quantum dot. They are the result of the Coulomb charging and quantum confinement effects. The discrete energy levels give rise to new transport characteristics; electrons can only flow from the source to the drain when an energy level inside the quantum dot is aligned with the source-drain Fermi level. This energy alignment is regulated by the gate voltage. Thus, instead of a linear relation between the drain current and the gate voltage as in a conventional transistor, a periodic oscillation would occur with each conductance peak representing the addition of a single electron to the quantum dot.

Extensive studies have been conducted in the past on electron quantum dots created in GaAs/AlGaAs material system by means of depleting the electrons using the patterned gates[20].[21]. A similar scheme has also been applied to silicon devices.[22,23] With this scheme, however, the size of the quantum dots is generally over 100 nm, often limited by the depletion width in the semiconductors. The energy due to quantum confinement is very small in such systems, and thus the discrete levels in the quantum dot are primarily due to the classical Coulomb interaction of the charges, or the charging effect of the dot.[4,5,24] As a consequence, most experimental studies were performed at temperatures much below the liquid helium temperature due to the small energy spacing. The transport properties of such structures can mostly be explained by the Coulomb blockade theory.[25]

To achieve quantum dots with smaller dimension, we have developed a scheme by using Silicon on Insulator (SOI) structure.[1,2] It utilized a very thin superficial silicon layer to provide vertical confinement. The quantum dot was first defined by means of nanolithography, and thermal oxide was grown to further reduce the dot size. Because of this innovative technology, one can obtain quantum dots with size beyond the limitation imposed by lithography. The confinement has also greatly improved since the Si dot is essentially confined by a hard wall potential due to the large band-offset between Si and SiO_2. As a result, the operation temperature has increased dramatically.[1,2]

The compatibility of this technology with conventional CMOS processing has also enabled us to demonstrate the first single-hole effect quantum dot transistors.[26] Although hole resonant tunneling in III-V resonant tunneling diode structures,[27] and hole Coulomb blockade in Si-SiGe diodes have been observed,[28,29] this is the first time that a three-terminal transistor device has been realized, which provides new possibilities for innovative circuits that utilize complementary pairs of quantum-dot transistors. Moreover, we have observed that the conductance oscillations for both the single-electron QDT and the single-hole QDT can persist at temperatures above liquid nitrogen temperature. This is the result of the large energy separations formed inside the quantum dot, with contributions from both Coulomb interaction and the quantum confinement.

7.2 THEORETICAL BACKGROUND

We consider a model system that consists of a quantum dot that is weakly coupled to a source and a drain through two tunnel junctions, and capacitively coupled to a gate electrode. The source and the drain are considered reservoirs for the electrons. The quantum dot is in thermal equilibrium with the reservoirs through the exchange of electrons. The probability of finding N electrons in the quantum dot is then given by the grand canonical distribution function

$$P(N) = \text{constant} \times \exp\left(-\left[F(N) - NE_F\right]/k_B T\right) \tag{7.1}$$

where $F(N)$ is the free energy of the dot. As T approaches 0 K, only a single value of N maximizes $P(N)$, namely an integer that minimizes the thermodynamic potential $\Omega(N) = F(N) - NE_F$; while $P(N)$ is exponentially small for other values of N. It is this condition that determines the stability of a quantum dot system at low temperature by the suppression of the charge fluctuations. Such stability is maintained up to a characteristic temperature T^* that is determined by $k_B T^* \sim \Delta E$, where ΔE is the typical energy level separation in the quantum dot.

Current can flow through the quantum dot when $P(N) = P(N+1)$.[30,xxv] In this case, charge tunnel through the quantum dot via intermediate state $N \rightarrow N+1 \rightarrow N \rightarrow N+1 \rightarrow N$ ... with one electron going through at a time, is called single-electron tunneling. As $T \rightarrow 0$ K, a necessary condition to have $P(N) = P(N+1)$ is $\Omega(N) = \Omega(N+1)$,

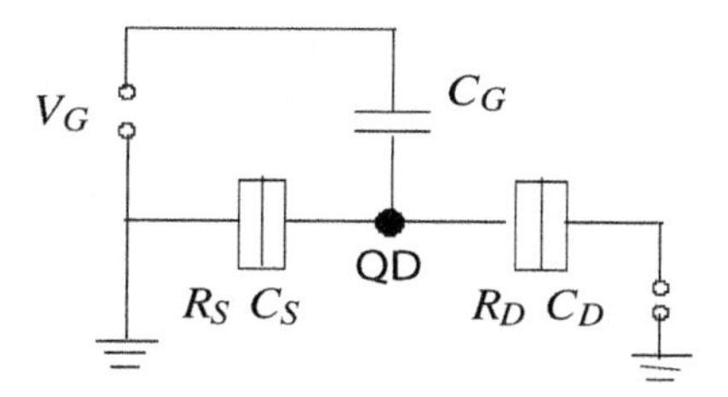

FIGURE 7.3 Equivalent circuit for a quantum dot system showing the capacitive coupling between the quantum dot and the source, the drain and the gate electrodes.

$$F(N+1)-F(N)=E_F. \tag{7.2}$$

$F(N+1) - F(N)$ is the chemical potential of the last electron added to the quantum dot, and E_F is the chemical potential of the reservoir. Equation (7.2) is exactly the equilibrium condition for the quantum dot and the reservoirs.

7.2.1 Energy of the Quantum Dot System

At $T = 0$ K, the free energy $F(N)$ is simply the energy of the N-particle system obtained by solving the many-particle Schrödinger equation $H(N)|N,l\rangle = E(N,l)|N,l\rangle$. Here an index l is added to stand for a set of quantum numbers that characterizes the N-particle states with energy E (N, l), starting from $l = 0$ to represent the ground state. Numerical solutions have been obtained for a number of systems,[31–34] but generally the energy spectrum is difficult to calculate for any real Si quantum dot system that will be discussed in this chapter.

Traditionally, the approach to such a problem has been simplified by treating the quantum dot as a confined 0-D (zero-dimensional) system that has quantized energy levels of ε_p, and by treating the Coulomb interaction between the charges in terms of the electrostatic energy of a capacitively coupled system.[5,9,35] The total energy of the system can be written as

$$E(N,l)=\sum_p \varepsilon_p + E_{el}(N) \tag{7.3}$$

where ε_p is the single particle level due to the energy quantization in the quantum dot, and E_{el} is the electrostatic energy that includes both the Coulomb interaction of the charges inside the dot and the interaction energy between the dot and the electrodes.

The electrostatic energy of the quantum dot system can be obtained with the help of the equivalent circuit shown in Figure 7.3. The excess charge on the dot, Q_0, can be related to its electrostatic potential ϕ, the drain voltage V_{DS} and the gate voltage V_G as

$$Q_0 = C_{tt}\phi - C_D V_{DS} - C_G V_G \tag{7.4}$$

Here we have used C_{tt} to represent the total capacitance of the dot, C_G, C_D and C_S to represent the dot to gate, dot to drain and dot to source capacitances, respectively. The total capacitance of the dot includes the self-capacitance C_0 and the sum of the mutual capacitances between the dot and all the electrodes ($C_\Sigma = C_G + C_D + C_S$), *that is,*

$$C_{tt} = C_0 + C_\Sigma = C_0 + C_G + C_D + C_S \tag{7.5}$$

Since the charge on the dot can only change by integer multiples of e, the electrostatic energy needed to add N electrons to the initially neutral dot island is

$$E_{el}(N, V_{DS}, V_G) = \int_0^{Ne} \phi(Q_0)\, dQ_0 = \frac{(Ne)^2}{2C_{tt}} + \frac{Ne}{C_{tt}}(C_D V_{DS} + C_G V_G) \tag{7.6}$$

We now use this expression for the energy of the system to understand the transport properties through the quantum dot channel. First we will consider the zero temperature situation where the system is in its ground state $E(N, l = 0)$. Using Equation (7.5), we can define

$$\begin{aligned} \mu(N, V_{DS}, V_G) &\equiv E(N, 0, \{V_{DS}, V_G\}) - E(N-1, 0, \{V_{DS}, V_G\}) \\ &= \varepsilon_N + (N - \frac{1}{2})\frac{e^2}{C_{tt}} + \frac{C_D}{C_{tt}} eV_{DS} + \frac{C_G}{C_{tt}} eV_G, \end{aligned} \tag{7.7}$$

where $\mu(N)$ can be regarded as the electrochemical potential of the last electron added to the dot. According to Equation (7.2), an electron can pass through the dot only if $\mu(N) = E_F$. So $\mu(N)$ can also be considered as the effective energy level of the *N*th electron. Then the condition $\mu(N) = E_F$ is the same as that of the resonant tunneling through the *N*th level. At fixed gate and drain voltage, the spacing between the effective energy levels is

$$\Delta E = \mu(N) - \mu(N-1) = (\varepsilon_N - \varepsilon_{N-1}) + \frac{e^2}{C_{tt}}, \tag{7.8}$$

which is the sum of the 0-D states level spacing plus the charging energy term e^2/C. It is this ΔE that determines the characteristic temperature T^* introduced before, above which the single-electron effect is obscured by thermal fluctuations.

Equation (7.7) says that the effective energy level $\mu(N)$ shifts linearly with the drain voltage V_{DS}, and the gate voltage V_G. This is consistent with the equivalent circuit model: the electrostatic potential on the dot depends linearly on the electrode voltages. This dependence is illustrated with the schematic band diagram of the quantum dot (Figure 7.4).

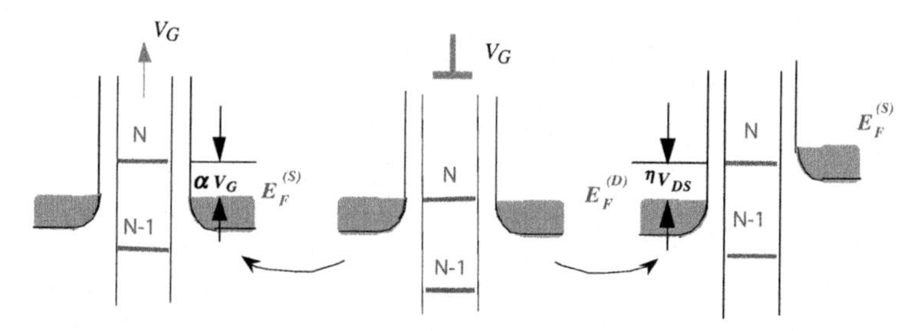

FIGURE 7.4 Schematic band diagram showing how the energy levels inside the quantum dot shift linearly with gate voltage V_G (left) and with drain voltage V_{DS} (right), (e.g., the level shifts by ηV_{DS} with respective to $E_F^{(S)}$ when a finite V_{DS} is applied).

We can introduce two coefficients $\alpha = C_G / C_{tt}$, and $\eta = C_D / C_{tt}$, which relate the shift of the energy level to the change of the gate and drain voltages:

$$\alpha = \frac{\Delta\mu}{e\Delta V_G}, \ \eta = \frac{\Delta\mu}{e\Delta V_{DS}} \tag{7.9}$$

7.2.2 Conductance Oscillation and Potential Fluctuation

Under small drain bias V_{DS}, a series of V_G can be found to satisfy $\mu(N) = E_F$. This leads to a series of oscillation peaks through the quantum dot in the linear conductance as a function of V_G. The spacing between the oscillation peaks can be found as follows. The change of E_F in the reservoir can be neglected due to the large 2-D (two-dimensional) density of states in the source and drain region. Thus, for the adjacent conductance peaks, we have

$$\mu(N-1, V_{DS}, V_G^{(N-1)}) = \mu(N, V_{DS}, V_G^{(N)}) = E_F, \tag{7.10}$$

By using the expression given in Equation (7.7), we get

$$\begin{aligned} \Delta V_G &= V_G^{(N)} - V_G^{(N-1)} \\ &= \frac{C_{tt}}{C_G}\left(\frac{\varepsilon_N - \varepsilon_{N-1}}{e} + \frac{e}{C_{tt}}\right) = \frac{1}{\alpha e}\left((\varepsilon_N - \varepsilon_{N-1}) + \frac{e^2}{C_{tt}}\right) \end{aligned} \tag{7.11}$$

Notice first that the term in the bracket is simply the energy level separation. By knowing α and peak spacing, the energy spacing can be obtained. Second, if the charging energy term e^2/C_{tt} is much larger than the single particle energy spacing, the conductance peaks will be regulated to a quasiperiodic function in gate voltage, even though the energy spectrum due to quantum confinement can be irregular. This is often called Coulomb blockade oscillations. By the same token, the deviation from periodic behavior in conductance versus V_G characteristics is an indication that the quantization energy of the quantum dot is also significant.

It is interesting to look at the fluctuation of the dot potential due to the tunneling of a single electron through the quantum dot. As discussed earlier, when Equation (7.2) is satisfied at the Nth conductance peak, the quantum dot undergoes an $N \rightarrow N+1 \rightarrow N$ transition. Although the gate voltage is fixed at this point, the electrostatic potential of the dot increases when the additional $N+1$th charge adds to the dot, and drops back when the $N+1$th charge tunnels out of the dot. According to Equation (7.4), this change in ϕ is given by

$$\Delta\phi = \frac{\Delta Q}{C_{tt}} = \frac{e}{C_{tt}} \tag{7.12}$$

Thus we see that the potential of the quantum dot fluctuates with an amplitude of e/C_{tt}, and at a rate determined by how frequently the charge traverses the dot. The time-averaged fluctuation of the dot potential between the two adjacent conductance peaks has indeed been verified experimentally.[36]

7.2.3 Transport under Finite Temperature and Finite Bias

Applying a drain bias V_{DS} will open an energy window between the source and the drain Fermi levels, *that is,* $E_F^{(S)} - E_F^{(D)} = eV_{DS}$. However, the change in the number of charges on the dot is still suppressed if [37]

$$\mu(N+1, V_{DS}, V_G) > E_F^{(S)} > E_F^{(D)} > \mu(N, V_{DS}, V_G). \tag{7.13}$$

On the other hand, current can flow if $\mu(N)$ lies between $E_F^{(S)}$ and $E_F^{(D)}$, especially, single-electron tunneling through the Nth level occurs if

$$\mu(N+1, V_{DS}, V_G) > E_F^{(S)} \geq \mu(N, V_{DS}, V_G) \geq E_F^{(D)} > \mu(N-1, V_{DS}, V_G) \tag{7.14}$$

Here the discussion is still restricted to the low temperature condition such that $E_F^{(S)} - \mu(N) >> k_B T$ and $\mu(N) - E_F^{(D)} >> k_B T$, so that the thermal activation is suppressed. We will show that a peak in the differential conductance separates the blockade and the tunneling regime.

The tunnel current through the quantum dot can be expressed as[38,39]

$$I_D = \frac{e}{h}\int \Gamma(E) D(E)\left[f(E - E_F^{(S)}) - f(E - E_F^{(D)})\right] dE \tag{7.15}$$

where $f(E - E_F)$ is the Fermi-Dirac distribution function; $\Gamma(E)$ is the energy dependent tunneling rate which in general also depends on V_{DS} and V_G; $D(E)$ is the density of states inside the dot. If we only examine the behavior of I_D in the neighborhood of the Nth level, we can approximate $D(E) = \delta(E - \mu_N)$, where $\mu_N \equiv \mu(N)$. At finite temperature T and vanishing drain bias ($V_{DS} \rightarrow 0$), the linear conductance can be calculated:

$$G_D = \frac{e}{h}\Gamma(\mu_N)f'[(E-\mu_N)/k_BT]$$
$$= \frac{1}{4k_BT}\frac{e^2}{h}\Gamma(\mu_N)\mathrm{sech}^2\left(\frac{\alpha[V_G - V_G^{(N)}]}{2k_BT}\right) \tag{7.16}$$

$f'(E/k_BT)$ is the derivative of Fermi-Dirac function, and describes a thermally broadened peak that has a width $\propto k_BT$ and height $\propto 1/T$. The full width at half maximum (*FWHM*) of the peak height can be related to temperature T as $\alpha(FWHM) = 3.5\ k_BT$ (see Figure 7.5). So the peak width provides a natural energy scale for analysis.

At finite V_{DS}, we can study the differential conductance which is defined and calculated as

$$g_D(\mu_N) = \left.\frac{\partial I_D(V_{DS}, V_G)}{\partial V_{DS}}\right|_{Vg,T}$$
$$= -\frac{e^2}{h}\Gamma(\mu_N)\times\left[\eta f'(\mu_N - E_F^{(S)}) + (1-\eta)f'(\mu_N - E_F^{(D)})\right] \tag{7.17}$$

Here we see that as long as $E_F^{(S)} \neq E_F^{(D)}$, i.e., $V_{DS} \neq 0$, two peaks will appear in the differential conductance. From the position of the two peaks in relation to the bias voltage, we can obtain the values of α and η. This can be done by using the explicit dependence of the effective energy level on the drain and gate bias, *that is,* $\mu(N, V_G, V_{DS}) = \mu_N^{(0)} - \alpha e V_G - \eta e V_{DS}$. Let us denote $V_G^{(N)}$ to be the position of the Nth conductance peak at zero bias, which is determined by $E_F^{(S)} = \mu(N, V_G^{(N)}, 0)$ or $E_F^{(S)} = \mu_N^{(0)} - \alpha e V_G^{(N)}$ according to Equation (7.2). If we write $V_G = V_G^{(N)} + \Delta V_G$, we get

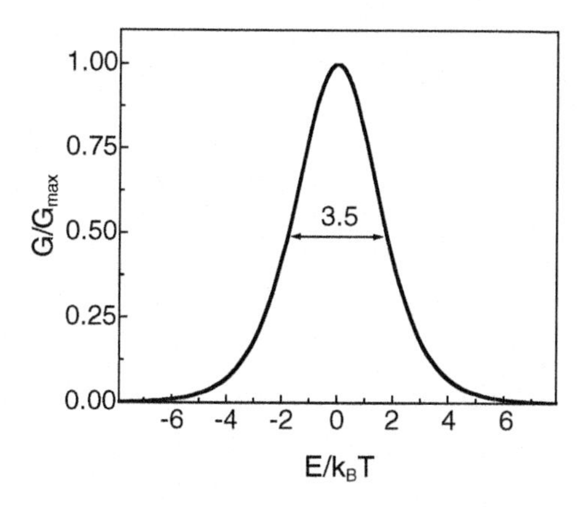

FIGURE 7.5 Line shape of a thermally broadened conductance peak. The *FWHM* is 3.5 k_BT.

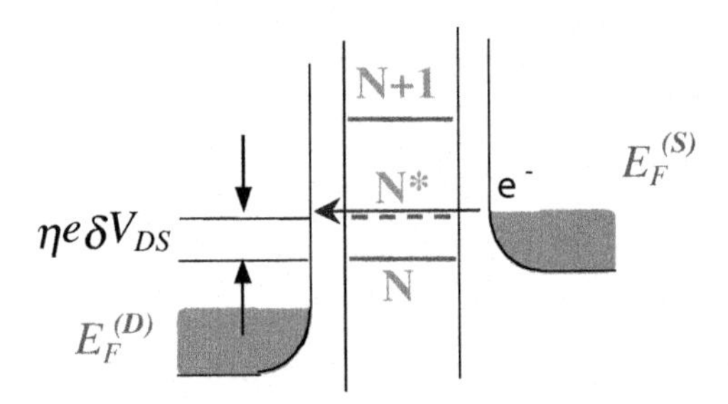

FIGURE 7.6 With sufficient high V_{DS}, both charge state N and its excited state N^* are available for tunneling, while the next charge state with N+1 electrons is still outside the energy window due to the large charging energy.

$$g_D(V_G,V_{DS},T)=\frac{e^2}{4hk_BT}\Gamma(V_G,V_{DS})\times\left\{\eta\,\mathrm{sech}^2\left(-\frac{\alpha e\Delta V_G+\eta eV_{DS}}{2k_BT}\right)\right. \\ \left.+\,(1-\eta)\,\mathrm{sech}^2\left(-\frac{\alpha e\Delta V_G-(1-\eta)eV_{DS}}{2k_BT}\right)\right\} \tag{7.18}$$

We can see that a single peak at $V_{DS} = 0$ splits in two with increasing V_{DS}. With respect to $V_G^{(N)}$, one located at $\Delta V_G^{(-)} = -\eta V_{DS}/\alpha$, and the other at $\Delta V_G^{(+)} = (1-\eta)V_{DS}/\alpha$. So we have

$$\frac{\Delta V_G^{(+)}}{\Delta V_G^{(-)}}=\frac{1-\eta}{\eta} \quad \text{and} \quad \Delta V_G=\Delta V_G^{(+)}-\Delta V_G^{(-)}=\frac{1}{\alpha}V_{DS} \tag{7.19}$$

where ΔV_G is the separation of two peaks in gate voltage. These two relations will be used in Section 4.3 to extract the values of η and α from the experiments.

Another effect that occurs under finite bias condition is that the transport through the quantum dot is no longer restricted to only the ground state.[40,41] Given additional energy, electrons can also occupy higher orbital states (i.e., excited 0-D states) in the confined quantum dot. Because of the opening of an energy window between $E_F^{(S)}$ and $E_F^{(D)}$, these excited states can also participate in the conduction though the quantum dot (Figure 7.6). The figure shows that at fixed V_G, increasing V_{DS} will successively bring the ground state N and the next excited state N^* to align with the source Fermi level. Note that N and N^* corresponds to the same number of electrons in the dot, but with different energies. With this we can find a direct measure of the energy spacing ($\Delta\varepsilon$) due to spatial quantization, if the bias voltage difference between the two successive alignments is δV_{DS}, $\Delta\varepsilon = \eta e\Delta V_{DS}$. This argument will be used later in Section 4.4.

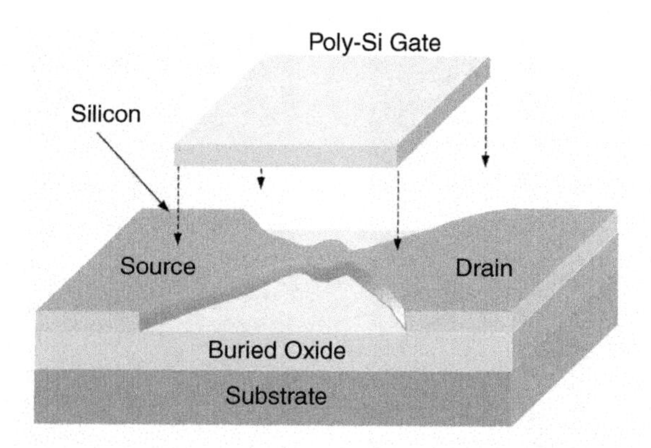

FIGURE 7.7 Device schematic of single-charge quantum-dot transistor on SOI.

7.3 DEVICE STRUCTURE AND FABRICATION

The schematic of our quantum dot transistor is shown in Figure 7.7. A quantum dot channel was formed in the top silicon layer of the wafer made by Separated by Implanted Oxygen (SIMOX). It has a dot-shaped island connected to the source and drain region by two narrow constrictions. A polysilicon gate covers the dot, the constrictions, and part of the source and drain region. When the gate voltage is applied to reach strong inversion, a 2-D electron gas (or hole gas in a hole quantum dot device) are formed outside the quantum dot, and act as the charge reservoirs as discussed in the previous section. The only difference between an electron quantum dot transistor and a hole quantum dot transistor is the doping of the source and drain region—n-type doping for the former and p-type for the latter. The quantum-dot channel is lightly doped with boron at a concentration of 3×10^{15} cm^{-3}.

The two constrictions in the silicon channel act as tunnel junctions. This is due to the rising of the conduction band at the narrow constrictions as compared with the wide region. For example, if the silicon wire has a thickness of 5 nm, the conduction band edge will increase by 100 meV.[42] It is this potential energy difference that forms the tunnel barrier.

In fabrication, the starting SIMOX wafer has a 60-nm-thick superficial silicon layer and 360-nm-thick buried oxide. A 30-nm-thick sacrificial oxide was grown as a masking layer, then the active area was patterned with optical lithography to isolate individual devices. An abacus bead-shaped pattern was first defined in the polymethyl methacrylate (PMMA) resist layer using electron beam lithography. The pattern was then transferred into the sacrificial oxide layer by reactive ion etching (RIE) with PMMA as the etching mask. Next, chlorine-based RIE was used to transfer the dot pattern into the silicon layer with oxide as a mask. Figure 7.8(a) shows an SEM picture of a dot channel etched into the silicon layer. The oxide mask was then removed using HF, and some of the buried oxide under the dot channel was also etched away in this process. This turned out to be beneficial for reducing the size of the quantum dot, because during the following gate oxide formation, not only the top and the side of the quantum dot but also the bottom of the dot were exposed to

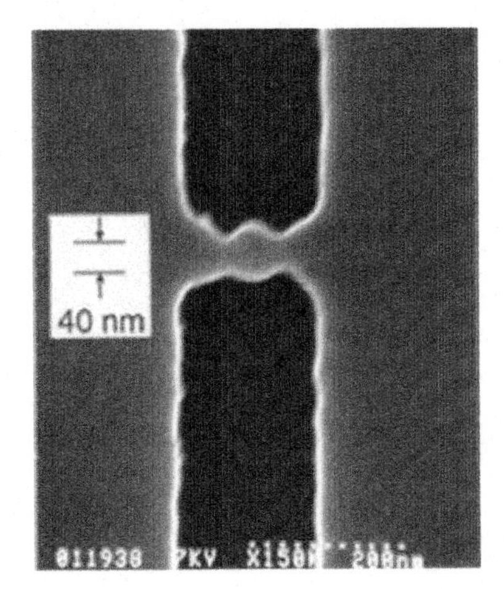

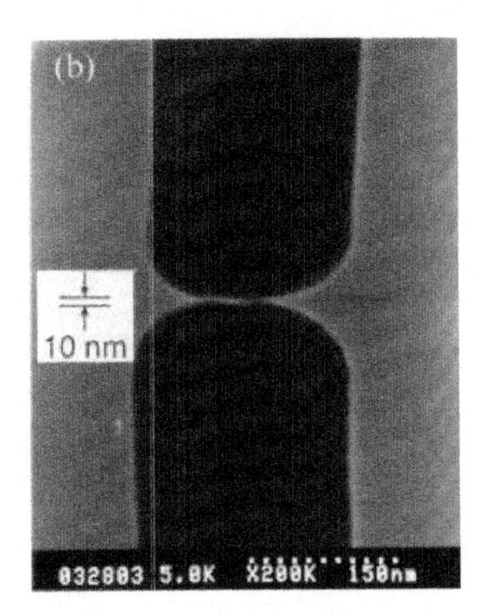

FIGURE 7.8 Scanning electron micrograph of a quantum-dot channel etched into the top silicon layer of SOI wafer (a) before, and (b) after oxidation. The width of the dot in (b) is about 10 nm. Reproduced with permission from the American Institute of Physics.[18,26]

oxidation. Since silicon was consumed during the process, after a 42-nm oxide growth we expect that the silicon dot was further reduced by about 35 nm from its initial size. Gate oxidation was done at a high temperature of 1000°C to anneal any damage caused by RIE. Polysilicon was then deposited and patterned by photolithography to cover the entire quantum-dot channel. This was followed by the self-aligned source/drain ion implantation, using phosphorous and boron, respectively, for the electron and hole QDTs. Dopant activation was achieved by annealing the sample at 950°C in a furnace. Aluminum with a titanium barrier layer was used in the final metalization. Finally, the sample was sintered in the forming gas (a mixture of hydrogen and nitrogen) to reduce the interface states. A detailed fabrication procedure can be found in Reference.[43]

It is helpful to take a careful look at the factors that limit the final size of the quantum dot. First, the vertical size of the final silicon dot was determined both by the initial top silicon thickness and the silicon consumption during the two oxidation processes, and it is estimated to be reduced to about 30 nm in the end. The lateral and transverse sizes were controlled by the e-beam lithography as well as by the oxidation process. Using e-beam lithography to define the dot pattern gives better control of the dot shape and the dot size as compared with the stress dependent oxidation method.[17] The SEM image of such a silicon dot after removing the oxide is shown in Figure 7.8(b), which is estimated to be 10 nm in width, 30 nm in length and 30 nm in depth. The dot has the shape of an ellipsoid rather than a sphere. In order to get a more symmetric dot, the distance between the constrictions must reduce further.

7.4 EXPERIMENTAL RESULTS AND ANALYSIS

The devices were tested in a Heliox cryostat, an He^3 sorption pumped refrigerator, with a variable temperature range from 0.5 to 300 K. DC measurements used an

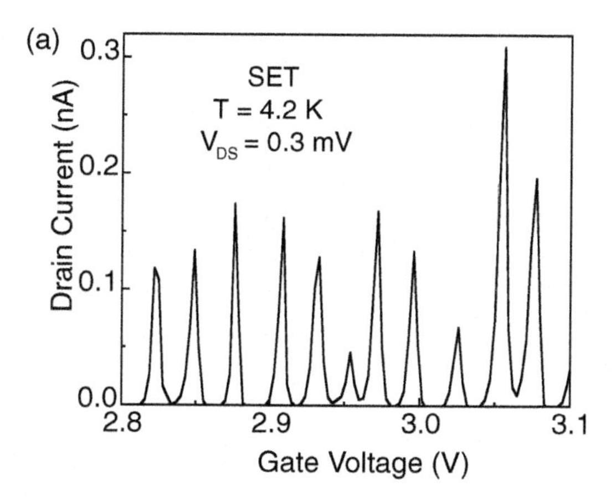

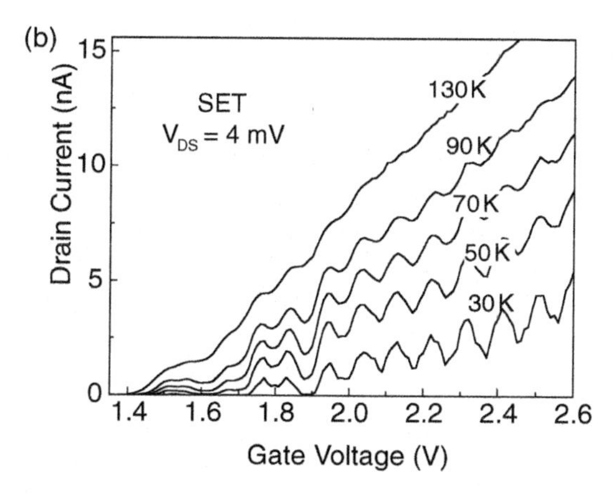

FIGURE 7.9 I – V characteristics of a single-electron QDT at (a) 4.2 K, and (b) from 30 K to 130 K.

HP4145 Semiconductor Analyzer. To reduce the influence of random noise, long integration times were used in order to average 256 values at each data point. To reduce the noise at low bias voltage, standard phase-sensitive detection was used with an excitation voltage of 20 to 50 μV, and a frequency of 13 to 20 Hz.

7.4.1 Single-Electron Quantum-Dot Transistor

The devices were first tested at 4.2 K under fixed low drain-source bias (V_{DS}) so that the transport is in the linear regime (i.e., the drain current I_D is linearly proportional to V_{DS}). The drain current vs. gate voltage characteristics for an electron quantum-dot transistor are shown in Figure 7.9(a), where distinct oscillations were observed in the drain current. This is the signature of discrete energy states inside the quantum dot; since the drain source bias is small, electrons can only tunnel through the energy state in the quantum dot that is in resonance with the Fermi level

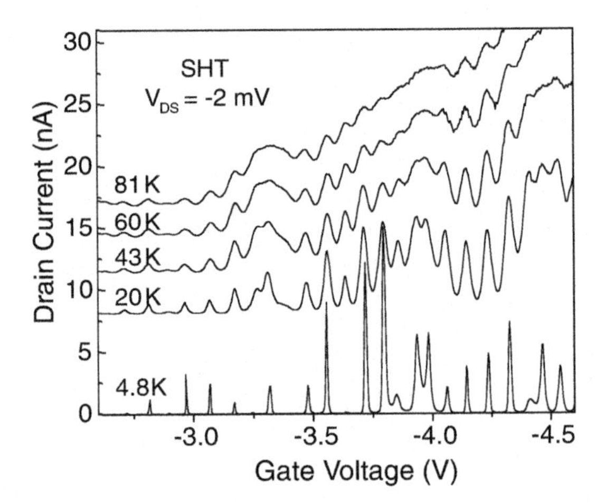

FIGURE 7.10 I-V characteristics of a single-hole QDT at temperatures from 4.8 to 81 K. The traces have been displaced for clarity.

of the source and drain. Under such condition, the current peaks. On the other hand, when the source/drain Fermi level is aligned in between two available states, the tunneling process is blocked by the presence of an energy gap, the so-called Coulomb blockade effect. Such oscillations can still be observed at a temperature above 100 K (Figure 7.9[b]). This is in contrast to the previous experiments on GaAs quantum dots, which were generally performed at temperatures in the milli-Kelvin range.[20,21,24]

7.4.2 Single-Hole Quantum-Dot Transistor

The I-V characteristics of a hole quantum dot transistor are shown in Figure 7.10 for temperatures from 4.2 up to 81 K. As in the case of the single-electron transistor, discrete current oscillations were clearly observed at low temperatures. Note that the gate voltage is negative so as to induce holes in the quantum dot, and the drain bias is also negative in order to drive the holes to flow from the source to the drain. As temperature increased, the oscillation peaks broadened and eventually smeared out at ~100 K. The peak broadening, as discussed in the previous section, is due to the thermal smearing of Fermi-Dirac distribution at the source/drain reservoirs.

Next we will use this single-hole quantum-dot transistor as an example to extract some important device parameters based on the discussions given in the previous sections. First we would like to know the energy level separations in the dot. We can make a rough estimate by using the energy scale provided by the width of the oscillation peak. Since we have shown that the peak has a width of 3.5 k_BT at temperature T, the fact that the oscillation smears out at ~100 K implies that the energy level spacing is about 30 meV. For a more accurate assessment, the conductance data of the device under a drain bias of 20 μV was obtained, shown in Figure 7.11.

In Figure 7.11, we notice that despite the overall broadening of the peaks, the height for individual peaks vary differently with temperature. Some increase as the temperature is raised and some decrease, while some stay approximately constant.

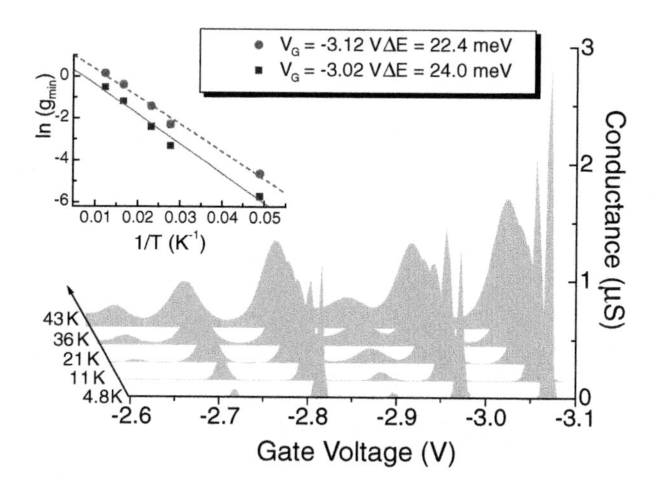

FIGURE 7.11 Peak width and height variation for the conductance oscillations at different temperatures (the same hole-QDT as shown in Figure 7.10). Insert shows that the conductance minimum is thermally activated.

At sufficiently high temperature, all the peaks show increases in height. This is very similar to the behavior of the oscillation reported before for the electrons tunneling through the quantum dot.[24,44] Explanation was provided in terms of the variation of the tunneling rate through different energy levels.[45]

Energy level separation can also be estimated from the activation energy obtained at the conductance minima. Because the conductance minimum corresponds to the condition that the source/drain Fermi level is aligned to the middle of two adjacent energy levels in the quantum dot, charge can only go through the quantum dot by thermal activation, and the conductance rises exponentially with increasing temperature. As shown in the insert of Figure 7.11, the conductance minima depends exponentially on k_BT, from which the activation energy can be obtained. The energy spacing is about twice the activation energy, and is found to be ~24 meV for the two chosen gate voltages. This value varies from one minimum to another, indicating a nonuniform energy level distribution inside the quantum dot.

7.4.3 Transport Characteristics under Finite Bias

With an increased drain bias, the transport through the quantum dot channel becomes highly nonlinear. This can be best seen by measuring the drain current I_D as a function of the drain bias V_{DS} while keeping the gate voltage V_G fixed. The result is shown in Figure 7.12, where I_D is plotted as a function of both the V_{DS} and V_G. Drain bias is from –50 to 50 mV, and the gate voltage is from –2.40 to –2.60 V. Observing this nonlinear characteristic, we see that the current is strongly suppressed in some regions due to the blockade of the hole tunneling. With increased V_{DS}, however, current starts to flow when V_{DS} is large enough to provide the hole with sufficient energy to overcome the energy gap between the adjacent energy levels.

In Figure 7.13 we present the differential conductance $g_D = \partial I_D/\partial V_{DS}$ (calculated from the experimental data), and plot g_D as a function of both V_{DS} and V_G in a gray

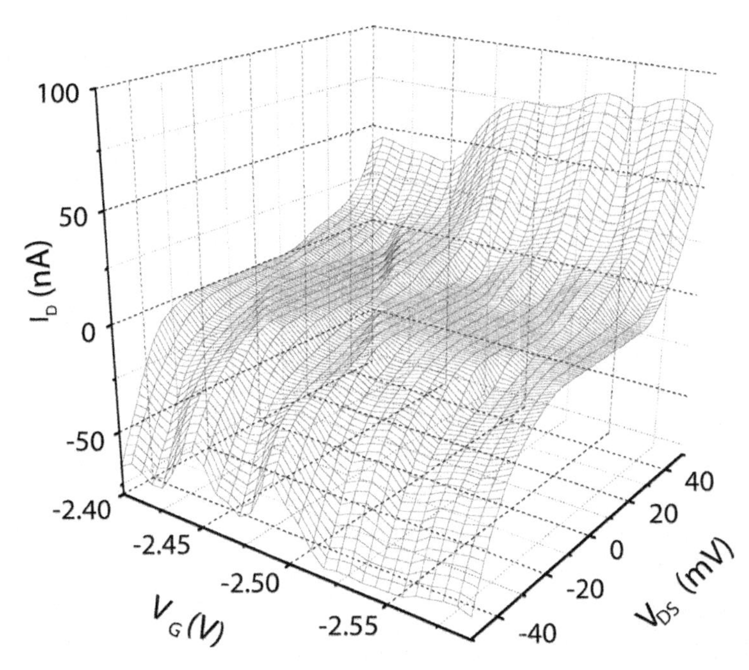

FIGURE 7.12 Drain current vs. drain bias characteristics of the hole QDT, gate voltage changes from –2.40 to –2.60 V in –0.01V step.

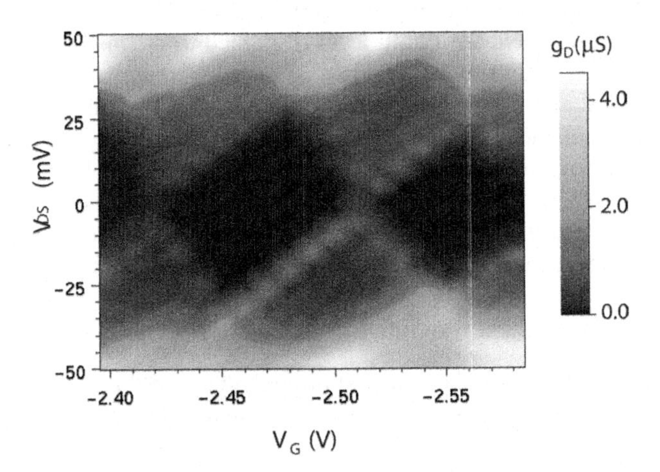

FIGURE 7.13 Gray scale representation of differential conductance g_D as a function of both drain bias and gate voltage.

scale scheme. The brighter the color, the larger the conductance. Regions of zero conductance shown in black are enclosed in the diamond-shaped boundaries. The bright boundaries separating the blockade from the tunneling regime are peaks in the differential conductance as discussed in Section 7.2.2. The single conductance

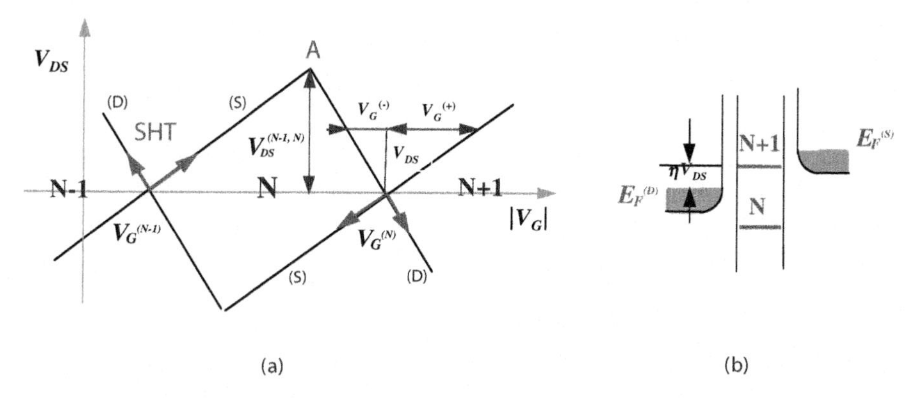

FIGURE 7.14 (a) A schematic showing the boundary between the tunneling and blockade region. (S) and (D) are labels for the boundary that corresponds to the alignment of the resonant level with the source and drain Fermi level, respectively. Other labeled variables are referred in the text. (b) A schematic band diagram.

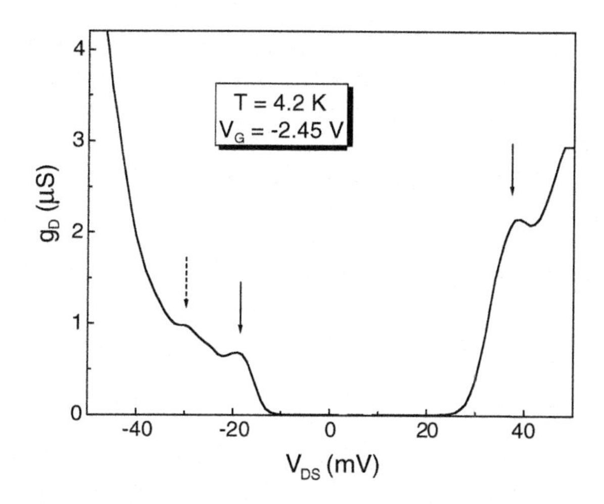

FIGURE 7.15 Differential conductance g_D plotted as a function of V_{DS} at fixed V_G. The two peaks indicated by the solid arrows separate the tunneling and the blockade region. The additional peak (indicated by the dashed arrow) is due to the transport through the excited 0-D channel.

peak observed at low V_{DS} splits into two with increasing V_{DS}, leaving two traces of conductance peaks that form the boundary. The boundary lines appear to be linear on the V_{DS}–V_G plane, which are shown schematically in Figure 7.14.

The physics for the appearance of the two peaks can be understood more clearly from Equation (7.17), which is the result of the opening of two conduction channels, one at the source side $\mu_N = E_F^{(S)}$, and the other at the drain side $\mu_N = E_F^{(D)}$. For gate voltages in between the two peaks, there is always a level within the energy window of $E_F^{(S)} - E_F^{(D)}$ that is available for tunneling.

The peak splitting effect, as we have discussed in Section 2.2, is essentially due to the bias-induced shifting of the energy level in the quantum dot, and the difference in $E_F^{(S)}$ and $E_F^{(D)}$.[46] The Nth effective energy level that produces the single conductance peak under the vanishing bias ($V_{DS} \to 0$) is originally aligned with $E_F^{(S)}$ and $E_F^{(D)}$, and is shifted by ηV_{DS} with respect to the $E_F^{(S)}$ when the drain bias is increased to V_{DS}. To bring it back to the resonant condition, we need to increase V_G by $V_G^{(+)} = (1\text{-}\eta)V_{DS}/\alpha$ to make it aligned with $E_F^{(D)}$, or decrease V_G by $V_G^{(-)} = \eta V_{DS}/\alpha$ to make it aligned with the $E_F^{(S)}$. Thus two conductance peaks would appear. From this simple analysis, we can immediately obtain the values for the two coefficients from Figure 7.13: $\alpha = 0.32$ and $\eta = 0.4$.

Using the diagram in Figure 7.14(a), we can also find the energy level separation. Let us look at the crossing point of two boundary lines at high drain bias (for example, A in Figure 7.14[a]). On one hand, it is on the boundary (D) where the Nth level is aligned with the drain Fermi level (i.e., $\mu[N, V_{DS}^{(N-1,N)}, V_G^{(N)}] = E_F^{(D)}$); on the other hand, it is also on the boundary (S) where the N–1th level is aligned with the source Fermi level (i.e., $\mu[N-1, V_{DS}^{(N-1,N)}, V_G^{(N)}] = E_F^{(S)}$). Using expression (7.7), we have $eV_{DS}^{(N-1,N)} = \varepsilon_N - \varepsilon_{N-1} + e^2/C_\Sigma = \mu_N - \mu_{N-1}$. So the drain bias (multiplied by e) at the crossing point is exactly the energy spacing between the N–1th and Nth level. Applying this result to the middle diamond shown in Figure 7.13, we can find that the energy level separation between the two neighboring states is $\Delta E \approx 30$ meV. On the other hand, since we already have determined α, we can get ΔE from the conductance peak separation: $\Delta E = \alpha \Delta V_G \approx 0.32 \times 92 = 29.7$ meV. The two values are very close. According to Equation (7.8), the ΔE just obtained is the sum of the charging energy and the quantum energy. Their values can be separately determined if we have knowledge about the energy spacing of the 0-D states. This information can be obtained by studying transport through the excited states.

7.4.4 Transport Through Excited States

Experimentally, the excitation spectrum of the 0-D states can be probed by using finite drain bias, which was discussed in Section 2.2. This is because transport through excited 0-D states is allowed due to the opening of the energy window between source and drain Fermi level. As shown in Figure 7.6, each time V_{DS} is increased to a point that a 0-D state is brought into the energy window, a new channel is opened for tunneling, resulting in a change in the drain current. This will be reflected as additional peaks in $\partial I_D/\partial V_{DS}$. Shown in Figure 7.15 is a typical result for the differential conductance at certain gate voltage: the two peaks indicated by the solid arrows correspond to the ground state transport, and the peak indicated by the dashed arrow at higher bias corresponds to the next available excited state. By measuring the drain bias difference between the peaks and multiplying by $e\eta$, we get the typical 0-D states energy spacing of about 5 meV.

Finally, by taking the single particle energy spacing of the 0-D states to be 5 meV, the charging energy is then ~25 meV, which corresponds to $C_{tt} \approx 6.4$ aF ($\because E_C \equiv e^2/C_{tt}$). By the definition of α and η, we can find the various capacitance values to be $C_G \approx 2.0$ aF, $C_D \approx 2.6$ aF, and $C_S + C_0 \approx 1.8$ aF, respectively. To further determine the values for C_S and C_0, we can exchange the source and drain electrodes

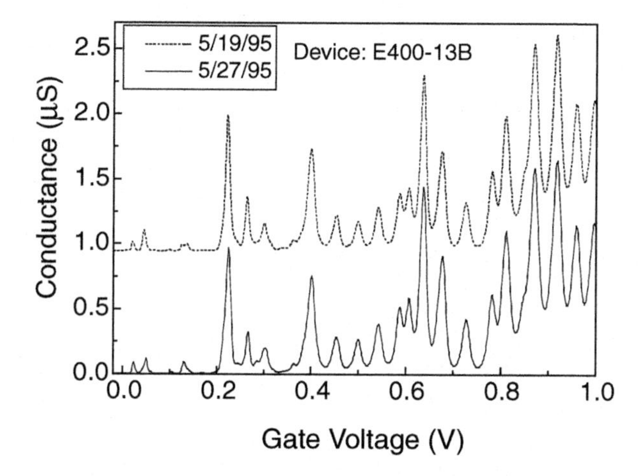

FIGURE 7.16 Conductance oscillations of a QDT measured at two different dates ($T = 4.2$ K), showing the repeatable structures after thermal cycling.

and perform another bias-dependent study to get a graph similar to Figure 7.13, from which we can get the ratio $\eta' = C_S / C_{tt}$, and C_S and C_0 can be obtained accordingly. Analysis of the single-electron transistor can be done in a similar fashion, and will not be repeated here.

7.5 ARTIFICIAL ATOM

Semiconductor quantum dots are often referred to as artificial atoms.[47] This term is very suitable to some of these silicon quantum dot devices, because their transport characteristics are exactly repeatable from time to time, and moreover, they are also strikingly the same even after thermal cyclings. Such behavior reflects the intrinsic properties of the quantum dot. Figure 7.16 shows the conductance oscillations of an electron QDT taken on two different dates. Not only has the threshold voltage of the device not changed, but also the positions and the amplitudes of the peaks are nearly the same.

This effect is believed primarily due to the absence of impurity atoms inside the silicon quantum dot. With a channel doping concentration of 3×10^{15} cm^{-3} in our sample, the probability of finding a single impurity atom inside a 10-nm-diameter quantum dot is less than 0.2%. Without impurity movement during thermal cycling, the transport characteristic is solely determined by the intrinsic quantum dot structure, that is, the confining potential, just as the case in a real atom. Moreover, the channel of a quantum-dot transistor need not have any doping at all, completely eliminating the fluctuations due to the presence of the impurity atoms.

The difference between a semiconductor quantum dot and a real atom is that coulomb interaction dominates in the former case (at least for a dot diameter greater than 10 nm), while it is a much weaker effect in a real atom as compared with the strong quantum confinement effect. However, the signature of the quantum states can still be revealed from the transport characteristics, such as the transport through

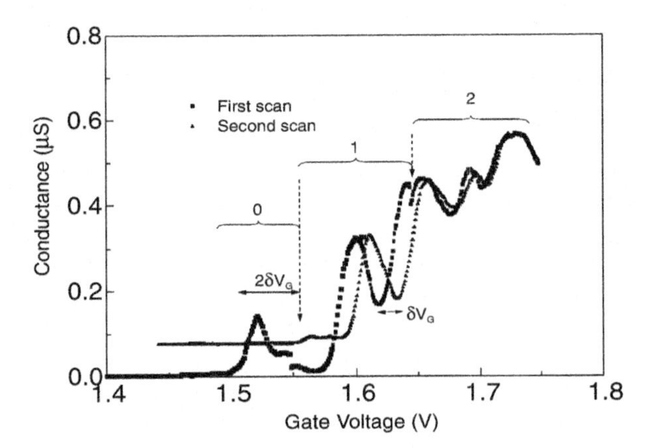

FIGURE 7.17 Discrete conductance switchings (indicated by the two dashed arrows) during the initial scan, which caused threshold shift during the subsequent scan.

the excited states as described in the previous section. In addition, it is the result of the quantized electron orbit inside a quantum dot that gives rise to the deviation from the periodic conductance oscillation as predicted by the coulomb blockade theory. Tarucha et al.[48] have studied vertical transport through a quantum dot that has cylindrical symmetry and lateral parabolic potential, and have identified a "shell" structure associated with the degeneracies of single-particle states. Moreover, the electron-electron interaction determines the filling of spin states, which can be predicted according to the Hund's rule in atomic physics. In this example, the artificial atom indeed deserves its name.

7.6 SINGLE CHARGE TRAPPING

In this section, we will describe another interesting effect encountered in the experiments on the quantum dot transistors — the bistability in the I_D–V_G characteristics associated with single-electron trapping. We will also discuss its possible cause and its implications.

The effect is shown in Figure 7.17, which was taken at 20 K for a QDT device. The two traces in the figure were obtained during two consecutive scans, with an interval of a few seconds between. Apart from the usual conductance oscillations, two noticeable characteristics can be seen. The first is the sudden drop of current at two gate voltages during the first scan (as shown inside the circles). The second is a memory effect—a shift in the threshold voltage after the initial scan. Such a memory effect only occurred if the interval between the two subsequent scans was short (from a few seconds up to ~5 min at 20 K). For longer intervals, repeated scans gave the same results as the first one, only the exact locations of the sudden current drop varied slightly.

We attribute each sudden drop of current to the trapping of a single electron.[49] Because the trapped charge lowers the local potential, each trapped electron can cause a shift of the threshold voltage in the conducting dot. As a result of the increase

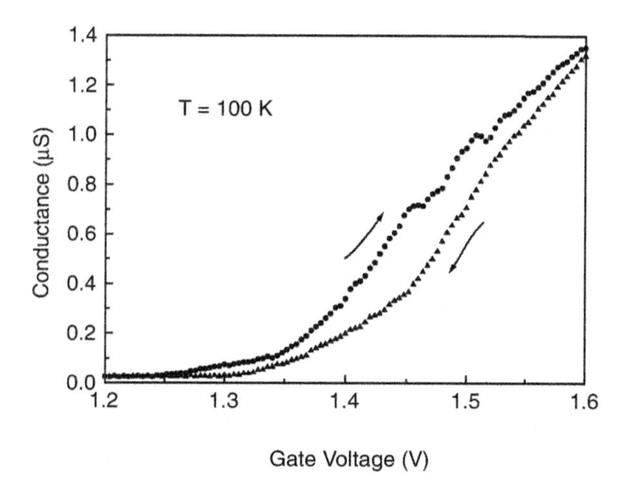

FIGURE 7.18 Hysteresis behavior observed at 100 K. The conductance oscillation already disappeared.

in the threshold voltage, the number of conduction electrons in the channel is reduced at given V_G, and hence the current drops. In Figure 7.17 we see that a total of two electrons had been trapped after the first scan. We can use the two discrete switchings to divide the total voltage range into three regions; each region is labeled by the number of electrons that has been trapped. Each trapped electron should create a discrete shift in the threshold voltage. Moreover, memory effect can occur if the subsequent scan is done within the charge holding time. Since the second scan in Figure 7.17 was taken within a few seconds of the first scan, the two trapped charges had not been released, and they remained trapped during the second scan. We can also notice that the amplitude of the first peak observed in the initial scan was reduced significantly during the second scan.

As compared with the first trace, the amount of threshold voltage shift during the second scan depends on the net difference in the number of trapped electrons, which is different for the three regions. Suppose each electron can cause a threshold shift of δV_G, then for region "0" the shift in the second trace is $2\delta V_G$; for region "1", the shift is δV_G; and for region "2", the shift is 0. This is because detrapping of charges occurred at higher gate voltage, and the number of *net* trapped electrons is 2, 1, and 0, respectively, for the three regions. From Figure 7.17, we can find δV_G ~ 22 mV. Noteworthy is that the hysteresis can persist to a much higher temperature — 100 K as shown in Figure 7.18. Even through the conductance oscillations had already smeared out at this temperature, the hysteresis could still be observed clearly between forward and backward scans.

There are two possible reasons for the electron trapping. One could be that the two electrons are trapped at two interface states that have different energies. The other possibility is that the electron tunnels from the conducting dot to a neighboring island dot that is not along the conduction path and are subsequently trapped inside this nonconducting dot. Similar behavior had been reported by Fujiwara et al. for an Si single-electron transistor with satellite Si islands.[50] The existence of the

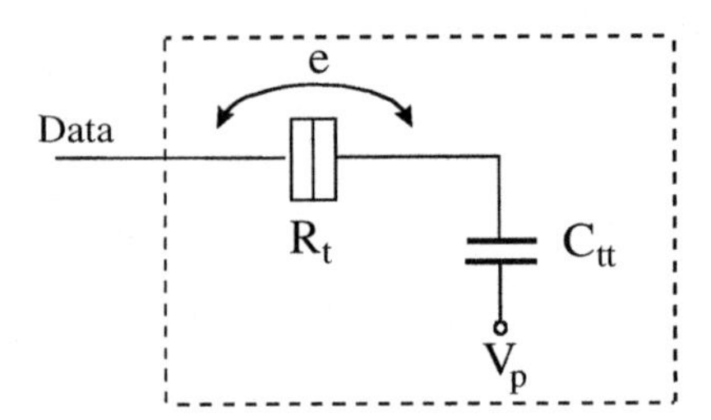

FIGURE 7.19 General scheme of semiconductor memory using capacitor for charge storage. The resistive element, R_t, can be a transistor as in a DRAM, or a tunnel oxide as in an E^2PROM.

hysteresis indicates that the process is irreversible; at the gate voltage where the current drops, the electron is trapped into the nonconducting dot and it stays there. No random telegraphic signal[1] was observed in this device. This is only possible, as suggested by Fujiwara et al., if at least two junctions are involved for the electron trapping. A likely case is that when an electron enters the conducting dot from the source, the Coulomb blockade between it and the neighboring dot is lifted simultaneously, which enables the electron to transfer to the nonconducting dot.

The dramatic difference in the amplitude of the first peak in the initial and the subsequent scans could be attributed to the change in the confinement potential caused by the trapped charge. It is known that the electron tunneling rate through the quantum dot depends critically on the potential inside the dot.[51] We think that the trapped electron not only affected the overall potential of the conducting channel, but also influenced the potential configuration inside the conducting dot.

Finally, we want to draw attention to the fact that a *single* electron can produce a measurable and discrete threshold voltage shift, leading to a distinct hysteresis behavior and memory effect. This implies that a *single-electron* memory can possibly be realized. The observation that the hysteresis exists even after the conductance oscillations disappear means that a quantum-dot channel is not necessarily needed for such an effect. It suggests that a structure with a quantum dot for storing charge, and coupled to a narrow straight channel, can be used to implement a single-electron memory. This will be the subject of the next section.[52]

7.7 INTRODUCTION TO MEMORY DEVICES

To increase the storage density of semiconductor memories, the size of each memory cell must be reduced. A smaller memory cell also leads to higher speed and lower power consumption. This is the incentive for studying the nanoscale semiconductor memory.

One of the general schemes for semiconductor data storage is by storing charges on a capacitor. The charged state and the uncharged state can be used to represent binary information 1 and 0, respectively. Usually charges are transferred to the capacitor through a resistive element as shown inFigure 7.19. The resistive element can be a transistor channel, as in the case of a Dynamic Random Access Memory (DRAM) cell, or a tunnel oxide in the case of a floating gate memory such as

Electrically Erased Programmable Read Only Memory (EEPROM). The presence or absence of the charges on the capacitor can be detected by using a transistor; in the case of a DRAM, it is represented by the bit line voltage difference when the transistor is selected; in the case of a floating gate memory, it is represented by the transistor's threshold shift.

The choice of capacitor depends on how well the channel or the junction could retain the stored charges on it. For example, it is known that a DRAM transistor has off-state leakage current through the channel, so its charge storage capacitor has to maintain a sufficient capacitance value when scaling down the memory cell. But for the floating gate memory, the tunnel oxide provides a very good energy barrier for the charge carriers. So it is very difficult for the charges to leak out from the floating gate, and hence the memory can be nonvolatile. In this case, the floating gate capacitor can be scaled down, so long as the charges stored on it can cause a significant and detectable change in the transistor's threshold voltage. This is the reason why we are interested in making a nanoscale memory based on the floating gate scheme.

The motivation for this work is to investigate the ultimate limit of a floating gate MOS memory. In a conventional floating gate memory, there are typically on the order of 10^4 electrons stored on the floating gate to represent one bit of information. The ultimate limit in scaling down the floating gate memory is to use only one electron for the same purpose, hence the name "single-electron MOS memory" (SEMM). The advantage of such a memory is that not only can it be very small, but also it can provide some unique characteristics that are not available in the conventional device, such characteristics as quantized threshold voltage shift and quantized charging voltage.

First, we will introduce the operation principle of the single-electron memory, and previous works in this area. Then we will present our study on the first room-temperature crystalline silicon single-electron MOS memory. We will also address the issues that are important to the functioning of such a device.

7.8 FLOATING GATE SCHEME

To make single-electron memory practical, both thermal fluctuation and quantum fluctuations of the stored charge have to be minimized. The former condition can be satisfied by making the floating gate small enough so that the energy level separation in the dot (ΔE) is greater than kT. The latter condition can be satisfied by making the tunnel resistance greater than the quantum resistance $h/e^2 \sim 26\ \mathrm{k\Omega}$, a condition which is easily satisfied by using oxide as a tunnel barrier. A floating gate memory scheme naturally fits both requirements.

The operation of the floating-gate-based semiconductor memory is illustrated in Figure 7.20. The floating-gate storage node is sandwiched between a conducting channel and a control gate. When the control gate is forward biased with respect to the source and the drain, electrons can be injected into the floating gate from the channel. The resulting stored electrons can screen the channel from the gate potential, and change the threshold voltage of the device.

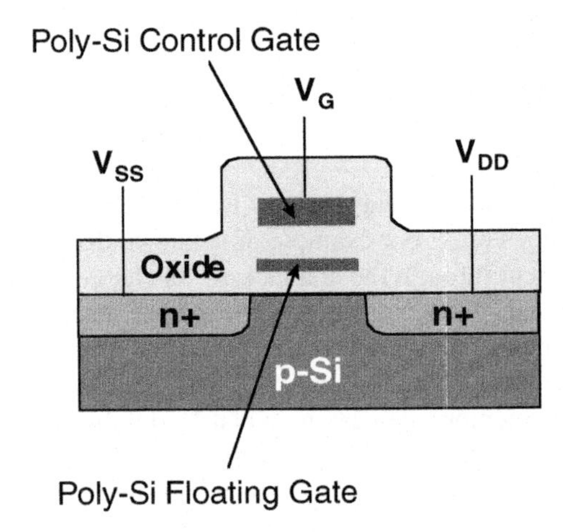

FIGURE 7.20 Schematic of a floating-gate memory device.

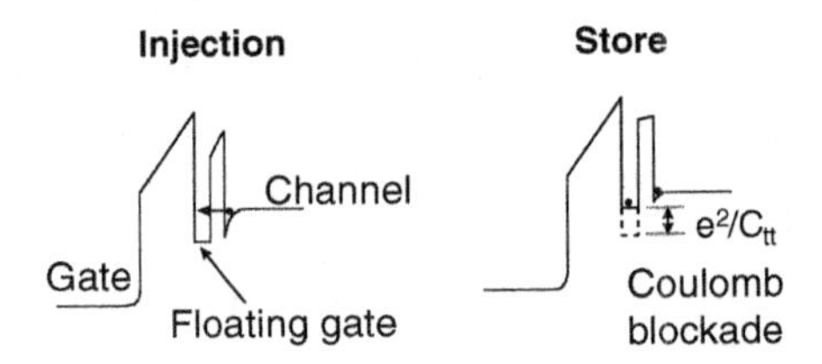

FIGURE 7.21 Schematic band diagram before and after an electron injection onto the dot. A single electron stored in the dot can raise its potential by e_2/C_{tt}, blocking other electrons in the channel from coming into the dot.

In a single-electron memory, the number of electrons that can be injected into the floating gate during each writing process can be exactly controlled by means of the energy constraint. For example, if the energy increase due to the storing of a single electron (charging energy plus quantization energy) on the floating gate is large compared with the potential energy difference between the channel and the floating gate, transfer of other electrons will be blocked. The schematic of this operation is shown in Figure 7.21. This makes it possible to store one bit of information by a single electron, and the information can be read by sensing the threshold difference between the two states.

One of the two previous approaches using the floating-gate concept was to build the device in a tiny polysilicon strip Figure 7.22(a).[53] An ultrathin (3 to 4 nm) polysilicon film was used for the active region of the device. Because of the potential fluctuations due to the nonuniform film thickness, an electron percolation path in the polysilicon strip can form a conducting channel, and one of polysilicon grains near the conduction path acts as the floating gate. Although single-electron memory effect had been observed in such a device, the structure itself intrinsically prevents

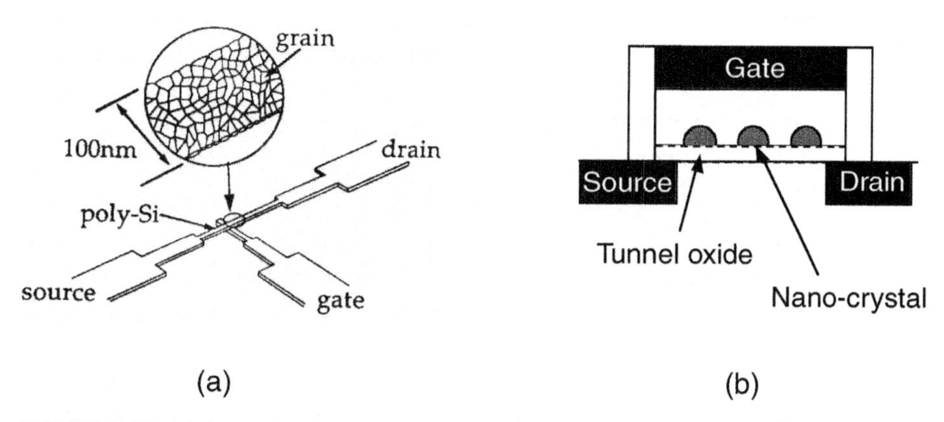

FIGURE 7.22 (a) A special polysilicon thin film transistor that demonstrates room temperature single-electron memory effect.[53] (b) A novel EEPROM structure that uses Si (or Ge) nanocrystal as charge storage nodes.[54]

a precise control of the channel size, the floating-gate dimension, and the tunnel barrier, making it difficult to be implemented on a large scale.

Another approach was to replace the floating gate of a conventional floating-gate transistor memory with nanocrystal silicon grains (Figure 7.22[b]) while keeping the rest of the device unaltered.[54] Fast charging and long retention time has been obtained with this structure. However, there was no clear observation of single-electron effect, due to the intrinsic broad distribution in the size of the silicon nanocrystals and the tunnel barrier thickness under each storage node. The statistical variations in both structures could lead to large fluctuations in the threshold voltage shift and in the charging voltage.

7.9 SINGLE-ELECTRON MOS MEMORY (SEMM)

7.9.1 Structure of SEMM

In order to reduce the variation in the device structure, we would like to build a single-electron memory device in crystalline silicon that has well-controlled dimensions. We defined the transistor channel and the floating gate by using lithography. We have observed that discrete charging of the floating gate leads, at room temperature, to a quantized threshold voltage shift and a staircase relation between the shift and the charging voltage. Furthermore, the charging process is self-limited.

Two key features of such SEMM are (as shown in Figure 7.23): (1) the width of silicon MOSFET channel is less than the Debye screening length, and (2) the floating gate is a nanoscale square (dot). Otherwise, the device is similar to an ordinary floating-gate MOS memory. The narrow channel ensures that storing a single electron on the floating gate is sufficient to screen the entire channel width from the potential on the control gate, leading to a significant threshold voltage shift. A small floating gate is used to significantly increase the quantum confinement energy (due to small size) and the electron charging energy (due to small capacitance), so that the

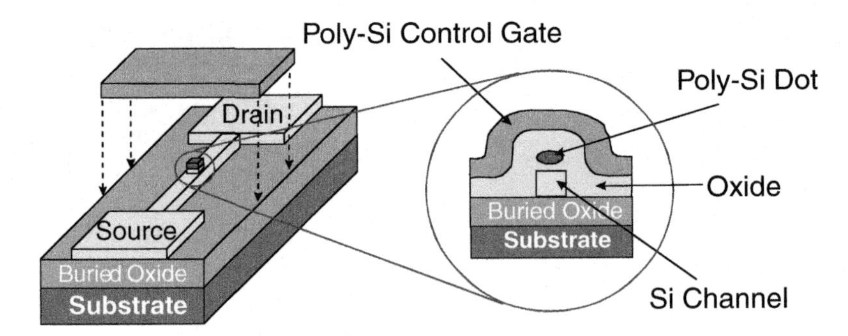

FIGURE 7.23 Schematic of a single-electron MOS memory that has a narrow silicon channel and a nanoscale polysilicon dot as the floating gate. The cross section view details the floating gate and the channel region. Reproduced with permission from the American Association for the Advancement of Science.[16]

energy level separation ΔE is greater than $k_B T$, suppressing the charge fluctuation on the floating gate. Note that the control gate in this device is very long, but the device threshold is determined by the section located at the floating gate.

7.9.2 Fabrication Procedure

In fabrication, we started with a separation-by-implanted-oxygen (SIMOX) wafer that has a top silicon layer thinned to 35 nm. A native oxide of ~1 nm on top of the silicon was used as the tunnel oxide. Then an 11-nm-thick polysilicon film to be used as the floating gate was deposited by low pressure chemical vapor deposition (LPCVD) (Figure 7.24[a]). The width of the channel and the floating gate were patterned in a self-aligned manner by electron beam lithography (EBL), a lift-off of Cr, and a chlorine-based reactive ion etching (RIE) (Figure 7.24[b]). The channel width before oxidation varies from 25 to 120 nm. The length of the floating gate was patterned by a second-level EBL, Cr lift-off and RIE, making it a square shape (Figure 7.24[c]). The native oxide acted as an etch-stop layer for the RIE. After this, an 18-nm-thick oxide was thermally grown, partially consuming silicon and reducing the thickness of the polysilicon dot by ~9 nm and the lateral size of the dot and the silicon channel width by ~18 nm. Then a 22-nm-thick oxide was deposited using plasma enhanced chemical vapor deposition (PECVD), making the total control-gate-oxide thickness 40 nm. Next, polysilicon was deposited by LPCVD and the control gate was patterned to a 3-μm length that covers the floating gate and part of the narrow channel (Figure 7.24[d]). After a self-aligned source/drain ion implantation, the final contacts were made, and the devices were sintered in a hydrogen and nitrogen forming gas to reduce the interface state density. Some fabrication procedures described here are similar to those used in our previous works.[55–57]

The scanning electron micrograph Figure 7.25) shows a 28-nm-diameter polysilicon dot defined on top of a 28-nm-wide silicon channel before they were reduced by oxidation. For easy alignment in the second-level EBL, a line was exposed instead of a small square. The line was etched in the buried oxide, and does not affect the electrical characteristics of the device.

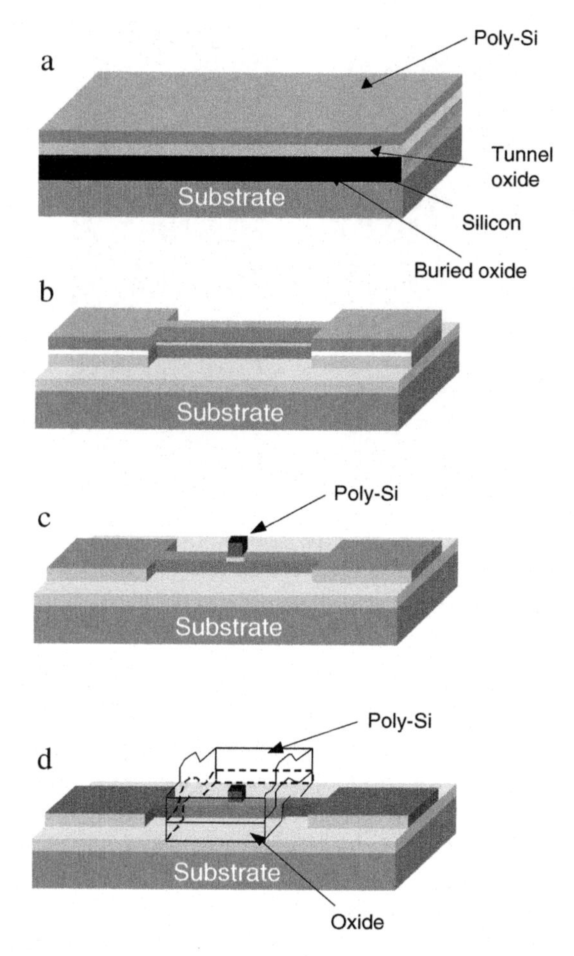

FIGURE 7.24 Fabrication procedures for the nanoscale floating gate memory. (a) Starting SIMOX wafer with multi-layer films. (b) First level EBL and RIE define the transistor channel. (c) Second level EBL and RIE define the polysilicon dot. (d) Formation of control oxide and control gate. Reproduced with permission from the American Institute of Physics.[52]

Note that in these devices, no tunnel oxide was intentionally grown between the channel and polysilicon floating gate. The reasons are twofold: (i) to allow fast charging and (ii) to minimize the potential difference between the channel and the floating dot during the charging process, so that for a given charging voltage, the coulomb blockade effect can regulate the number of electrons that will be stored on the floating gate. In these devices, a potential barrier still exists between the channel and the floating gate, because of the grain boundary in polysilicon and the thin native oxide.

7.9.3 Experimental Observations

The devices were characterized at room temperature in a two-step process. First a positive voltage pulse (2 to 14 V) relative to the grounded source was applied to the

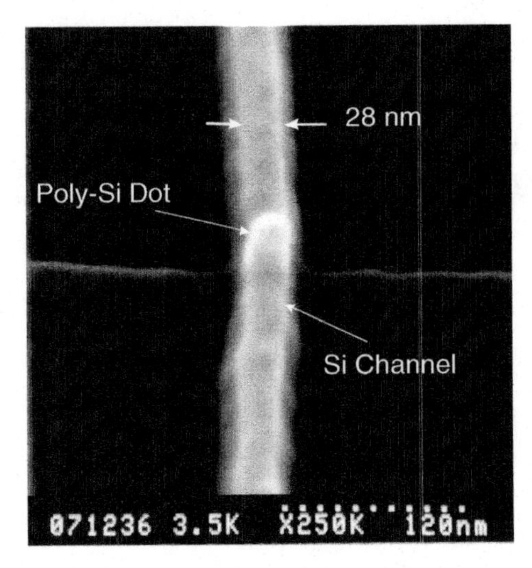

FIGURE 7.25 Scanning electron micrograph of the narrow silicon channel and the polysilicon dot on top, before size reduction by thermal oxidation. The width of the channel and the size of the dot are both 28 nm. Reproduced with permission from the American Institute of Physics.[52]

control gate, while the drain voltage was maintained at 50 mV. This process causes the electrons to tunnel from the channel to the floating gate. The drain current of the transistor was then measured as a function of the gate voltage with a 50 mV source drain voltage, from which the threshold voltage (V_{th}) shift was obtained. A simple switching circuit was used to allow the measurement of the I-V characteristics to be taken within 1 sec after the charging process was completed.

A SEMM that has an ~10-nm-wide channel and a floating gate of ~7 × 7 nm square and 2 nm thick — the smallest in our fabrication was characterized under different charging voltages. The device dimension was estimated from SEM measurement and the oxidation rate. However, self-limiting oxidation might occur,[58] making it difficult to assess the exact size.

Figure 7.26 shows the *I-V* characteristics of the device after the control gate was pulsed by different voltages from 2 to 14 V. Although the charging voltage was varied continuously, the threshold voltage of the device (defined as the gate voltage at which the drain current reaches 100 pA) always makes a discrete shift, each with an increment of 55 mV. Each threshold shift requires a charging voltage increment of ~4 V. Figure 7.27(a) clearly shows a staircase relation between the threshold voltage shift (ΔV_{th}) and the charging voltage applied on the control gate. Moreover, for a given charging voltage, the threshold voltage shift is independent of the time it takes to charge the floating gate, indicating that the charging process is self-limited (Figure 7.27[b]).

Because there is no intentional tunnel oxide, the charge stored at the floating gate can be held for ~5 sec after the control gate potential was set back to zero, and the threshold voltage of the device would then return to its original value (the leftmost trace in Figure 7.26). However, if the gate voltage is biased just below the threshold, the charge can be retained for a much longer time.

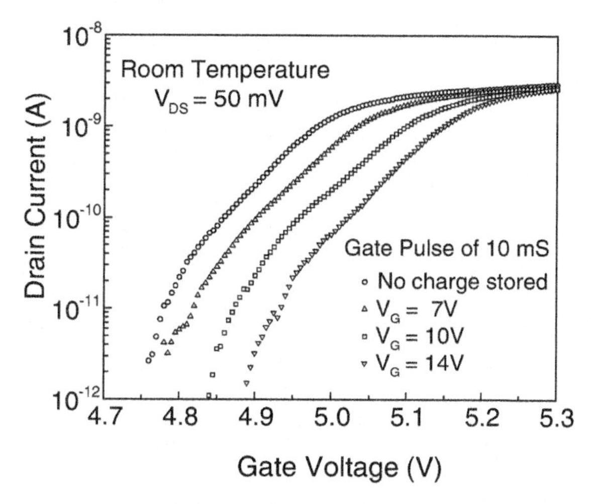

FIGURE 7.26 The I-V characteristics of a SEMM device before and after the charges being stored in the floating dot. For a charging voltage pulse from 2 to 14 V, the threshold voltage shift is quantized with an increment of ~55 mV. The trace with (o) symbol represents the case where no charge was stored in the dot, and the other three traces show the results after positive gate pulses had been applied with progressively larger magnitude. Reproduced with permission from the American Association for the Advancement of Science.[16]

For comparison, we have also fabricated devices that have no floating gate on top of the channel. These devices did not exhibit any memory effect regardless of what voltage pulse was applied to the control gate. This observation indicates that, as expected, it is the electrons stored inside the floating gate that cause the memory effect.

7.9.4 Analysis

The behavior of the device can be explained by the single-electron charging effect. First, the voltage required for adding a single electron to the floating gate can be calculated from the fact that the charging voltage appears primarily between the control gate and the floating gate because of the use of very thin tunnel oxide between the channel and the floating gate. To add one electron to the floating gate requires a voltage increment of e/Cfg to be applied to the control gate, where Cfg is the capacitance between the floating gate and the control gate (Figure 7.28). The capacitance Cfg for the 7 × 7 nm floating gate and a 40-nm control oxide is about 4.4 × 10–20 F, resulting in a single-electron charging voltage of 3.6 V, close to the experimental value of 4 V.

The discrete shift in threshold voltage (ΔVth) due to a single electron stored in the floating gate can be calculated using the following model. The charge induced in the channel (directly under the floating gate) is determined by both the floating gate voltage (V_F) and the control gate voltage (V_G), because the control gate partially wraps around the channel. The induced channel charge under the floating gate can be approximated as

$$Q_{ch} = C_{fc}V_F + C_{cg}V_G \tag{7.20}$$

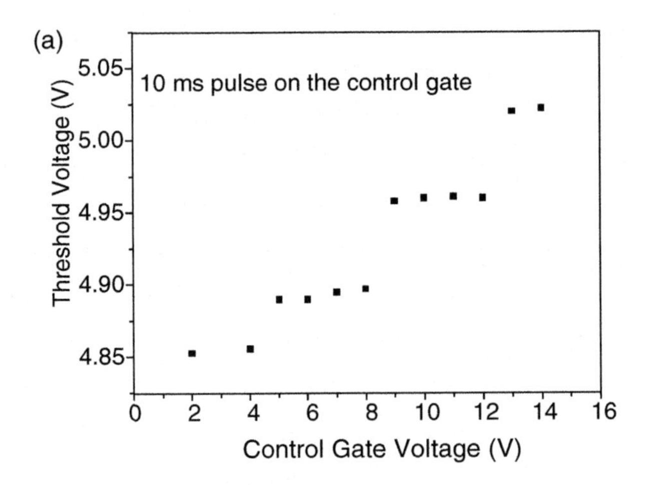

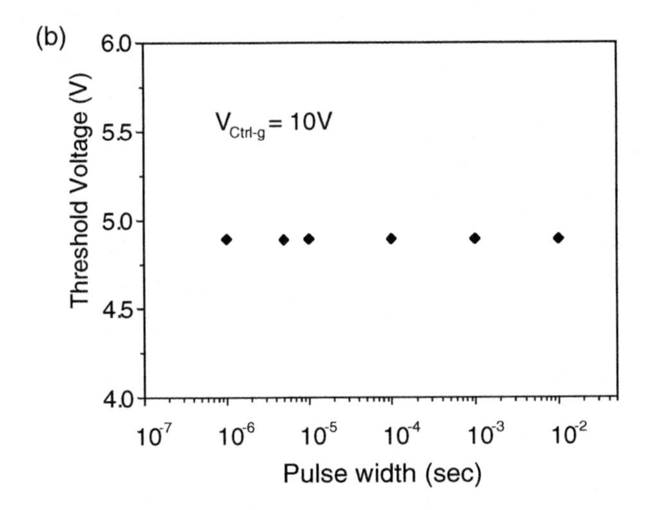

FIGURE 7.27 Threshold voltage of the SEMM as a function of (a) the charging voltage on the control gate, showing a staircase relation with an interval of ~4 V; and (b) the charging time while the charging voltage pulse is fixed at 10 V, indicating a self-limited process.

where C_{fc} is the mutual capacitance between the floating gate and the channel, and C_{cg} is the mutual capacitance between the channel and the control gate (Figure 7.28). The floating gate potential can be related to the floating gate charge (Q_F) and the control gate voltage as

$$V_F = \frac{Q_F}{C_\Sigma} + \frac{C_{fg}}{C_\Sigma} V_G \,. \tag{7.21}$$

Here $C_\Sigma = C_{fc} + C_{fg}$. Putting Equation (7.21) into Equation (7.20) yields an expression for the channel charge:

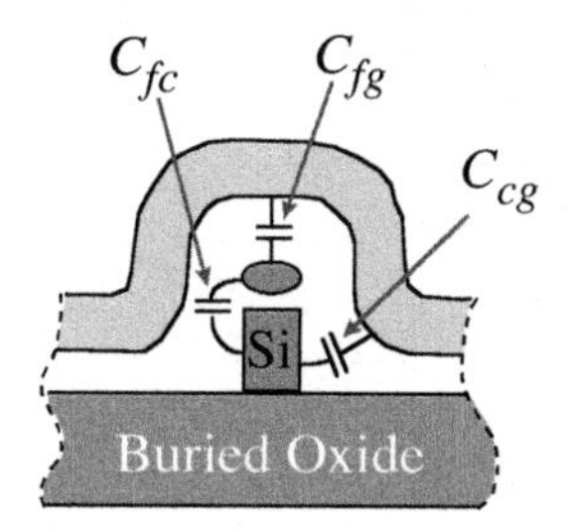

FIGURE 7.28 Schematic cross section of SEMM showing the capacitive coupling between the various elements.

$$Q_{ch} = \frac{C_{fc}}{C_\Sigma} Q_F + (C_{cg} + \frac{C_{fc}}{C_\Sigma} C_{fg}) V_G \tag{7.22}$$

When one more electron is added to the floating gate (i.e., $\Delta Q_F = e$), to maintain the same channel charge Q_{ch}, the threshold voltage has to increase by

$$\Delta V_{th} = \frac{e}{C_{fg} + \frac{C_\Sigma}{C_{fc}} C_{cg}} \tag{7.23}$$

Because the polySi dot is very close to the silicon channel, but far away from the control gate, we can take $C_\Sigma \approx C_{fc}$. So the threshold shift due to addition of single electron to the floating gate can be evaluated as

$$\Delta V_{th} = \frac{e}{C_{fg} + C_{cg}} \tag{7.24}$$

In a conventional floating gate memory $C_{cg} = 0$, and the threshold shift is simply $\Delta V_{th} = e/C_{fg}$. But in our case, the threshold shift is reduced because the charge in the floating gate can only partially screen the gate potential. The channel-to-control gate capacitance (C_{cg}) can be estimated in the following way: The channel length that can be controlled by a charge on the floating gate is roughly equal to the Debye screen length, which is ~70 nm for the given channel doping at room temperature. The channel thickness for the device is ~26 nm. For the control oxide thickness of 40 nm and the area of 70-nm × 26-nm × 2 nm, C_{cg} is about 2.5×10^{-18} F ($>>C_{fg}$), and hence $\Delta V_{th} \approx 64$ mV, which is again consistent with the experimental value.

Finally, the self-limiting charging process can be explained by three factors: (i) The energy level spacing in the floating gate is large compared with kBT at room temperature. For a 7-nm-by-7-nm silicon square embedded in SiO2 the quantum energy spacing is ~50 meV and the Coulomb energy spacing is 60 meV (assuming the oxide between the dot and channel is 1 nm thick). An electron has to overcome

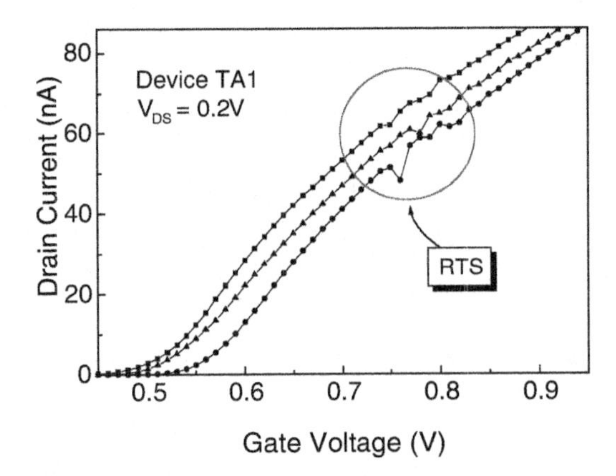

FIGURE 7.29 Transfer characteristics of a memory device that exhibit random telegraphic signals.

this energy difference to get into the floating gate. (ii) The barrier layer is thin, the voltage drop between the channel and the floating gate is very small; and (iii) once an electron is added into the floating gate, the potential of the floating gate will raise, further reducing the voltage difference between the channel and the floating gate and preventing another electron from tunneling into the floating gate. Therefore, for a fixed charging voltage, the charging process is self-regulated and stops once the floating gate is charged with a fixed number of electrons, leading to a threshold shift independent of charging time and a staircase relation between the charging voltage and the threshold shift.

7.9.5 Effects of Trap States

We have established experimentally that the memory effect was only observed in devices that have nanoscale floating gates. However, there is still the concern of charge trapping in the oxide or in the interface states, which could also cause local threshold voltage shift. But these different effects can be distinguished, because the threshold shift due to traps would not give multiple equally spaced threshold shifts (since the charge will be trapped at different locations of the channel) and would have a time-dependent charging process.[59] Experimentally, the effect of charges captured by the trap states could also be distinguished from the threshold shift caused by the charges stored in the floating gate as will be discussed next.

It is well known that in very small MOSFET devices, the alternate capture and emission of carriers by the individual trap state generates discrete switching signals in the drain current, referred to as the random telegraphic signal (RTS).[60] Occasionally, we also observe RTS in some of our memory devices. The I-V characteristic of a memory device that exhibits such switching behavior is shown in Figure 7.29. The circled region shows strong current fluctuation; the sudden drop of current is a consequence of an increase of the local threshold voltage of the device due to charge trapping (the detail will be discussed later). The three traces correspond to different amounts of charges injected into the floating gate (ΔV_{th} was not quantized in this

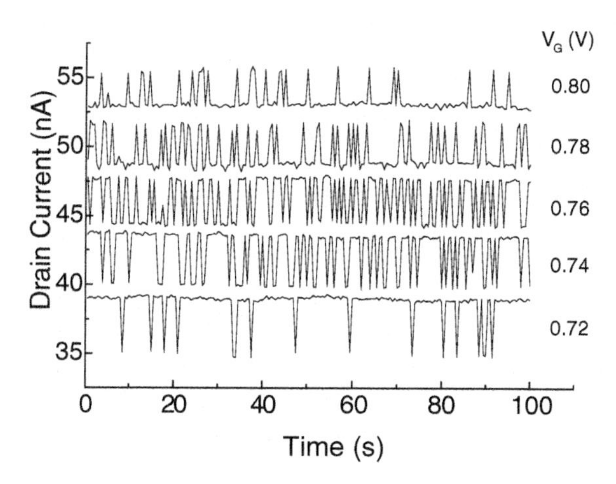

FIGURE 7.30 Time domain measurement of the RTS signal.

device because the floating gate size was larger than 20 nm). So we see the effect of charge capture by the trap state is independent of charge storage in the floating gate.

In addition, because the energy level of the trap state is fixed with respect to the conduction band edge, the trapping and detrapping characteristics depend solely on the position of the trap level relative to the Fermi level. That is why the RTS always fall into the same gate voltage range in Figure 7.29. Such dependence is more clearly seen in Figure 7.30 which is a time domain measurement for the RTS. Assuming the electron capture by an initially neutral trap site, then the high current level corresponds to the trap-empty state (neutral), and the low current level corresponds to the trap-filled state (negatively charged). Voltage-dependent time constants for the capture and emission of electrons by the trap states can also be obtained from Figure 7.30.[lxi]

7.10 EFFECT OF THICKER TUNNEL OXIDE

We have also fabricated devices with similar structures but with thicker tunnel oxide. However, the self-limiting charging process was not observed for the devices that have 2 nm or 5 nm tunnel oxide. Part of the reason is that the polysilicon dot is not small enough (typical size greater than 30 nm), and so the potential energy drop across the tunnel barrier during the charging process is greater than the charging energy. A typical I-V characteristics is shown in Figure 7.31. For these devices, the threshold shift cannot always be measured from the initial uncharged state, because of the difficulty in removing the electrons from the floating gate (and this will be discussed later). We found that the threshold voltage shifted continuously with charging voltage. At given charging voltage, the threshold shift increased with charging time as seen in Figure 7.32, which is measured separately for another device. We also found that for charging voltage below 9 V, the threshold voltage hardly shifted, even with charging time greater than 1 sec. The devices generally

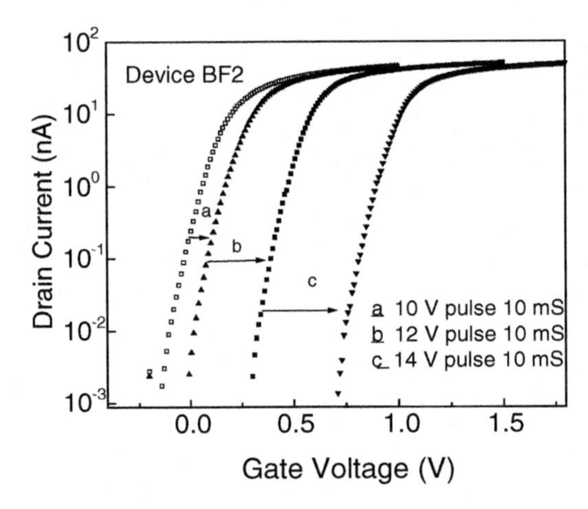

FIGURE 7.31 I-V characteristics of a memory device with 2 nm tunnel oxide under different charging conditions.

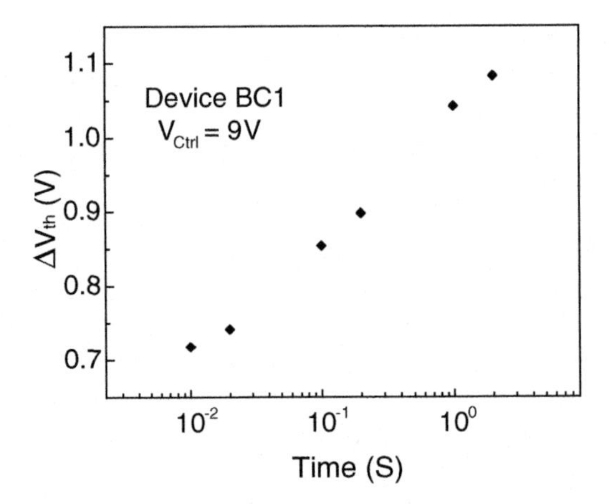

FIGURE 7.32 Threshold shift as a function of charging time for a device with 2-nm-thick tunnel oxide.

have much longer charge retention time (during several weeks of observation), and are expected to behave like nonvolatile memories.

The behavior can be understood by using a simplified model developed for conventional EEPROM. By assuming the tunneling through the tunnel oxide is of *Fowler-Nordheim* (*FN*) type, we will show that a closed form can be obtained which relates the threshold voltage shift to the charging voltage and charging time. However, this is only an approximation, which should be modified to take into account the direct tunneling through the thin tunnel oxide (when oxide thickness is below 40 Å). Because of this and the lack of knowledge of some precise device parameters, we do not intend to make quantitative comparison with the experimental results.

Rather, the equations will be used to understand qualitatively the device behavior, especially the presence of the apparent threshold charging voltage.

From the discussion in the last section, we see that the threshold voltage change due to storing of charge Q_F on the floating gate is given by

$$\Delta V_{th} = \frac{-Q_F}{C_{fg} + C_{cg}} \tag{7.25}$$

which is the collection of the electrons tunneled from the channel during the charging process,

$$-\frac{dQ_F}{dt} = AJ(E) \tag{7.26}$$

where A is the effective area of the tunnel junction, and $J(E)$ is the tunnel current density of the assumed FN type:

$$J(E) = \alpha E^2 \exp(-\beta/E) \tag{7.27}$$

Here E is the electric field in the tunnel oxide, which can be approximated as

$$E = \frac{\phi_F}{d_0} \tag{7.28}$$

where ϕ_F is the floating gate potential, which can be related to the floating gate charge by

$$\phi_F = \frac{Q_F}{C_\Sigma} + \frac{C_{fg}}{C_\Sigma} V_G \tag{7.29}$$

A differential equation for $E(t)$ can be obtained by combining Equations (7.26) through (7.29):

$$-C_\Sigma \frac{dE}{dt} = A\alpha E^2 \exp\left(-\beta/E\right) \tag{7.30}$$

which can be solved to give

$$E(t) = \beta \ln^{-1}\left(\frac{A\alpha\beta t}{d_0 C_\Sigma} + \exp\left(\beta/E(0)\right)\right) \tag{7.31}$$

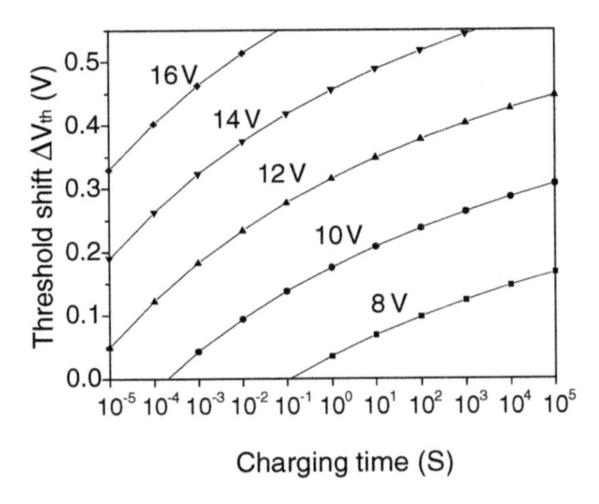

FIGURE 7.33 Calculated threshold shift as a function of charging time at different charging voltages for a nano-floating gate memory of 50 nm by 50 nm size and has 2 nm tunnel oxide.

Now the threshold voltage shift from Equation (7.25) can be written as

$$\Delta V_{th} = \frac{1}{C_{fg} + C_{cg}} \left(C_{fg} V_G - \frac{C_\Sigma d_0 \beta}{\ln P(t)} \right) \tag{7.32}$$

where

$$P(t) = \frac{A\alpha\beta t}{C_\Sigma d_0} + \exp\left(\frac{\beta C_\Sigma d_0}{C_{fg} V_G} \right) \tag{7.33}$$

Comparing Equation (7.32) with the threshold shift obtained for a conventional floating-gate EEPROM, we find that if the control voltage increases by ΔV_G, instead of having $\Delta V_{th} = \Delta V_G$ as in the conventional memory,[61] the threshold voltage changes only by a fraction of ΔV_g: $\Delta V_{th} = \Delta V_G \, C_{fg}/(C_{fg} + C_{cg})$. This is again because the control gate is partially wrapping around the channel giving a *non*-zero C_{cg}. From the experimental data, this fraction can be obtained as $C_{fg}/(C_{fg} + C_{cg}) = \Delta V_{th}/\Delta V_G \approx$ 0.07.

A plot of the threshold shift from Equation (7.32) is shown in Figure 7.33 in the long time limit. The following parameters were used: $\alpha = 9.625 \times 10^{-7}$ A/V^2, $\beta = 2.765 \times 10^8$ V/cm,[62] A = 30 nm × 30 nm, d_0 = 2 nm, $C_{fg} = 2.2 \times 10^{-18}$ F, and $C_{fg}/(C_{fg} + C_{cg})$ = 0.07, and $C_\Sigma \approx C_{fc}$. In this example, there is a charging threshold voltage of ~ 8 V, below which a charging time greater than 1 sec is required to cause a sizable shift in threshold voltage. In the experiment, this voltage threshold is ~ 9 V.

7.11 DISCUSSION

Now we would like to discuss a number of other issues. First, despite the extremely small floating gate and the very low channel doping concentration, the device has a good subthreshold slope of 108 mV/decade. This is because the inversion layer induced by the control gate effectively acts as an ultrashallow source and drain for the device. This also suggests that a double-gated MOSFET could be a solution to the future ultrashort channel transistors. This is because the surface potential is essentially constant under strong inversion, so that the effect of drain induced barrier lowering (DIBL) can be better managed. Moreover, there is no lateral depletion involved in such a structure, because the channel and effective source/drain have the same type of doping, so no PN junction will be formed. This could help to minimize the punch-through effect which is due to the touching of depletion boundaries in the conventional MOSFET under high drain bias. Previous work also supported the effectiveness of using an inversion layer as the shallow source/drain junction.[63,64]

Second, the threshold voltage can be much larger than the present 55 mV, if the control-gate-oxide thickness is increased or the fringing gate capacitance (Cfg) can be reduced. Third, if a thicker tunnel oxide is used between the channel and the floating gate the charge on the floating gate can be held much longer than the current 5 sec. And fourth, clearly the SEMM should be investigated more thoroughly, especially the effect of variations in device size and stray charges on the threshold voltage.

Finally, the single-electron memory potentially has a number of advantages over conventional memories: (1) the quantized characteristics of the device make it immune to the noise from the environment—unless the noise level reaches a certain threshold, it will not affect the memory state. The immunity to noise is especially important for the future terabits integration, simply because of the sheer large number of devices present on a single small chip area. (2) the inherent quantized nature of the SEMM makes it possible to easily implement multilevel logic storage in a single memory cell; (3) the device can operate at a higher speed due to the use of only one or few electrons during writing and erasing; (4) for the same reason, the device can also have ultralow power consumption.

In conclusion, researches on single-electron effects in Si quantum dots can help to build a solid knowledge base for the exploration of future Si nanoelectronics. Single-electron transistors can be used as ultrasensitive charge sensors in numerous applications. Single-electron effects observed in Si nanodots can be a potential route to build very high density memory chips with added functionalities.

REFERENCES

1. Rossum, M.V., The future of microelectronics: Evolution or revolution? *Microelectron. Eng.*, 34, 125–134, 1996.
2. See, for example, *Single Charge Tunneling: Coulomb Blockade Phenomena in Nanostructures*, ed. by H. Grabet and M. H. Devoret, Plenum, New York, 1992.
3. Giaever, I. and Zeller, H.R., *Phys. Rev. Lett.*, 20, p.1504, 1968.
4. Fulton, T.A. and Dolan, G.J., Observation of single-electron charging effects in small tunnel junctions, *Phys. Rev. Lett.*, Vol. 59, pp. 109–112, 1987.
5. Geerligs, L.J., Anderegg, V.F., Holweg, P.A.M., and Mooij, J.E., Frequency-locked turnstile device for single electrons, *Phys. Rev. Lett.*, Vol. 64, pp. 2691–2694, 1990.

6. Kouwenhoven, L.P., Quantized current in a quantum dot turnstile, *Physica Scripta*, vol. T42, pp. 133–135, 1992.
7. Nakazato, K.,Blaikie, R.J., and Ahmed, H., Single-electron memory, *J. Appl. Phys.*, Vol. 75, pp. 5123–5134, 1994.
8. Ji, L.,Dresselhaus, P.D., Han, S., Lin, K., Zheng, W., and Lukens, J.E., Fabrication and characterization of single-electron transistors and traps, *J. Vac. Sci. Technol.*, Vol. B12, pp. 3619–3622, 1994.
9. Averin, D.A., Odintsov, A.A., Vyshenskii, S.V., Ultimate accuracy of single-elecron dc current standards, *J. Appl. Phys.*, Vol. 73, pp. 1297–1308, 1993.
10. Hirvi, K.P., Kauppinen, J.P., Korotkov, A.N., Paalanen, M.A., and Pekola, J.P., Coulomb blockade thermometry: 1d arrays and solitary tunnel junctions in the $E_c < kT$ regime, *Meeting abstract of 190th Electrochemical Society meeting*, vol. 96-2, p. 547, 1996.
11. Tucker, J.R., Complementary digital logic based on the 'Coulomb blockade', *J. Appl. Phys.*, Vol. 72, pp. 4399–4413, 1992.
12. Chen, R.H., Korotknov, A.N., and Likharev, K.K., Single-electron transistor logic, *Appl. Phys. Lett.*, Vol. 68, pp. 1954–1956, 1996.
13. Electron imaging performed at Bell labs, *Semiconductor International*, p. 84, June 1997.
14. Takahashi, Y., Nagase, M., Namatsu, H., Kurihara, K., Iwdate, K., Nakajima, Y., Horiguchi, S., Murase, K., and Tabe, M., *IEDM Tech. Dig.*, 938, 1994.
15. Yano, K., Ishii, T., Hashimoto, T., Kobayashi, T., et al., A Room Temperature Single-Electron Memory Device Using Fine-Grain Polycrystalline Silicon, *IEDM Tech. Dig.*, pp. 541–544, 1993.
16. Guo, L.J., Leobandung, E., and Chou, S.Y., A single-electron transistor memory operating at room temperature, *Science*, Vol. 275, 649, 1997.
17. Takahashi, Y., Nagase, M., Namatsu, H., Kurihara, K., Iwdate, K., Nakajima, Y., Horiguchi, S., Murase, K., and Tabe, M., Fabrication technique for Si single-electron transistor operating at room temperature, *Electronics Lett.*, Vol. 31, pp.136–137, 1994.
18. Leobandung, E., Guo, L.J., Wang, Y., and Chou, S.Y., Observation of quantum effects and Coulomb blockade in silicon quantum-dot transistor at temperature over 100 K, *Appl. Phys. Lett.*, Vol. 67, 938, 1995.
19. Ali, D. and Ahmed, H., *Appl. Phys. Lett.*, Vol. 64, p. 2119, 1994.
20. Meirav, U., Kastner, M.A., and Wind, S.J., Single-electron charging and periodic conductance resonances in GaAs nanostructures, *Phys. Rev. Lett.*, Vol. 65, pp. 771–774, 1990.
21. Kouwenhoven, L.P., van der Vaart, N.C., Johnson, A.T., Kool, W., Harmans, C.J.P.M., Williamson, G., Staring, A.A.M., and Foxon, C.T., *Phys. B - Condensed Matter*, 85, 367, 1991.
22. Matsuoka, H., Ichiguchi, T., Yoshiyuki, T., and Takeda, E., *Appl. Phys. Lett.*, 64, 586, 1993.
23. Matsuoka H. and Kimura, S., *Appl. Phys. Lett.*, 65, 613, 1995.
24. Fields, S.B., Kastner, M.A., Meirav, U., Scott-Thomas, J.H.F., Antoniadis, D.A., Smith, H.I., and Wind, S.J., Conductance oscillations periodic in the density of one-dimensional electron gases, *Phys. Rev. B*, 42, pp. 3523–3536, 1990.
25. Beenakker, C.W.J., Theory of Coulomb-blockade oscillations in the conductance of a quantum dot, *Phys. Rev. B*, 44, pp. 16461656, 1991.
26. Leobandung, E., Guo, L.J., and Chou, S.Y., Single hole quantum dot transistors in silicon, *Appl. Phys. Lett.*, Vol. 67, pp. 2338–2340, 1995.

27. Mendez, E.E., Wang, W.I., Ricco, W.I., and Esaki, L., Resonant tunneling of holes in AlAs-GaAs-AlAs heterostructures, *Appl. Phys. Lett.*, Vol. 47, pp. 415–417, 1985.
28. Yakimov, A.I., Markov, V.A., and Dvurechenskii, A.V., *Philos. Mag. B*, 65, 701, 1992.
29. Paul, D.J., Cleaver, J.R.A., and Ahmed, H., *Phys. Rev. B*, 49, 16514, 1994.
30. Glazman, L.I. and Shekhter, R.I., Coulomb oscillations of the conductance in a literally confined heterostructure, *J. Phys. Condend. Matter 1*, pp. 5811–5813, 1989.
31. Maksym, P.A. and Chakaraborty, T., *Phys. Rev. Lett.*, 65, 108, 1990.
32. Häusler, W.H. and Kramer, B., *Phys. Rev. B*, 47, 16353, 1993.
33. Johnson, N.F. and Payne, M.C., Exactly solvable model of interacting particles in a quantum dot, *Phys. Rev. Lett.*, Vol. 67, pp. 1157–1140, 1991.
34. Johnson, N.F. and Quiroga, L., Analytic results for N particles with $1/r^2$ interaction in two dimensions and an external magnetic field, *Phys. Rev. Lett.*, Vol. 74, pp. 4277–4280, 1995.
35. Weis, J., Haug, R.J., Klitzing, K.V., and Ploog, K., Transport spectroscopy of a confined electron system under a gate tip, *Phys. Rev. B*, 46, pp. 12837–12840, 1992.
36. Field, M., Smith, C.G., Ritchie, S.A., Frost, J.E.F., Jones, G.A.C., and Hasko, D.G., Measurements of Coulomb blockade with a noninvasive voltage probe, *Phys. Rev. Lett.*, Vol. 70, pp. 1311–1314, 1993.
37. Weis, J., Haug, R.J., Von Klitzing, K., and Ploog, K., Transport spectroscopy on a single quantum dot, *Semicond. Sci. Technol.*, Vol. 9, pp. 1890–1896, 1994.
38. Meir, Y., Wingreen, N.S., and Lee, P.A., Transport through a strongly interacting electron systems: Theory of periodic conductance oscillations, *Phys. Rev. Lett.*, Vol. 66, pp. 3048–3052, 1991.
39. Wang, Y. and Chou, S.Y., Observation of bias-induced resonant tunneling peak splitting in a quantum dot, *Appl. Phys. Lett.*, Vol. 64, pp. 309–311, 1994.
40. Foxman, E.B., McEuen, P.L., Meirav, U., Wingreen, N.S., Meir, Y., Belk, P.A., Belk, N.R., Kastner, M.A., and Wind, S.J., Effects of quantum levels on transport through a Coulomb island, *Phys. Rev. B.*, Vol. 47, pp. 10020–10023, 1993.
41. Johnson, A.T., Kouwenhoven, L.P., de Jong, W., van der Vaart, N.C., Harmans, C.J.P.M., and Foxon, C.T., Zero-dimensional states and single electron charging in quantum dots, *Phys. Rev. Lett.*, Vol. 69, pp. 1592–1595, 1992.
42. Needs, R.J., Bhattacharjee, S., Qteish, A., Read, A.J., and Canham, L.T., First-principle calculations of band-edge electronic states of silicon quantum wires, *Phys. Rev. B.*, Vol. 50, pp. 14,223–14,227, 1994.
43. Leobandung, E., Nanoscale MOSFETs and Single Charge Transistors on SOI, Ph.D. thesis, University of Minnesota, Duluth, June, 1996.
44. Weis, J., Haug, R.J., Klitzing, K.V., and Ploog, K., *Phys. Rev. B*, 46, 12837, 1992.
45. Jalabert, R.A., Stone, A.D., and Alhassid, Y., *Phys. Rev. Lett.,* 68, 3468, 1992.
46. Nicholls, J.T., Frost, J.E.F., Pepper, M., Ritchie, D.A., Grimshaw, M.P., and Jones, G.A.C., *Phys. Rev. B*, 48, 8866, 1993.
47. Kastner, M.A., Artificial atoms, *Physics Today*, Vol. 46, No. 1, p. 24, 1993.
48. Tarucha, S., Vertical Transport Through a Few-Electron Quantum Dot, Meeting Abstract of 190th Electrochemical Society meeting, Vol. 96-2, p. 551, 1996.
49. Kirton, M.J., Uren, M.J., Collins, S., Schulz, M., Karmann, A., and Scheffer, K., Individual defects at the Si:SiO_2 interface, *Semicond. Sci. Technol.,* Vol. 4, pp. 1116–1126, 1989.
50. Fujiwara, A., Takahashi, Y., Murase, K/. and Tabe, M., Time-resolved measurement of single-electron tunneling in a Si single-electron transistor with satellite Si islands, *Appl. Phys. Lett.*, Vol. 67, pp. 2957–2959, 1995.

51. Jalabert, R.A., Stone, A.D., and Alhassid, Y., *Phys. Rev. Lett.*, 68, 3468, 1992.
52. Guo, L.J., Leobandung, E., and Chou, S.Y., Fabrication and characterization of room temperature silicon single-electron memories, *J. Vac. Sci. Technol.*, Vol. B15, pp. 2840–2843, 1997.
53. Yano, K., Ishii, T., Hashimoto, T., Kobayashi, T., Murai, F., and Seki, K., Room-temperature single-electron memory, *IEEE Transaction on Electron Devices*, Vol. 41, pp. 1628–1637, 1994.
54. Tiwari, S., Rana, F., Hanafi, H., Hartstein, A., Crabbe, E.F., and Chen, K., A silicon nanocrystals based memory, *Appl. Phys. Lett.*, Vol. 68, pp. 1377–1379, 1996.
55. Fischer, P.B. and Chou, S.Y., 10 nm electron beam lithography and sub-50 nm overlay using a modified scanning electron microscope, *Appl. Phys. Lett.*, Vol. 62, pp. 2989–2991, 1993.
56. Fischer, P.B., Dai, K., Chen, E., and Chou, S.Y., 10 nm Si pillars fabricated using electron-beam lithography, reactive ion etching, and HF etching, *J. Vac. Sci. & Technol.*, Vol. B11, pp. 2524–2527, 1993.
57. Leobandung, E., Guo, L., and Chou, S.Y., Single electron and hole quantum dot transistors operating above 110 K, *J. Vac. Sci. & Technol.*, Vol. B13, pp. 2865–2868, 1995.
58. Liu, H.I., Biegelsen, D.K., Ponce, F.A., Johnson, N.M., and Pease, R.F.W., *Appl. Phys. Lett.*, 64, 1383, 1994.
59. See, for example, Davis, J.R., *Instabilities in MOS Devices*, Gordon and Breach Science Publishers, New York, 1980.
60. Kirton, M.J. and uren, M.J., Noise in solid-state microstructures: A new perspective on individual defects, interface states and low-frequency (1/f) noise, *Advance in Physics*, Vol. 38, pp. 367–468, 1989.
61. Kolodyn, A., Nieh, S.T.K., Eitan, B., and Shappir, J., Analysis and modeling of floating-gate EEPROM cells, *IEEE Trans. Electron Devices*, Vol. ED-33, pp. 835–844, 1986.
62. Yang, E.S., *Microelectronics Devices*, chap. 13, McGraw-Hill, New York, 1988.
63. Harstein, A., Albert, N.F., Bright, A.A., Kaplan, S.B., Robinson, B., and Tornello, J.A., A metal-oxide-semiconductor field-effect transistor with a 20-nm channel length, *J. Appl. Phys.*, Vol. 68, pp. 2493.
64. Wong, H.-S., Gate Current Injection and Surface Impact Ionization in MOSFET's with a Gate Induced Virtue Drain, *IEDM Tech. Dig.*, pp. 151–154, 1992.
65. Hartstein, A.; Albert, N.F.; Bright, A.A.; Kaplan, S.B.; Robinson, B.; Tornello, J.A., A metal-oxide-semiconductor field-effect transistor with a 20-nm channel length. *J. Appl. Phys.*, Vol. 68, pp. 2493–2495, 1990.
66. Gate current injection and surface impact ionization in MOSFETs with a gate induced virtual drain, *IEDM Tech. Dig.*, pp. 151–154, 1992.

8 Silicon Memories Using Quantum and Single-Electron Effects

Sandip Tiwari

8.1 INTRODUCTION

Nearly four decades ago, Neugebauer and Webb[1] recognized that when capacitance is reduced, the electrostatic energy required to charge the capacitance ($e^2/2C$) can be made to be of the order of magnitude of thermal voltage (10s of meV) at dimensions in the 10-nm range which results in a capacitance in the aF (10^{-18} F) range. This implies that discrete single-electron transmission or storage events can be observed, and unless the electrostatic energy is available to the electron from an energy source such as a power supply, the transition is prohibited. This is known as Coulomb blockade. Figure 8.1 shows an illustrative schematic, with band-diagrams for such a confined system. For a confined system in a semiconductor, quantum effects can also be significant due to a reduced number of states. In three-dimensionally confined semiconductor structures, the order of magnitude of the confinement energy can be similar to that due to Coulomb blockade. Fulton and Dolan,[2] in 1987, demonstrated the first single-electron transistor by showing the modulation of the Coulomb blockade region by an applied bias voltage, and in recent times, there has been tremendous interest in understanding of this mesoscopic system,[3] From a device point of view, an important consequence of the use of single-electron and quantum effects is the reduction in the number of charged carriers.

Reduction of the number of charged carriers used in the operation of devices results in low power consumption and for some device examples increased reliability by reducing the total energy per operation. However, the reduction of the number of electrons is not without changes in properties whose reproducibility microelectronics has taken advantage of. One is a reduction in the "collective phenomena" that we have relied on to achieve reproducibility in microelectronics. Examples of such collective phenomena are the number of electrons flowing through the channel, the number of electrons transferred during a CMOS switching event, and the number of dopants used to control the threshold voltage. A smaller number of electrons leads to larger fluctuations in the current, a smaller number of electrons transferred during switching leads to larger fluctuations in the switching voltage levels or time constants, and a smaller number of dopants leads to larger fluctuations in the threshold voltage or equivalent parameters. Small dimensions are not easy to achieve, and

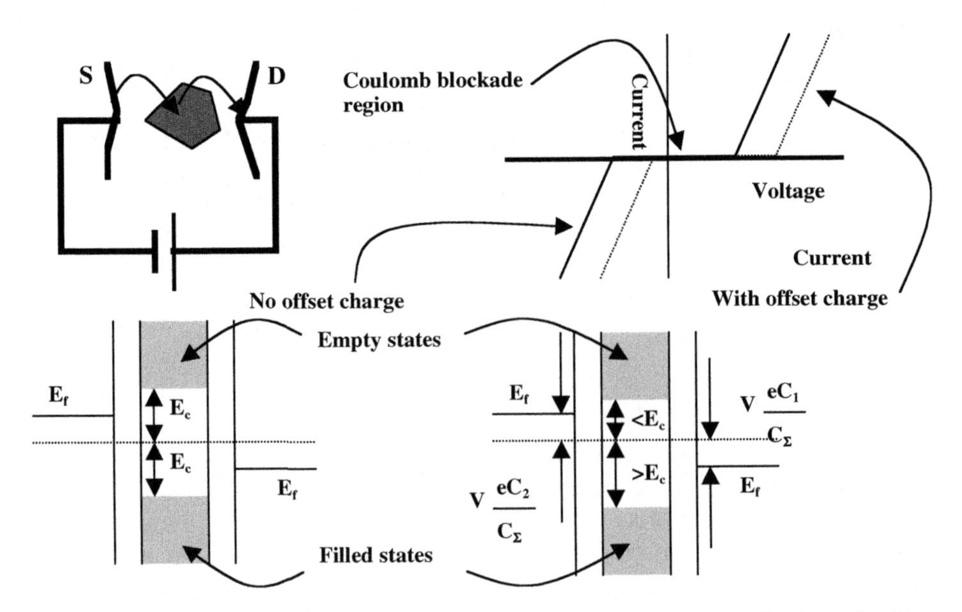

FIGURE 8.1 Schematic of the transfer process of an electron upon application of a bias voltage V between two electrodes that the particle is confined in between and resulting conduction characteristics. The band diagrams show Coulomb blockade condition, one in the absence of offset charge and another in presence of offset charge. The capacitances of the two junctions are C_1 and C_2 and total capacitance C_Σ.

hence many implementations of single-electron devices suffer from reduced gain, voltages needed, and temperatures at which the phenomena can be observed unambiguously. We will discuss these in the context of the operation of these devices and limitations of CMOS with which they usually need to be integrated. And, in particular, we will focus on the attributes of devices that are formed as flash memory structures.

8.2 SINGLE-ELECTRON EFFECT

Figure 8.1 shows a schematic representation of the process of charge transfer through a small confined volume of a particle with high density of states at nanometer scale. For the single-electron events to be clearly observable,[4] a number of requirements must be met. The state that the electron occupies in the particle is confined, and for the event to be observable the change in system energy upon transmission of an electron is larger than thermal energy. The uncertainty principle tells us that the width of the eigenstate is $\Delta E \approx h/2\pi\tau$ where $\tau = 1/\Gamma$ is the lifetime (related inversely with the tunneling rate). The change in system energy ($\Delta U = QV = e\,I\,R_T$ (where Q is the charge, V is the voltage, I is the current ($e\Gamma$), and R_T is the tunnel resistance), upon transition of an electron, is $\Delta U = e^2\,\Gamma R_T$. The energy width of the eigenstate is, therefore, $\Delta E = h/\,2\pi\,\Gamma = h/2\pi\,\Delta U\,/\,(e^2\,R_T)$. A clear observation of Coulomb blockade requires $\Delta U >> \Delta E$, which is the condition

$$R_T \gg \frac{(h/2\pi)^2}{e^2} = 4.1\ k\Omega$$

Note that this resistance is different from that of quantum resistance,[5] usually alluded to in superconducting tunneling of Cooper pairs, of $R_q = h/4e^2$, which has the magnitude of 6.4 kΩ. So, the first condition is that the resistance of the barrier be larger than 4.1 kΩ. This places an impedence constraint that limits the gains that one can obtain from a transistor that uses the Coulomb blockade effect as an intrinsic phenomenon. The second condition for observation for Coulomb blockade is that the energy $(e^2/2C)$ be larger than kT. The overriding time constant for the transmission of this electron is an RC time constant, which is greater than 100 fsec and can, in practical structures, be very large. This time constant dominates since it is larger than the time constant from uncertainty principle, that is, of the certainty of observation, of $h/(e^2/2C)$, which is of the order of 10 fsec, as well as of the transmission of a wave packet through the barrier, of $h/2\pi\ d(ln(T(E))/dE$ at $E=E_f$, which is very small. This large time constant also implies that a small current flows (1 electron/100 fs is ~ 1.6 μA). In most experiments, this current is typically a nA. However, coupling the effect of stored electrons to field effect of a transistor allows a larger current because carriers are more mobile in the barrier-free channel and do not have to be limited by the barrier impedance effects. Now, let us consider a small particle which has these requisite properties. For the moment, we assume that there is no spurious charge (electrically neutral; no trapped electron at interfaces or in the bulk) and hence electric field terminations occur only between the island and the electrodes, when a power supply is connected at the electrodes. Because only discrete tunneling events are allowed for the flow of charge and hence the change in electrochemical potential, the electrochemical potential of the particle a nanocrystal follows the inequality $\mu_{nxtl} \le C_\Sigma \left(|\mu_{nxtl}|/e\right) \le (e/2)$, where C_Σ is the total capacitance between the particle and its surroundings. This is equivalent to having polarization charge, or an offset charge Q_{ofs}, of

$$Q_{ofs} \approx C_\Sigma \frac{|\mu_{nxtl}|}{e} \le \frac{e}{2}$$

For the situation where the offset charge is zero, and C_1 and C_2 are the coupling capacitances, the Coulomb charging energy of E_c is accounted for in the energy diagram (see Figure 8.1) by raising the energy of the unoccupied states by E_c. Tunneling to the particle occurs when the energy of states in the lead align with the E_c -shifted unoccupied states of the particle. Tunneling from the occupied states of the particle also occur with a change in energy of E_c; the energy diagram accounts for this by shifting the energy of occupied states by E_c. So, a barrier E_c exists for flow of an electron whether it is onto the particle or off the particle. Under favorable conditions, an electron tunnels from the left onto the particle with the expenditure of energy E_c by the power supply. The number of electrons in the occupied states has now increased by one and the electrochemical potential of the particle (aligned

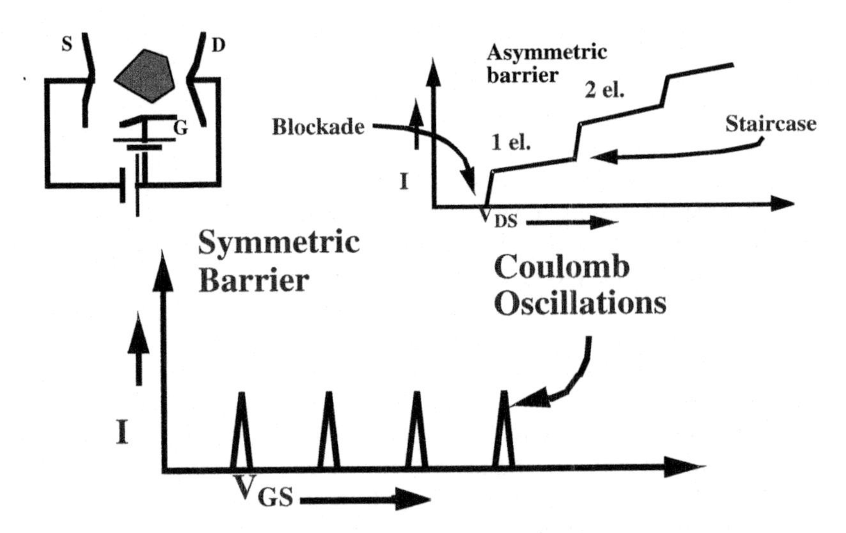

FIGURE 8.2 Schematic of the transfer process of an electron upon application of a bias voltage V between two electrodes that the particle is confined in between. Both diagrams show Coulomb blockade conditions, one in absence of the offset charge and another in the presence of offset charge. The capacitances of the two junctions are identified by the subscript 1 and 2. In the case of symmetric barriers, current is blocked over a range. In the case of asymmetric barriers, the current-voltage characteristics appear as a staircase corresponding to the transfer of 0,1,2,... electrons.

with the first electrode) is higher than the second electrode. Tunneling can now occur off the island to the second electrode because it is energetically favorable, and the system returns back to its initial state. Similar arguments hold when the first tunneling event is from the particle to the second electrode. In this case, the first tunneling event occurs upon alignment of the energies between the particle and the second electrode. This leads to the lowering of the energy of the particle, and now empty states are available for tunneling from the first electrode. In both cases the Coulomb blockade energy is still $e^2/2C_\Sigma$. When the barriers are identical, in the absence of spurious charge, one observes flow of current only above a certain voltage, the Coulomb blockade voltage. When barriers are not identical one observes a Coulomb staircase as shown in Figure 8.2.

The offset charge represents the polarization of the particle due to electric fields terminated on the small quantum-dot. This offset may be intentional, such as from an electric field due to a gate nearby (as shown in Figure 8.2) with potential that can be varied as in a single-electron transistor, or it may be due to unintentional causes, such as an electron trapped on a defect or an interface state in the enclosing matrix of barriers. In the event of transfer of an electron from the reservoir onto the island under application of a bias V and in the presence of a polarization charge Q_{ofs}, the change in energy, for $C_1 >> C_2$, is

$$eV = \frac{\left(e + Q_{ofs}\right)^2}{2C_2} - \frac{Q_{ofs}^2}{2C_2} = \frac{e}{2C_2}\left(1 + \frac{2Q_{ofs}}{e}\right)$$

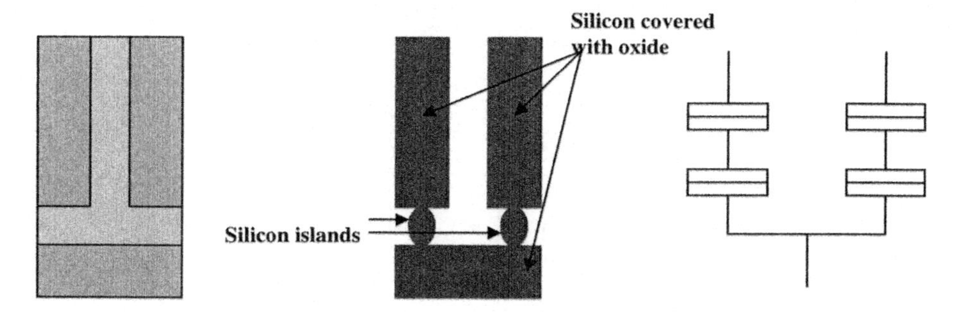

FIGURE 8.3 Pattern-dependent oxidation that leads to formation of small silicon island in silicon-on-insulator together with connection regions. Oxidation leads to formation of silicon islands because of decreased oxidation at corners due to stress. The island results from the lithographically defined thin region. (after Takahashi et al.)

that is, if the offset charge is *-e/2*, conduction is allowed. And, the largest Coulomb blockade occurs for e/2, and is e/C_2. The offset can appear due to polarization induced by a gate. Thus, under gate modulation, it is possible to have a condition where there is no blockade, so the conduction can be modulated from off (blockade) to on condition. Thus, if one has an external gate, changing it leads to current flow at precise voltages, while no current flows in between. And, one can determine the sum capacitance for the system by observing the control voltage difference between the current peaks. In the first single-electron transistor work of Fulton and Dolan,[2] performed on aluminum junctions, both the existence of this offset charge and the tunability of this conduction was demonstrated. Likharev[6] provides a very complete description of the transport properties of the system.

8.3 SINGLE-ELECTRON TRANSISTORS AND THEIR MEMORIES

The silicon-on-insulator substrate, with its buried oxide, is an excellent medium for the implementation of the single-electron transistor and its employment in a memory[7–9] since the small islands can be formed in the thin silicon layer on top of a low permittivity medium. Pattern dependent oxidation, such as at the intersection of two defined regions (silicon and a dielectric mask, for example, or narrow silicon regions of different thicknesses) leads to single-electron island regions formed through nonuniform stress-dependent oxidation (see Figure 8.3) without directly depending on lithographic patterning of the island. Since the confined regions occur in a matrix of silicon dioxide, which has a small permittivity, the effects can be observed at temperatures closer to room temperature, unlike the early experiments which used depletion regions of semiconductors and were performed near liquid He temperatures. Single-electron gated transfer is clearly observable in such structures and a variety of demonstrations, such as turnstiles, single-electron charged-coupled devices (CCD),[10] have been made. Similar to the pattern-dependent oxidation technique, a narrow wire formed in the silicon, at small dimensions, leads to confined regions due to random distribution of dopants which form local potential wells. With the introduction of a laterally controlling gate, tunneling through multiples of such

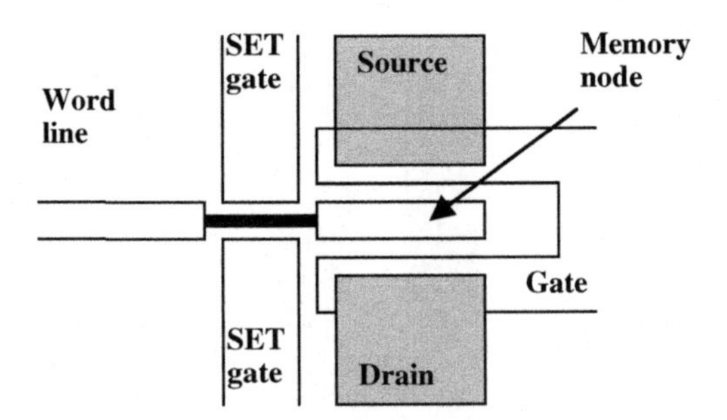

FIGURE 8.4 A coupled single-electron transistor – field-effect transistor structure. Electrons are stored on the memory node by transferring from the wordline using the SET gate that controls the multiple tunnel junctions along the nanowire. The field-effect transistor's conduction measures the presence of the electrons (after Smith et al.).

islands (multiple tunnel junctions) can be controlled and single-electron effects observed. Impurity potential wells exist wherever impurities are employed, that is, in most microelectronic geometries, but their charging energy is so small that one does not observe them at room temperature. CMOS structures, at small dimensions, are similar to the narrow wire geometries, and the Coulomb oscillations are observed in them,[11] so long as measurements are made at temperatures corresponding to the ~2 meV charging energies.

The blockade feature of the single-electron effect allows several interesting memory arrangements to be implemented. Figure 8.4 shows a schematic of an example[8] from Smith et al. in the form of a gain cell where the storage of electrons occurs on a memory node by transferring through a multiple-tunnel-junction single-electron transistor. The presence of electrons on the memory node, formed in the silicon layer of a silicon-on-insulator substrate, changes the characteristics of a field-effect transistor formed using the buried oxide as a gate oxide and a channel in the underlying silicon substrate, thus allowing detection of stored electrons and hence the state of the memory. Another example of the use of multiple tunnel junctions is described by Nakazato et al.[12]

All these structures depend on small dimension lithography to eventually translate into formation of structures that have Coulomb blockade and hence the ability to rectify and store by a suitable application of bias. Dimensional variations, temperature, and voltages that are dependent on the magnitude of coupling through the laterally defined lithographic dimensions have an effect on reproducibility that we will discuss later.

8.3.2 Memories by Scaling Floating Gates of Flash Structures

Coupling the discrete electron effects to the channel of a field-effect transistor has resulted in a number of successes in using single-electron effects at room temperature. There are a number of interesting examples of this approach. Yano et al.[13] used a thin polygrain silicon film, in which current flows through a series of connected grains with conduction constrained by presence of electrons on a grain that is separated from, but in close proximity to, the channel. Another example of such a

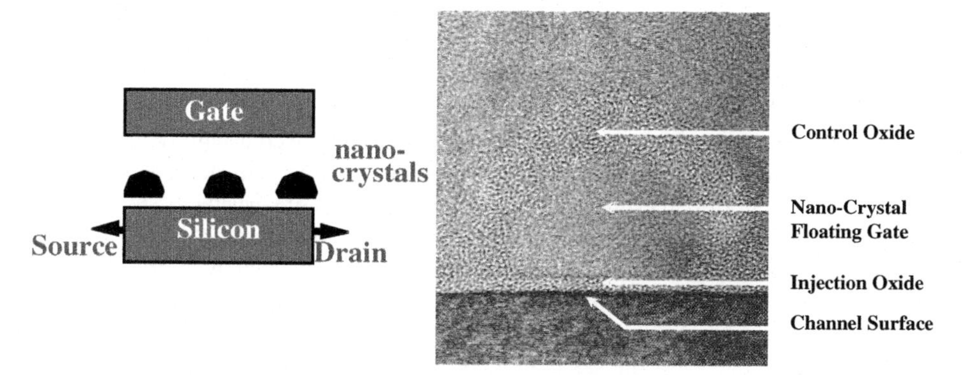

FIGURE 8.5 A cross section of nano-crystal memory (after Tiwari et al.).

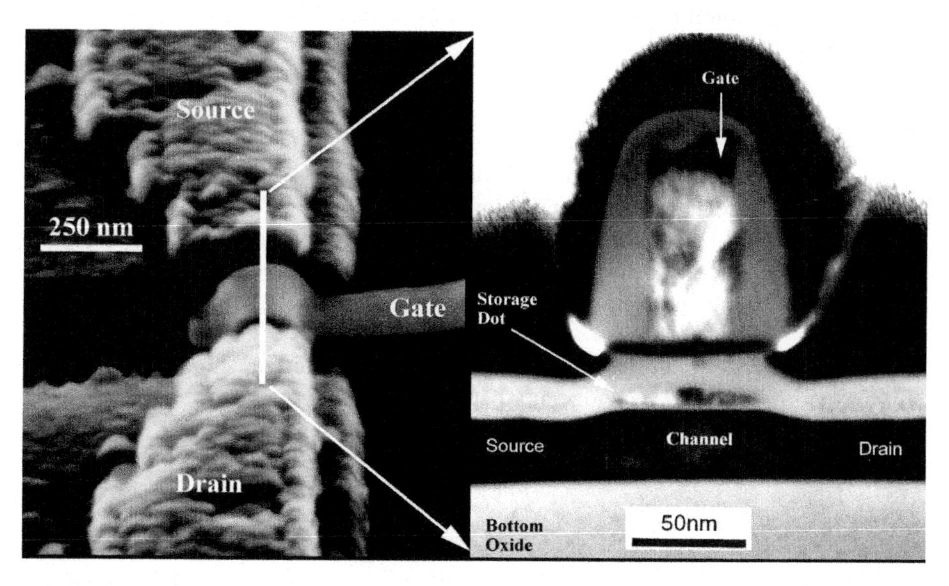

FIGURE 8.6 A cross section of quantum-dot memory (after Welser et al.).

coupled-channel approach is scaling of the floating gate region of a flash memory to dimensions where the discreteness effects become strong. Nanocrystal memories[14–17] (Figure 8.5) accomplish this through a distribution of quantum dots, while quantum-dot memories[18,19] (Figure 8.6) accomplish this by scaling the entire device geometry to the ultrasmall dimensions. Both are examples of flash memories that utilize quantum dot(s) between the gate and the channel of a field-effect transistor to store electrons, which screen the mobile charge in the channel, thus inducing a change in the threshold voltage or conductivity of the underlying channel. These quantum dots are transmissively coupled to the channel and isolated from the gate, and their processing can be accomplished together with CMOS processing. Their reduced dimension and confinement brings forth two important features that are absent in the conventional floating gate structures: a reduced density of states that restricts the states available for electrons and holes to tunnel, and the Coulomb blockade effect that arises from a larger electrostatic energy associated with placing a charged particle onto a smaller capacitance.

In a floating-gate memory, when the gate energy is lowered with regard to that of the source and the drain, electrons transfer to the floating gate storage nodes. For nanocrystal memories, these storage nodes are small single-crystal silicon islands (nanocrystals of an areal density in the 10^{11} cm^{-2} range) that are chemical vapor or aerogel deposited on the injection oxide. For the quantum-dot memory, this is a polycrystalline island patterned at the intersection of the gate and the channel line.

Electrons stored on the island screen the charge in the channel and hence lead to less channel charge for the same applied gate-to-channel potential. This is effectively a change in the threshold voltage. The biggest implication of the scaling of the floating gate region is that the number of electrons used in the device structure is discrete and small. But, this discreteness, or quantization, is not used directly in the device operation. Instead, it is coupled to the channel of the device, thus forming a gain cell. That is, these electrons trapped on the islands influence the conduction of a channel underneath them, and thus the conduction of the channel is a measure of the storage of the electrons. Barriers, used for storage of the electrons, are thus important to the write, erase, and the refresh conditions. But, read of the device, and the amount of signal delivered by the device, are related to the field-effect. The single-electron effects, or the quantum effects, provide a perturbation to it that are detected through the influence of immobile charge on mobile charge. The device behaves as a gain cell and it is limited in size by field-effect. So, limits of field-effect devices are equally important to limits of the floating-gate memory implementations.

For a nanocrystal density of ν_{nxt} of size t_{nxt}, a tunneling injection oxide of t_{inj}, a control oxide of t_{cntl}, and ν average number of electrons per nanocrystal, the threshold voltage is approximately given by:

$$\Delta V_T = \frac{e\overline{\nu n_{mxt}}}{\varepsilon_{ox}}\left(t_{cntl} + \frac{1}{2}\frac{\varepsilon_{ox}}{\varepsilon_{nxt}}t_{nxt}\right)$$

The spacing between the nanocrystals should be less than the screening length in order to minimize percolative transport in the channel underneath.

Figure 8.7 shows, low bias charged conditions (away from the completely discharged and charged branches), that are evidence of this percolation. The oxide in between the nanocrystals is kept large enough to suppress transport directly between the nanocrystals since leakage and subsequent loss to source and drain regions is one of the major methods for loss of charge in floating-gate structures. The control oxide is designed to be thick enough (7 to 15 nm) so that the only path for electron transport to and from the nanocrystals is from the silicon underneath-inversion channel region during injection and depletion region during ejection. The barrier height of Si/SiO_2 interface is large (~3.15 eV). Oxide thicknesses in the 1- to 10-nm range controls the transmission efficiency over nearly 20 decades. This

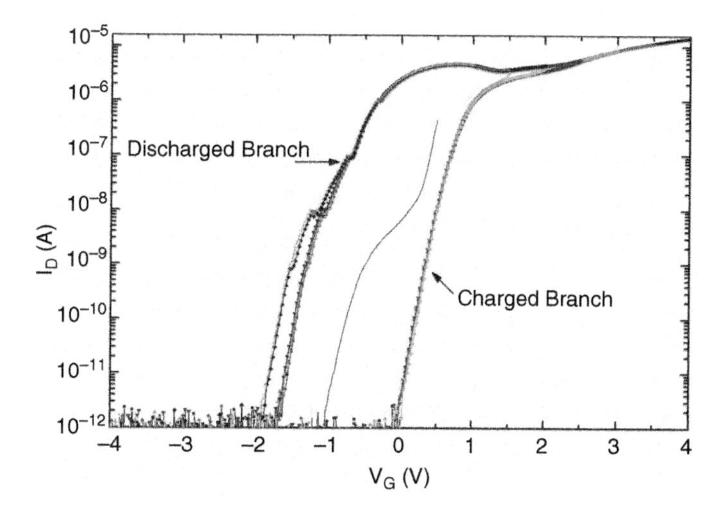

FIGURE 8.7 Current, at slow sweep, as a function of gate voltages for charging and discharging at VDS = 0.1 V. The discharged branch is obtained by sweeping from -4V to 0.5, 1.0, 1.5, ..., 4.0 V. The discharged branch is obtained by charging for extended time at 0.5, 1.0, 1.5, ..., 4.0 V and sweeping to −4 V. Charge movement can occur between the nanocrystals and the channel during the measurements for certain ranges of voltages and leads to the efficient effects observed at large voltages and hysteresis at low voltages. The extrapolated differences between the two branches allows calculation of the charge stored in the nanocrystals. In the logarithmic curves, at low voltages (0.5 V, middle curve), with less charging of nanocrystals, percolation transport can occur in the channel.

TABLE 8.1
Spherical Charge in SiO_2 Gate Stack with a 7-nm Gate Control Oxide

Diameter (nm)	C_Σ (aF)	E_c (eV)	C_{cntl} (aF)	E_0 (eV)	Single-electron V_T shift (V)
30	6.68	0.011	5.27	~ 0.003	0.03
20	4.45	0.018	2.57	0.007	0.062
10	2.23	0.036	0.71	0.03	0.225
5	1.11	0.072	0.19	0.104	0.84
3	0.68	0.118	0.069	0.29	2.31
2	0.45	0.178	0.031	0.65	> 5
1	0.22	0.364	0.008	2.6	> 10

allows the memories to be made volatile and high speed using small injection oxide thickness or nonvolatile and slower speed using large injection oxide thickness.

Table 8.1 summarizes a number of characteristics for the polysilicon/control oxide/silicon island/injection oxide/silicon gate stack. This is an approximate calculation meant to point out the main characteristics of the system. The capacitance

(C_Σ) is the self-capacitance of the silicon dot used in the calculation of the charging energy (E_c). These vary linearly (for capacitance) and inverse linearly (for energy) with dimension. The quantization effect of confinement in energy (E_0) varies as the inverse square of the dimension. At dimensions below 10 nm, the capacitance is small enough that it requires energy of the order of room temperature thermal energy to place an electron on the silicon island. As the dimension decreases further, the eigenenergy of the confined states allowed become larger than the single-electron electrostatic charging energy. When a single electron is stored on the island, it causes the channel threshold voltage to shift by a magnitude that is inversely controlled by the capacitance C_{cntl}. The shift should not exceed the operating voltages of the structure. Thus, dimensions of between 10 and 3 nm are usable and provide a substantial and observable effect.

In the case of nanocrystal and quantum-dot memories, the coupling capacitance to the channel is made to be significantly larger, with additional coupling to the other nanocrystals in the vicinity and, in particular, for the end nanocrystals there is stronger coupling to the source and drain reservoirs. The latter is not surprising; it is one of the major mechanisms for leakage of charge in floating-gate memories. While these results are secondarily geometry specific (e.g., box, sphere, and hemisphere shapes) because of the different degree of confinement, the estimates of capacitance described are within 20% of the more sophisticated calculations. The coupling capacitances to the channel and source and drain regions, therefore, have a stronger influence on the characteristics.

8.4 MODELING OF TRANSPORT: TUNNELING

Figure 8.8 shows distribution of the subband states in a three-dimensionally confined box of silicon coupled to the inversion layer. Confinement in silicon, with the different longitudinal and transverse masses, and six valleys along the six coordinate axes result in multiple ladders that need to be included. Note from Figure 8.8 that the number of states available to receive injection is small in the quantum-box for the lower energy range. That is, direct tunneling is certainly limited by the sink function for tunneling. Clearly in these structures, tunneling (from inversion layers into confined structures and removal) is the important intrinsic feature that needs to be understood mathematically. In reality, in addition to these features, there will exist unusual effects arising from traps strongly localized structures such as those due to interface states and intrinsic and extrinsic defects. Tunneling in oxides and tunneling between coupled systems are central to this, and we will next study these.

8.4.1 TUNNELING IN OXIDE

At small dimensions, two regions of tunneling are important: oxide tunneling (injection and control oxides) which determines the competing balance between injection and ejection, and between source and drain which determines the smallest limit of the device. As oxide thickness decreases, a large current density can be obtained through thin barriers. Indeed, a 0.15-nm change in oxide thickness leads to a current density change by a factor of 10. Note that a larger roughness arises from the top

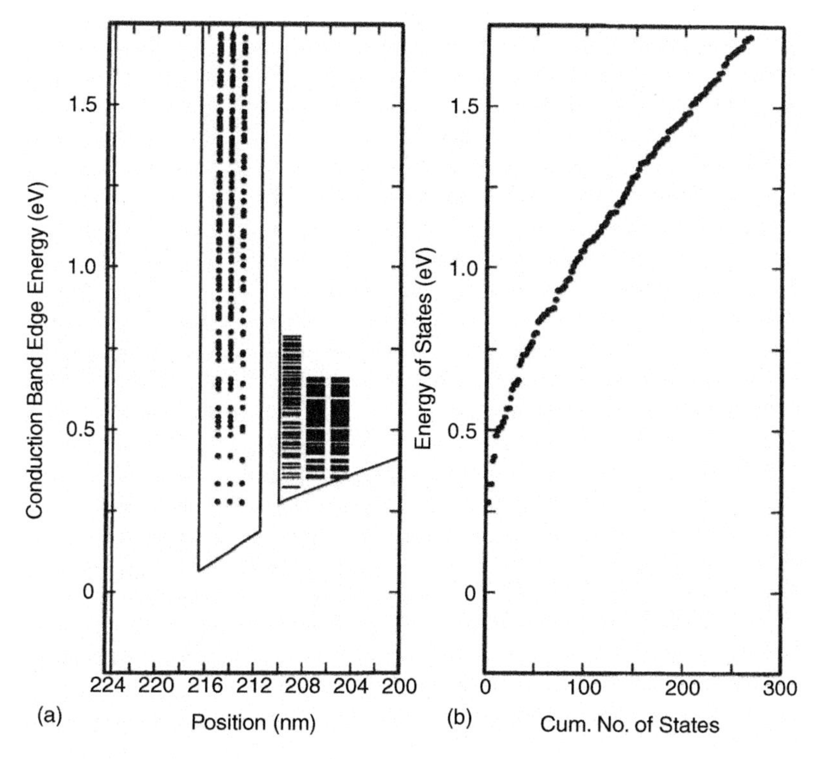

FIGURE 8.8 (a) Shows the band edge and discrete and subband-edge energies in a nanocrystal (<100> orientation of silicon) and inversion layer of silicon (<100> surface) at a symmetrical cross section through the device. The dots show the position in energy (not space) in the quantum dot and the lines show the subband-edge energy in energy (not space) in the two-dimensional electron gas. The triple ladder structure represents the effect of confinement in the three directions. (b) shows the corresponding cumulative number of states in the nanocrystal as a function of energy. The device structure is under 2.5 V bias between the n^+ gate and the substrate.

polysilicon gate/SiO_2 interface whereas the bottom surface is nearly atomically smooth. Random effects should be expected from the random gate control capacitance and dopant activation in this polysilicon/oxide region. In order to model the injection processes in the coupled quantum-dot systems, we will utilize quantum kinetic modeling.

8.4.2 Quantum Kinetic Equation

The calculation of charging and discharging in the coupled system[20] requires quite rigorous calculations together with approximations to make the calculation tractable. Details are given in the reference; here we summarize only the salient steps of this modeling, and then discuss the implications. The Hamiltonian for the coupled system consisting of the channel region (in inversion: two-dimensional electron gas) coupled to the quantum-dot is:

$$H = H_{2\deg} + H_{qd} + H_T \text{ or } H = H_0 + H_T, \text{ with } H_0 = H_{2\deg} + H_{qd}$$

and

$$H_{2\deg} = \sum_n \left(\varepsilon_n + eV\right) a_n^\dagger a_n$$

$$H_{qd} = \sum_m \varepsilon_m b_m^\dagger b_m + E_s \nu$$

where Es(n) is the electrostatic energy and m's and n's identify the quantum-dot and the channel states.

$$H_T = T_{nm} a_n^\dagger b_m + c.c.$$

The state of the system is $|n_n,n_m>$, where n_n and n_m represent the occupation number in the channel and the quantum-dot. It is useful to write time evolution of the density matrix in the Heisenberg representation:

$$i\frac{h}{2\pi}\frac{\partial}{\partial t}\hat{P}_H(t) = \left[H, \hat{P}_H(t)\right],$$

which yields the equation of motion in interaction representation:

$$\frac{\partial}{\partial t}\hat{P}_I(t) = -i\frac{2\pi}{h}\left[H_T, \hat{P}_I(t_0)\right] + \left(i\frac{2\pi}{h}\right)^2 \int_{t0}^{t}\left[H_T(t), H_T(t'), \hat{P}_I(t')\right]dt$$

and which can be formulated as the Rate/Master equation:

$$\frac{\partial}{\partial t}\vec{P}(t) = \vec{W} \cdot \vec{P}(t)$$

To calculate the time-dependence of the charging and discharging, we first calculate self-consistently for all eigenstates in the channel for a given number of electrons stored in the quantum dot for all gate voltages, repeat it for the different number of electrons allowed, determine the transition rates, and then determine the time-dependence from the rate equations. Probability of quantum-dot occupation number $\{n_m\}$ is:

$$p(\{n_m\})(t) = \sum_{\{nn\}} \left\langle \{n_n\}, \{n_m\} \middle| \hat{P}_I(t) \middle| \{n_n\}, \{n_m\} \right\rangle$$

and probability of having ν electrons in the quantum dot

$$\sum_m n_m = \nu \ ,$$

with the transition rates

$$P_\nu(t) = \sum_{n_m}\sum_{n_n}\left\langle \{n_n\},\{n_m\}\right|\hat{P}_I(t)\left|\{n_n\},\{n_m\}\right\rangle \sigma\left(\sum_m n_m, \nu\right)$$

determined using the coupling constants and the occupation statistics. The average and variance of electrons follow from this.

This now gives us the information from which many of the parameters of interest can be calculated.

8.4.3 Carrier Statistics and Charge Fluctuations

From the master equation, the stationary solution follows from setting the rate to zero. The transition matrix W is of dimensions $n_{max} \times n_{max}$ where n_{max} is the number of electrons in the quantum-dot set for computational tractability, and the vector P(t) = [po(t), p1(t), p2(t), ... pnmax(t)] is the probability of having 0, 1, 2, ... n_{max} electrons in the quantum dot.

This now allows determination of the time-dependence to relate to the classical expressions for current, threshold voltage, and also derive the spectrum for single quantum dot. Figure 8.9 shows an evolution of electrons in a dot due to injection from the inversion layer under three different bias conditions.

At the 2-V bias condition, it takes nearly 100 nsec before the average reaches one electron. The transition rates are too low because of the large oxide barrier height and small overlap. But, it changes rapidly with bias so that less than 10 nsec is needed at 4 V bias. The saturation in number of electrons between 100sec of nsec to 10sec of μs for the differing bias voltages represents the effect of reduced dimensions. As the charging process nears flat-band conditions, the injection process begins to slow down for the same reasons that slow the process at low bias voltages. Now consider the same structure during erasure (Figure 8.9) when a negative potential is applied at the gate to eject the electrons into the substrate. A number of starting electrons are considered for two differing voltages. The behavior does not have the detailed features of the injection process; the injection process reveals more of the details of the states being tunneled into. The time-constants of ejection are, however, quite similar to that of injection. At 2 V, not shown, the process has very appreciably slowed down. The lifetime in the dot has become very large.

Figure 8.10 shows mean and variance in the number of electrons for a calculation in which a maximum of three electrons is allowed. The variance is ½ at gate voltages where the mean number of electrons is integer + 1. The actual number of electrons in the quantum dot can take only integer values. A mean number of integer + ½ implies that the actual number of electrons is fluctuating rapidly between integer and integer + 1. The fluctuations in current, etc., follow from these calculations by coupling to the classical transport equations of the field effect.

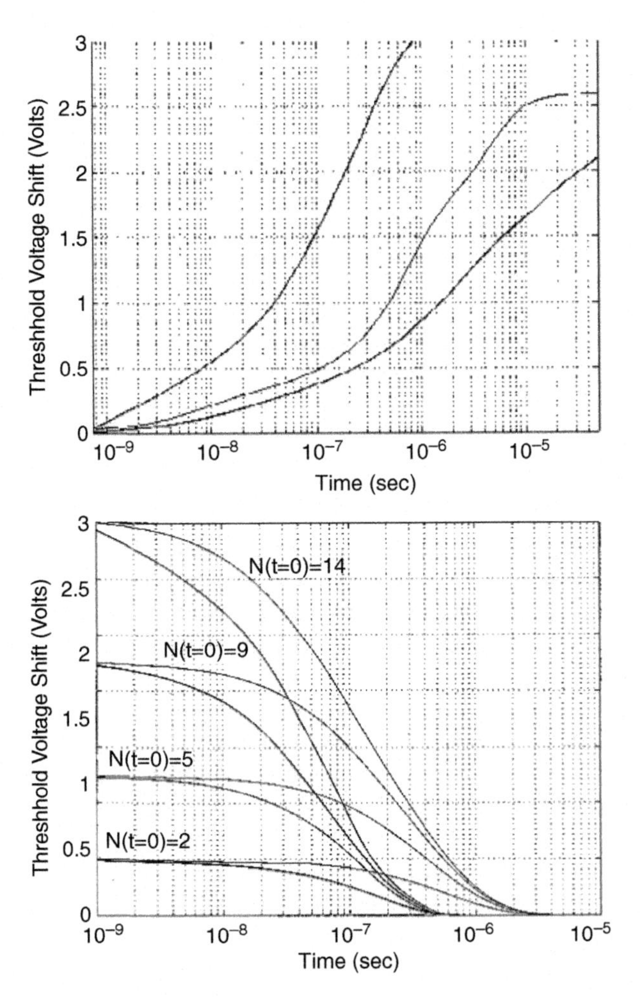

FIGURE 8.9 Evolution of the mean number of electrons in a silicon quantum dot (10 nm × 10 nm × 10 nm) from an inversion layer due to direct tunneling from an injection oxide of 1.5 nm at three different gate-to-inversion layer potentials. The evolution is also shown.

8.5 EXPERIMENTAL BEHAVIOR OF MEMORIES

Use of laterally uncoupled nanocrystals allows suppression of leakage mechanisms that limit the scaling of injection oxide in conventional floating-gate memory structures by leakage to other device regions. Thus, injection oxide can be reduced in thickness together with the use of a small number of electrons. For nanocrystals 5 nm in dimension, the Coulombic charging energy is ~0.07 eV and the subband energies ~0.1 eV, both larger than thermal energy at room temperature. For such nanocrystals that are 5 nm apart, that is, a nanocrystal density of 1×10^{12} cm^{-2}, with a control oxide thickness of 7 nm, the threshold shift is nearly 0.36 V for one electron per nanocrystal. Figure 8.11 shows examples of operational characteristics, measured dynamically, that confirm these expectations. The devices exhibit convergence since,

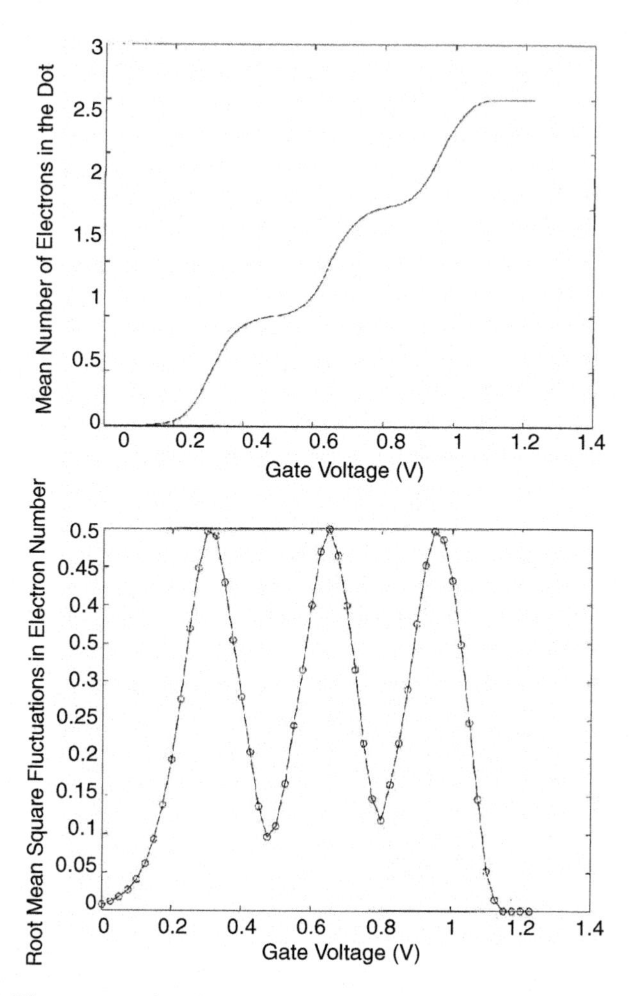

FIGURE 8.10 The mean and variance as a function of gate voltage for occupation in the quantum-dot for a model example of quantum box coupled to the channel and controlled by gate.

for any applied voltage, only a finite number of carriers can be accommodated. The figure also shows these threshold shifts at three different temperatures. At low temperatures, steps in the shifts can be discerned corresponding to the storage, on an average, of 1, 2, ... electrons. While this effect is still present at room temperature, it is masked by the variability in nanocrystal size.

The write times are considerably better than standard flash memory structures, and the voltages are low. This tunability of operation, in power and speed, and operation at small voltages, show the desirability of use of small dimensions in microelectronics because of their voltage, power, and compatibility with present practice of CMOS. Additional references[21,22] decribe recent progress in this area. We now focus on the experimental properties of the nanocrystal memories that are related to the use of small dimensions, particularly to understand the role of surfaces and smaller statistics in the presence of confinement and finite charge effects.

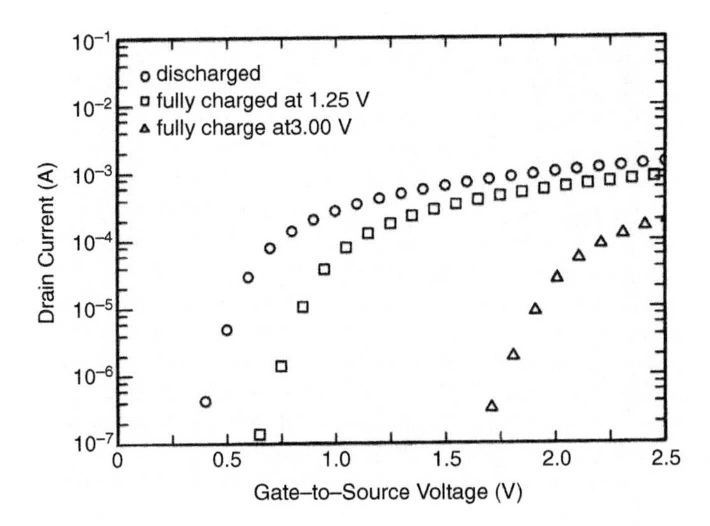

FIGURE 8.11 Bistability in nanocrystal memories resulting from reduced charge storage in the floating gate region. Injection oxide in this structure is 1.5 nm and it results in volatile operation. The lower graph shows threshold voltage shift as a function of applied gate voltage for 300, 77 and 40 K.

Similar behavior in observable in the single quantum-dot device structures. Figure 8.12 shows device characteristics that emphasize this behavior.

A short and a long time-constant are observed in measurements of charging and discharging such as through capacitance as also through measurements of random telegraph signal, that is, after an initial rapid change, such as the loss of the excess electrons of the charged branch, the electronic state of the structure changes slowly. The slow time-constant changes are most discernible in the voltage range of ± 1 V around flat-band condition, a voltage range where the coupling between the nanocrystal and the channel (inverted or depleted) is the weakest. However, the existence of two time-constants is indicative of a possible role for interface states. Note, however, that the preferred injection is still through an efficient transmission into the nanocrystal which has a large capture cross section, and forms a path to possible localization at the interface defect.

The evolution of charging and discharging can be observed through time-dependence. The charging follows an exponential relationship at the estimated tunnel oxide thickness of 1.8 to 2.0 nm, with a control oxide of 7 nm, of this structure and indicates time-constants of ~1 msec at 1.5 V and ~1 μsec at 3.5 V. The exponential relationship is indicative that the coupling between the nanocrystals and the inversion region does follow a field-dependence that is linked to the total oxide thickness of the structure.

Logarithmic transfer characteristics during the charging and discharging are shown in Figure 8.7 for a 3-nm tunnel oxide. The characteristics of the discharged branch are obtained by sweeping to a number of different gate voltages after charging at –4 V. The charged branch characteristics are similarly obtained following charging at different gate voltages and then sweeping the gate voltage to –4 V. In the discharged branch, injection begins to occur at the > 1 msec time-constant of the sweep

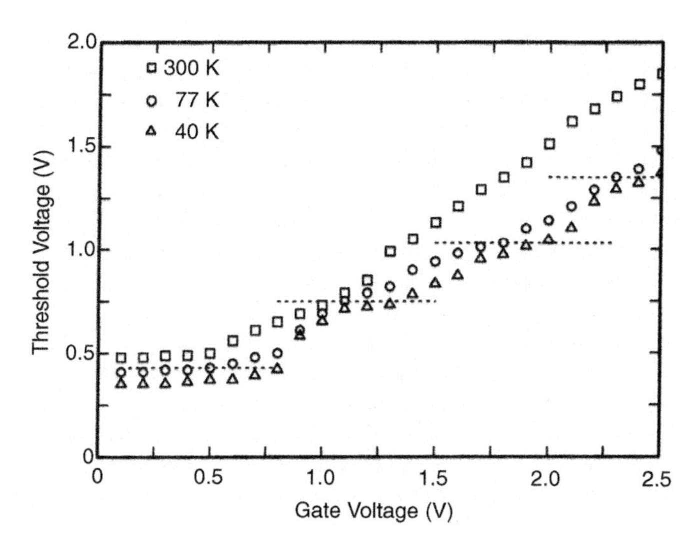

FIGURE 8.12 Single electron events lead to the discreteness observed in the current voltage characteristics of the top figure for a quantum-dot memory.

at about 0.5 V leading to a dynamic change in the threshold voltage that leads to the decrease in observed drain current. At greater than 2.5 V, the threshold voltage shift saturates at ~1.6 V. In the charged branch, the dynamic changes (~1 msec time-constant) occur in the 0.5- to 2.5-V range. All the charging curves exhibit the presence of inversion layer at their initial applied bias and down to 0.5 to 1 V gate bias. The charging at 0.5 V shows a smaller threshold voltage shift and perhaps percolative transport due to insufficient charging of nanocrystals. These characteristics reinforce a feature of the characteristics of Figure 8.7, that is, charge injection into the nanocrystals still allows the inversion region to persist, and charge exchange takes place over the ms time-constants in the 1.5- to 2.5-V range, and below that, it is extremely slow.

Figure 8.13 plots the capacitance as a function of bias for a large-area transistor structure. The measurement is made at a slow sweep rate with a bias sweep starting from a discharged condition. The dynamic threshold-voltage shift that occurs within 0.5 V of flat band again appears in these characteristics as a lowering of the net apparent capacitance. At high frequencies, both channel transmission-line and channel nanocrystal charging time effects reduce the measured capacitance. The two plateaus in capacitances near 2.0 V result from response of the inversion region as well as of nanocrystals that are now efficiently transmissively coupled to the inversion region. At bias voltages of 2.0 V, there exists an inversion region in the structures as expected from Figure 8.13 characteristics also, and as the bias voltage is increased above it, the charge density in the nanocrystals is sufficiently large and coupled efficiently enough to show as an additional increase in capacitance. As the bias is swept from 3 to 4 V, this excess capacitance corresponds to the storage of mid-10^{11} cm^{-2} electron density in the nanocrystals, or one or two additional electrons stored per nanocrystal. The high frequency behavior shows that the transmission-line effects dominate the nanocrystal coupling effects because of the large channel length.

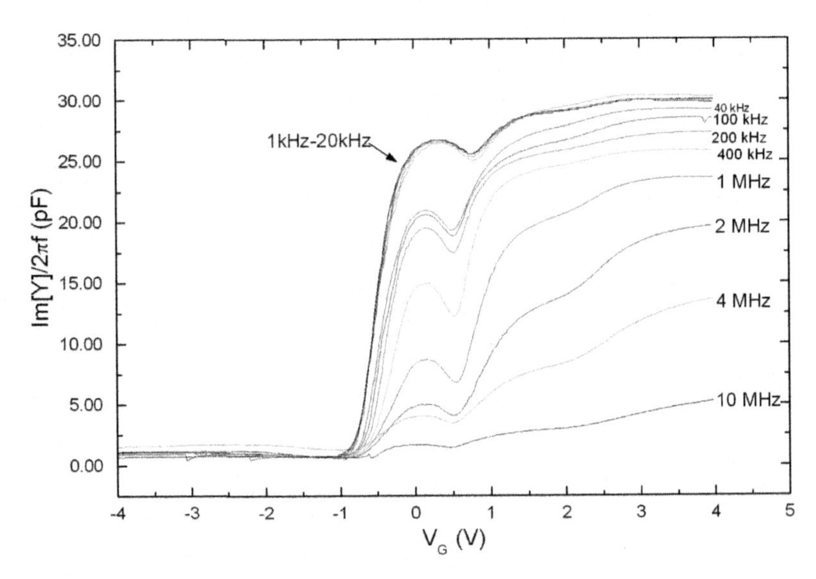

FIGURE 8.13 C-V characteristics of a large-area transistor structure for a range of frequencies at a slow sweep rate starting from discharged condition.

8.5.1 Percolation Effects

The nanocrystal memory operates by screening of the channel from the gate by stored electrons in the nanocrystal. The nanocrystals have a disordered distribution on the oxide surface. The occupation of the nanocrystals by electrons is modeled to the first order by the rate equation derived before. When only a fraction of the nanocrystals are occupied, and this is a function of applied bias and time, transport takes place underneath through the unscreened areas. This is similar to the classic problem of bond percolation. The conductance of this area, for example, measured at low drain-to-source voltages, therefore shows percolative behavior with a criticality that is dependent on time and voltage. We can model this system approximately by assuming a square lattice whose conductance we calculate between the edges. The occupation is determined using the rates, and for this calculation we do not consider any two-dimensional effects along the channel.

Percolation is clearly observable in these simulations, and a minimum necessary nanocrystal density, voltage and time are needed for reproducible operation. Note, however, that not all nanocrystals have to be filled with electrons for conduction to be shut off. Thus, time scales smaller than those required to fill all nanocrystals are sufficient, and probabilities significantly smaller than unity still allow operation.

8.5.2 Limitations in Use of Field Effect

The floating gate memory, since it employs field effect to achieve the gain cell, is still limited ultimately by dimensions, the same constraints that exist for field effect. We discuss some of these limits that arise in field effect (also, see e.g., Wong et al.[23]), as well as due to the practice of floating-gate structures, to gauge the limits of nanocrystal and quantum-dot memories.

8.5.3 Confinement and Random Effects in Semiconductors

Single charge tunneling, limited by the electrostatic energy argument, shows up best in metal systems, where the density of states is enormous and, hence, confinement does not place severe restrictions on the states occupied by the electron in the island. In practice, we work with semiconductor systems where the density of states is many orders of magnitude lower. A consequence of this is an additional energy conservation term related to the energy of the confined state occupied by the electron. Hence the arguments of the required bias are modified by the subband energy term. It increases the energy requirement by E_0 for transit of one electron.

Interface states are a major source of the offset charge in the floating-gate structure. If we assume that surface states on the nanocrystals are the largest source of the offset charge, for a cubic quantum-dot the mean threshold voltage and the standard deviation of the threshold-voltage are:

$$\overline{\Delta V_T} = \frac{e}{\varepsilon_{ox}}\left(t_{cntl} + \frac{1}{2}\frac{\varepsilon_{ox}}{\varepsilon_{Si}}t_{dot}\right)\frac{1}{4t_{dot}^2}\overline{vv}_{nxt},$$

$$\sigma(V_T) = \frac{e}{\varepsilon_{ox}}\left(t_{cntl} + \frac{1}{2}\frac{\varepsilon_{ox}}{\varepsilon_{Si}}t_{dot}\right)\frac{1}{t_{dot}}(6N_{\square T})^{0.5},$$

where $N_{\square T}$ is the interface state density. For oxide/silicon interfaces with thermally grown oxides the interface trap density is typically ~ 5×10^{10} cm^{-2}eV^{-1} or less. For comparison, the effect of random dopants is given by:[24,25]

$$\sigma(V_T) = \frac{e}{C_{gate}}\left(\frac{N_A z_d}{3WL}\right)^{0.5}\left(1 - \frac{z_{undop}}{z_d}\right)^{1.5}$$

where z_d is the depletion region of the retrograde implant and z_{undop} is the lightly doped region width. Figure 8.14 compares these fluctuation effects. As dimensions decrease, the magnitude of the variations increases.

8.5.4 Variances due to Dimensions

Stray charge has already been seen to have a significant effect on the characteristics. Since the charging energy varies inversely with capacitance the charging energy varies linearly with dimensional variance, that is,

$$\frac{\Delta E_c}{E_c} = \frac{\Delta L}{L}$$

Relative dimensional tolerances are similar to the relative voltage tolerances if the effect is an intrinsic part of device operation. Current, $(e\Gamma)$, is more severely affected by the dimensions and energy of the confining barrier due to the exponential

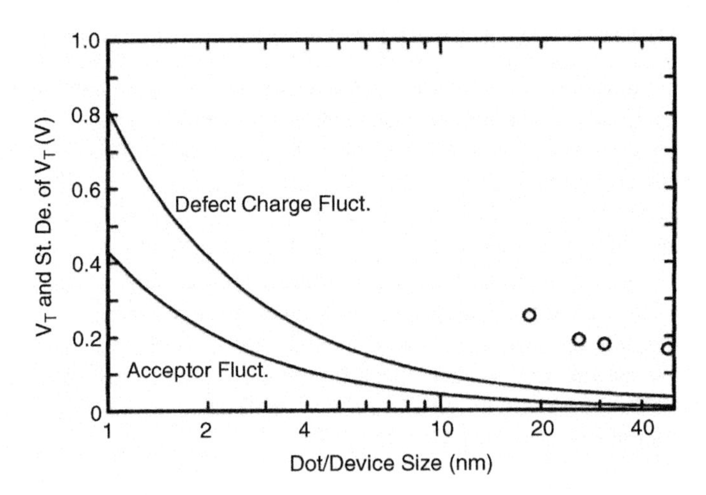

FIGURE 8.14 The threshold voltage and its standard deviation as a function of quantum-dot size for a quantum-dot memory. For comparison the expectation from acceptor fluctuations is also included.

dependence of quantum tunneling on the barrier height. Subband energies vary as the inverse square of the dimensions. Thus, the relative variation of the energy dependence is

$$\frac{\Delta E_c}{E_c} = 2\frac{\Delta L}{L}$$

a requirement nearly twice as strong as due to the Coulomb effects. This places a stronger constraint on device structures when dimensions are below ~7 nm.

The variance in threshold voltage of bulk MOSFETs increases with decreasing dimensions partly because of the random (Poisson) distribution of dopants and the limited number of dopants used to achieve the threshold voltage. A plausible solution to this is elimination of dopants, such as in double-gate and back-plane structures.[26,27] In the former, the electrostatics of the structure, through control of silicon channel thickness, allow for a normally-off device whose threshold voltage is determined by the material parameters. This threshold voltage cannot be made very high. In the latter, a back-gate is used that provides the back-barrier and controls the threshold voltage through an applied potential. In either case, the contribution of channel random doping is eliminated. But now, the variance in threshold voltage is determined by the lateral distribution of the dopants in the shallow contact regions, and can be interpreted as an effective channel length variation across the width of the device. Theoretically, such a variance is in the 1- to 5-mV range for 25 nm junction spacing, instead of the retrograde channel doping contribution of the order of 20 to 40 mV. Instead of channel doping, in the back-plane geometry (and with some modifications in the double-gate geometry), the thickness variations of silicon, through the linear electrostatic potential change ($\psi \sim t$ instead of $\psi \sim t^2$ for doped channels), now contributes to threshold voltage variation. Current SOI structures have a thickness variance of ~ 0.4 nm over 10 μm² areas. Assuming that this can be improved to 0.3 nm, a 10 mV threshold voltage variance leads to a limit in usable silicon thickness of > 10 nm. Confinement introduces an inverse square dependence on the subband energy. This is worse than the linear electrostatic potential dependence. Figure 8.15 shows the variation of this energy as a function of the thickness of the silicon channel. Below 5 nm in thickness, the subband energies change very rapidly.

8.5.5 Limits due to Tunneling

8.5.5.1 Tunneling in Oxide

At small dimensions, two regions of tunneling are important: oxide tunneling (injection and control oxides) which determines the competing balance between injection and ejection, and between source and drain which determines the smallest limit of the device. Figure 8.16 shows the calculated tunneling current.[28] As oxide thickness decreases, a large current density can be obtained through thin barriers. Indeed, a 0.15 nm change in oxide thickness leads to a current density change by a factor of 10. Note that a larger roughness arises from the top polysilicon gate/SiO_2 interface whereas the bottom surface is nearly atomically smooth. Random effects should be expected from the random gate control capacitance and dopant activation in this polysilicon/oxide region.

8.5.5.2 Tunneling in Silicon

With a bandgap of 1.1 eV and a low transverse effective mass (0.19 m_0, Bohr radius ~ 3 to 4 nm), tunneling between the source/drain regions and the substrate, and from the source to the drain, becomes significant when the distance scales are ~10 nm. For bulk nMOSFET structures, interband tunneling appears when acceptor doping

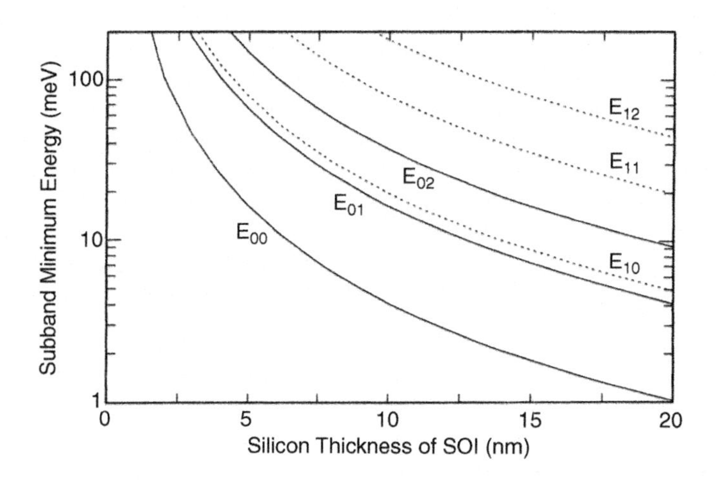

FIGURE 8.15 Subband minimum energy as a function of the silicon thickness of an SOI structure.

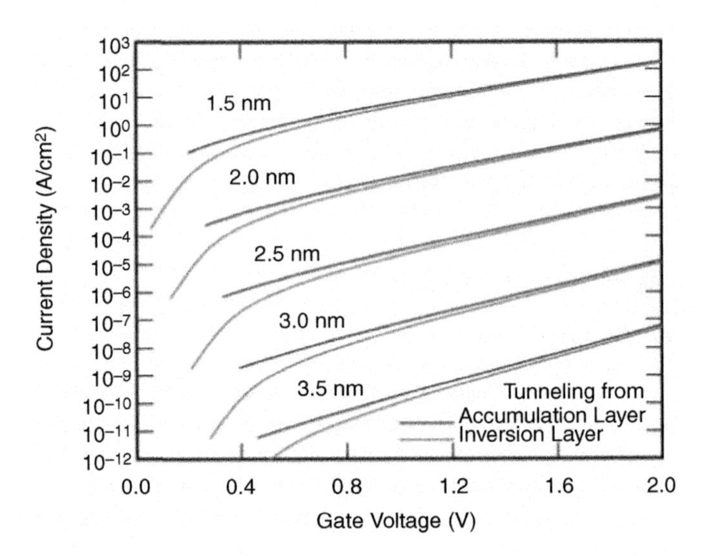

FIGURE 8.16 A cross section of a thin gate oxide and the calculated gate tunneling current from accumulation regions and inversion region.

in channel- or halo-doped regions approaches 2×10^{19} cm^{-3}. This tunneling occurs at the source junction and at the reverse-biased drain-substrate junction, and is a standby power constraint similar in nature to that from oxide thickness. In thin silicon structures, interband tunneling (conduction to valence band and back) is avoided if the threshold voltage of the device allows $V_D+(V_G- V_T)<E_g/e$, a condition that prevents tunneling at the drain end, and is equivalent to the threshold voltage not exceeding ~ 0.55 eV for 0.5-V drain bias. This threshold voltage is consistent with requirements of low subthreshold leakage currents for designs with good subthreshold swing. So, intraband tunneling between source and drain through the channel barrier is a fundamental constraint for adequate field-effect operation and needs to be satisfied in the quantum-dot memory. Figure 8.17 shows tunneling current between source and drain, for a 5-nm-thick sliver of silicon (junctions box doped 5×10^{18} cm^{-3} and 10 nm apart) for a quasi two-dimensional self-consistent Schrödinger-Poisson calculation. The longitudinal masses of the doubly degenerate ellipsoids form the lowest energy ladder with the smaller transverse mass available for tunneling. The fourfold degenerate in-plane ellipsoids cause tunneling through the longitudinal mass. Tunneling from the doubly degenerate states dominates. This tunneling current, between source and drain, establishes a fundamental constraint of 10 nm for channel length, where the field effect is not subsumed by tunnel effect, and is a practical constraint at which the standby current caused by tunneling leads to too high a standby power drain during chip operation. Geometries such as the straddle-gate structure[29] can satisfy this constraint while allowing for a small enough quantum dot.

Table 8.2 summarizes some of these constraints in the design of field-effect structures. In addition to the transistors, these issues also constrain the floating-gate-based memory structures.

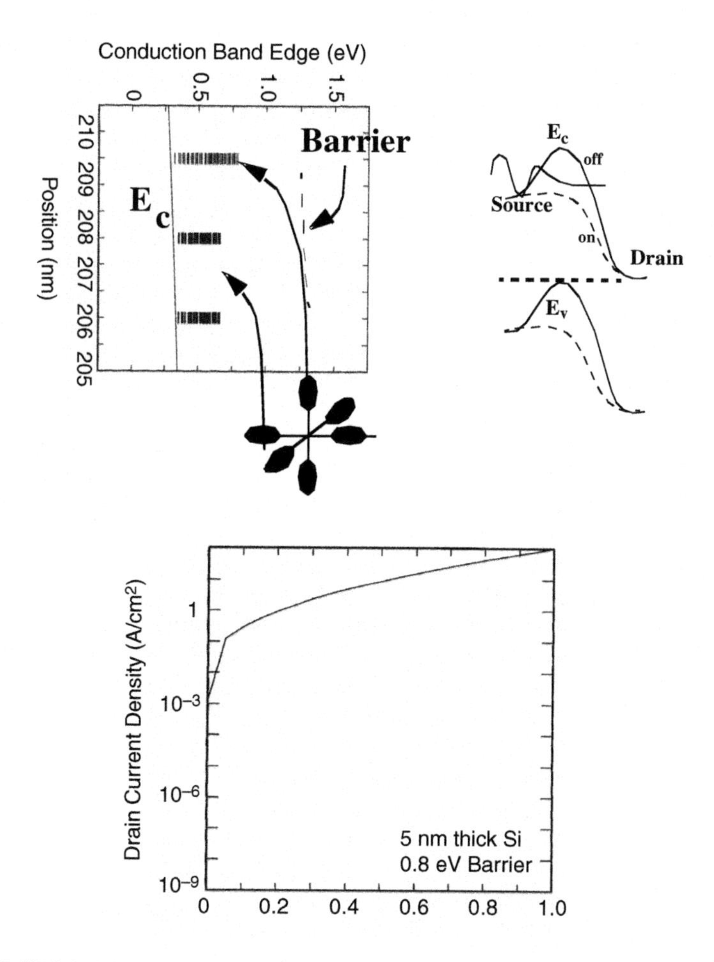

FIGURE 8.17 Model tunneling current calculation between source and drain for a gate bias causing 0.8-eV barrier. Largest tunneling occurs from the low transverse mass valleys.

8.6 CAN WE AVOID USE OF COLLECTIVE PHENOMENA?

Digital microelectronics optimizes device and interconnection technology with the circuit design to obtain the necessary functions at the characteristics desired (usually a combination of several attributes: speed, power dissipation, voltages of operation, density, reliability, noise immunity, cost, etc.). It requires a judicious blend, one of whose important components has been the reduction of dimensions. Smaller dimensions lead to faster devices, lesser area, and lower power — all desirable properties. High yields can be maintained only by improvements in technology and use of device designs that minimize variations arising from technology. CMOS, with the use of rail-to-rail circuits, restores levels and is less sensitive to variations so long as devices maintain reasonably active power gain. The use of collective phenomena has been implicit in this progress because the number of electrons used has always

TABLE 8.2
Constraints in the Design of Field-Effect Structures

Cause	Constraint	Comments
Random dopant distribution with reduced volume of depletion region	s_{vT} ~ 20–40 mV at L_{ch} ~ 30 nm	In bulk structures where channel doping profiles are optimized for short-channel control. Can possibly be alleviated by eliminating use of channel doping in thin silicon-based double-gate or back-plane structures
Lateral Random Dopant Distribution from source-drain extension regions	s_{vT} ~ 15 mV at L_{ch} ~ 10 nm	Lateral random distribution of doping leading to channel-length fluctuations. Also present in undoped channel structures, but can be controlled by use of inversion layers as extensions such as in Straddle geometries
Band-to-band tunneling in junction regions	Peak acceptor dopings ~ 2×10^{19} cm^{-3}	Leakage current limitations to channel and halo dopings constrain the short-channel control in bulk structures
Intra-band tunneling between the source-drain extension regions	L_{ch} ~ 10 nm	Maximum barrier height in silicon is approximately the bandgap; for useful V_T's tunneling leakage between the source-drain reservoirs constrains spacing to ~10 nm
Random silicon thickness distribution in thin-silicon channel structures	t ~ 10 nm	Surface roughness ~ 0.4 nm over 10 mm^2 results in threshold variations in thin-silicon geometries similar to that from random doping.
Gate SiO_2 thickness	t_{ox} ~ 1.5 nm	Leakage current through oxide dominated by current through accumulation regions in the source and drain extensions, and also from inversion region

been large enough. It reduces the variances arising from random effects and usually also takes care of variations arising from systematic effects (dimensions, resistances, etc.) arising from the practice of technology. Reduction of dimensions, while reducing the power (usually), also increases the variance due to random effects that appear in statistical fluctuations in voltages (threshold, threshold shifts, etc.) and in currents (magnitude, time and phase). So, implicit in this is a worsening of the noise margin for operation of devices, in logic and in memory, and minimum voltages that can be tolerated and that can compensate for voltage noise margins. For a CMOS inverter gate driving another inverter gate, the number of electrons that are transferred during a low loaded switching event is nearly 100 electrons at dimensions of ~10 nm. Deviations in threshold voltages change the amount of drive available approximately linearly with an average transconductance as the multiplication factor. A large consequence of any changes in drive current is a proportional change in switching times and hence problems in timing and clocking. For logic, therefore, the consequence of scaling size is serious; however, device design (through width) and circuit design (through careful timing analysis and design) can compensate and allow a functioning design.

For memories, the issues are more difficult. Static random access memory (SRAMs) are flip-flops made using CMOS gates. The previous comments apply to it; SRAMs also use lower voltages and are much more sensitive to variances in threshold voltages due to the need to minimize imbalances in the flip-flop. Dynamic memories use a large number of electrons on the capacitors (~40 fC of charge, equivalent to ~250,000 electrons). Fitting such a large number of electrons, as dimensions are scaled, is an increasingly difficult task because we are demanding that the capacitance not be scaled. An increase in the third dimension is needed and higher and higher aspect ratios are more and more difficult to achieve. An equally big problem is that of retention time. Transistors are designed to have fA of off-state current so that the electrons do not leak easily and refresh cycles are slow. However, changes in threshold voltages change the off-state leakage current exponentially for barrier-modulated transport, as is the case for electron diffusion. So, the leakage current changes by approximately a factor of 10 for every ~60 mV change in V_T. This is very difficult to design around through a worst-case design. Quantum-dot memories and the nanocrystal memories attempt to work around these issues by using the oxide barrier for storage of charge and using a gain cell (conversion of the change in electrostatic potential into the current carried by the device) and do it with low power by reducing the number of electrons needed to obtain the memory effect. Two problems arise: one is the need to work with low voltages, and the second is the need to work with the smaller numbers and their consequent Poisson variance problem.

First consider the issue of voltages. Smaller size increases the eigenenergies and the electrostatic charging energy as seen in Table 8.1. The voltages needed for the charging are increased by the lever effect ($\sim t_{cntl}/t_{inj}$) approximately a factor of 3 to 5. To work with voltages of ~ 3 V, the charging energies should not exceed ~ 0.6 to 1 eV. To compensate for random variations, in a quantum-dot memory one needs to work with ~ 5 electrons. So, this implies a size in between 10 and 3 nm. These dimensions still maintain a charging energy larger than thermal voltages. For nanocrystal memories, the large number of quantum dots help in minimizing the variance. A similar size range in dot size is still necessary. The improvement in power-speed trade-off is significant compared to alternative semiconductor memory structures. It is therefore quite likely that, if we are willing to trade speed for lower power, we will be able to work with smaller numbers of electrons and still achieve the control on electrical variations desired for microelectronics.

8.7 SUMMARY

As field-effect devices reach their operationalfundamental and practicallimits, and assimilate semiclassical and quantum-mechanical effects in their operation, increased sensitivity in static and dynamic operational fluctuations are inevitable. Thickness and length control of barriers and channels are clearly a very essential requirement and they have an increasing variation due to quantum effects. And, since tunneling currents vary exponentially, the consequences of leakage can also be quite significant. Small silicon memories, such as the nanocrystal and quantum dot, combine the field effect with the discreteness that comes from use of small dimensions. For control of electrical variations, nanodevices have to solve similar issues as CMOS.

REFERENCES

1. Neugebauer, C.A., and Webb, M.B., Electrical conduction mechanism in ultrathin, evaporated metal films. *J. Appl. Phys.*, 33, 74–82, 1962.
2. Fulton, T.A. and Dolan, G.J., Observation of single-electron charging effects in small tunnel junctions, *Phys. Rev. Lett.*, 59, 109–112, 1987.
3. Grabert, H. and Devoret, M.H., *Single Charge Tunneling*, Plenum Press, New York, 1992.
4. Devoret, M.H., in *Single Charge Tunneling,,* Grabert, H. and Devoret, M.H., Eds., Plenum Press, New York, 1992.
5. Tinkham, M., *Introduction to Superconductivit*y, 2nd ed., McGraw-Hill, New York, 1996.
6. K.K. Likharev, Single-electron devices and their applications, *IEEE Proc.*, 87, 606–632, 1999.
7. Takahashi, Y., Namatsu, H., Kurihara, K., Iwadate, K., Nagas, M., and Murase, K., Size dependence of the characteristics of Si single-electron transistors on SIMOX substrates, *IEEE Trans. on Electron Dev.*, 43, 1213–1217, 1996.
8. Smith, R.A. and Ahmed, H.A., Gate controlled Coulomb blockade effects in the conduction of a silicon quantum wire, *J. Appl. Phys.*, 81, 2669–2703, 1997.
9. Ahmed, H., Single electron and few electron memory cells, *Tech. Dig. of IEDM*, 363–366, 1999.
10. Fujiwara, A. and Takahashi Y., Manipulation of elementary charge in a silicon charge-coupled device, *Nature*, 410, 560–562, 2001.
11. Specht, M., Sanquer, M., Caillat, C., Guegan, G., and Deleonibus, S., Coulomb oscillations in 100 nm and 50 nm CMOS devices, *Tech. Dig. of IEDM*, pp. 383–385, 1999.
12. Nakazato, K., Blaikie, R.J., and Ahmed, H., Single-electron memory, *J. Appl. Phys.*, 75, 5123–5124, 1994.
13. Yano, K., Ishii, T., Hashimoto, T., Kobayashi, T., Murai, F., and Seki, K., A room-temperature single-electron memory device using fine-grain polycrystalline silicon, *Tech. Dig. of IEDM*, 541–544, 1993.
14. Tiwari, S., Rana, F., Chan, K., Hanafi, H., Chan, W., and Buchanan, D., Volatile and non-volatile memories in silicon with nano-crystal storage, *Tech. Dig. of IEDM*, 521–524, 1995.
15. Shi, Y., Saito, K., Ishikuro, H., and Hiramoto, T., Effects of traps on charge storage characteristics in metal-oxide-semiconductor memory structures based on silicon nanocrystals, *J. Appl. Phys.*, 84, 2358–2360, 1998.
16. Kim, I., Han, S., Kim, H., Lee, J., Choi, B., Hwang, S., Ahn, D., and Shin, H., Room temperature single electron effects in si quantum dot memory with oxide-nitride tunneling dielectrics, *Tech. Dig. of IEDM*, 111–114, 1998.
17. King, Y.C., King, T.J., and Hu, C., MOS memory using germanium nanocrystals formed by thermal oxidation of $Si_{1-x}Ge_x$, *Tech. Dig. of IEDM*, 115–118, 1998.

18. Welser, J.J., Tiwari, S., Rishton, S., Lee, K.Y., and Lee, Y., Room temperature operation of a quantum-dot flash memory, *IEEE Electron Device Letters*, 278–280, 1997.
19. Guo, J., Leobandung, E., and Chou, S., A silicon single-electron transistor memory operating at room temperature, *Science*, 175, 649, 1997.
20. Rana, F., Tiwari, S., and Welser, J.J., Kinetic modelling of electron tunneling processes in quantum dots coupled to field-effect transistors, *Superlattices and Microstructures*, 23, 757–770, 1998.
21. De Salvo, B., Gerardi, C., Lombardo, S., Barno, T., Perniola, L., Mariolle, D., Mur, P., Toffoli, A., Gely, M., Semeria, M.N., Deleonibus, S., Ammendola, G., Ancarani, V., Melanotte, M., Bez, R., Baldi, L., Corso, D., Crupi, I., Puglisi, R.A., Nicotra, G., Rimini, E., Mazen, F., Ghibaudo, G., Pananakakis, G., Monzio Compagnoni, C., Ielmini, D., Cacaita, A., Spinelli, A., Wan, Y.M., and van der Jeugd, K., How far will silicon nanocrystals push the scaling limits of NVMs technologies? *Tech. Dig. of IEDM*, 597, 26.1.1–26.1.4, 2003.
22. Muralidhar, R., Steimle, R.F., Sadd, M., Rao, R., Swift, C.T., Prinz, E.J., Yater, J., Grieve, L., Harber, K., Hradsky, B., Straub, S., Acred, B., Paulson, W., Chen, W., Parker, L., Anderson, S.G.H., Rossow, M., Merchant, T., Paransky, M. Huynh, T., Hadad, D., Chang, K.-M., and White, B.E., A 6 V embedded 90 nm silicon nanocrystal nonvolatile memory, *Tech. Dig. of IEDM*, 601, 26.2.1–26.2.4, 2003.
23. Wong, H.S.P., Frank, D.J., Solomon, P., and Welser, J.J., Nanoscale CMOS, *IEEE Proc.*, 87, 537–570, 1999.
24. Burnett D. and Sun, S.W., Statistical threshold-voltage variation and its impact on supply-voltage scaling, *Proc. SPIE*, 2636, 83–90, 1995.
25. Taur, Y. and Ning, T.H., *Introduction to VLSI Devices*, Oxford Univ. Press, Oxford, 1998.
26. Frank, D.J., Laux, S.E., and Fischetti, M.V., Monte Carlo simulation of a 30 nm dual-gate MOSFET: How short can Si go? *Tech. Dig. IEDM*, 553–556, 1992.
27. Yang, I.C., Vieri, C., Chandrakasan, A., and Antoniadis, D.A., Back gated CMOS on SOIAS for dynamic threshold voltage control, *Tech. Dig. IEDM*, 877–880, 1995.
28. Rana, F., Tiwari, S., and Buchanan, D.A., Self-consistent modeling of accumulation layers and tunneling currents through very thin oxides, *Appl. Phys. Lett.*, 69, 1104–1106, 1996.
29. Tiwari, S., Welser, J.J., and Solomon, P., Straddle-gate transistor: changing MOSFET Channel length between off- and on-state towards achieving tunneling-defined limit of field-effect, *Tech. Dig. IEDM*, 737–740, 1996.

9 SESO Memory Devices

Kazuo Yano

9.1 INTRODUCTION

9.1.1 How Nanotechnologies Solve Real Problems

The integrated electronics technologies have reached the turning point at around 100 nm for various reasons. Especially, CMOS devices suffer subthreshold leakage and gate leakage problems, which inevitably change the conventional miniaturization and performance enhancement trend of silicon integrated circuits. Although the gate leakage might be surmounted by high-k insulators, which are under intensive investigation, subthreshold leakage is directly related to fundamental Boltzmann statistical physics, and is becoming a very high hurdle to overcome.

Under these circumstances, nanodevices, which might go beyond CMOS devices, attract anticipation. However, when one takes "nano" literally, there is a large gap between the anticipation and reality. If we extrapolate the conventional miniaturization trend along "Moore's Law" down to nanometer level, one has to wait until 2030. Moore's law states the integration level of a chip increases by a factor of four every 3 years, which corresponds to 30 percent shrink of minimum feature size.

Here, I believe that nano should not be used to form all the features sizes, rather, partial introduction of nanostructures in the conventional device structures gives much impact on the limitations of conventional CMOS devices (Figure 9.1.)

SESO (Single-Electron Shut-Off) memory introduced here is conceived based on this grand vision.

9.1.2 New Direction of Electronics

Not only does the technical aspect of the pictures drive the change, but also does the market. The volume market that drives the electronics industry has moved from PC to mobile devices. Worldwide cellular phone unit shipment is four times larger than that of PCs, and new demand is been actively added; for example, 2003 was the year of camera on the phone. In the future, ubiquitous computing and networking technologies will flood computers in the entire life and business environments.

This mobile-to-ubiquitous direction drives strong requirements for advanced multimedia/broadband processing capabilities and human interfacing capabilities. However, on the other hand, these heavy workloads have to be handled with very limited power budget because of the battery-life restrictions. Particularly, subthreshold, drain-induced barrier lowering, and gate leakages impose serious power problems in the sub-100-nm region.

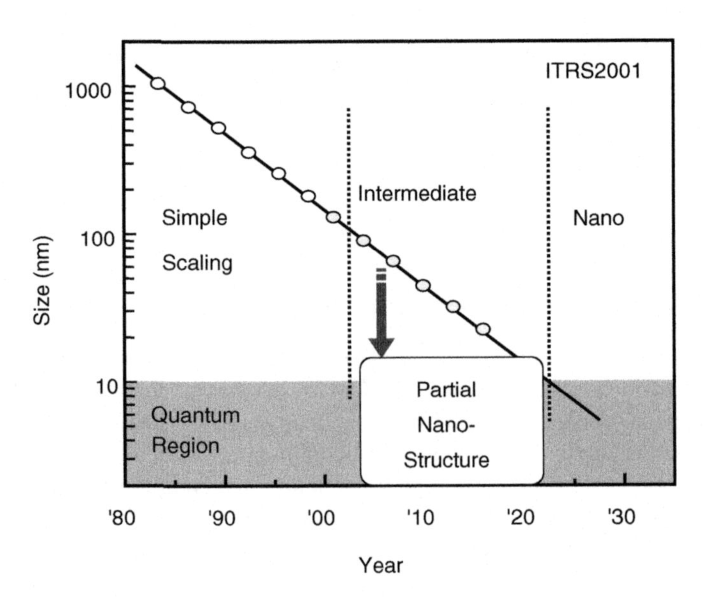

FIGURE 9.1 Miniaturization trend and the importance of partial nanostructure.

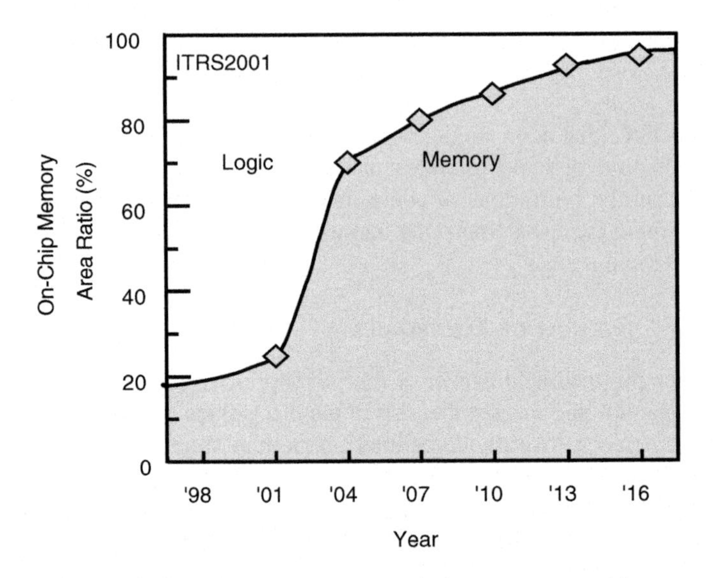

FIGURE 9.2 The trend of on-chip memory area ratio in system-on-a-chip.

The bottleneck to overcome the power problem is memory. Even in a processor LSI or a system-on-a-chip (SoC), a majority of transistors are used for memory. The memory-transistor ratio is estimated to increase as we cram more functionality on a chip as found in the ITRS (International Technology Roadmap for Semiconductors) roadmap, in which the memory ratio in a typical SoC will reach as high as 80 percent in 2007 as shown in Figure 9.2. Those massive memory bits will hold large programs

TABLE 9.1
Memory Classification for Mobile Application

	Application	Memory Type	Features
Volatile	Note PC/PDA Main Memory	DRAM	Large capacity Low cost
	Mobile Phone Main Memory	SRAM	Low power (~μA)
Non-Volatile	Program Storage	NOR Flash	High speed
	Data Storage	AND,NAND Flash	Large capacity Low cost

for complex software functions and large multimedia data, such as video and audio data. Memory is and will continue to be the dominant cost, power, and performance factors in processors and SoCs.

SESO memory is conceived to improve total cost, power, and performance of the sub-100-nm system under the above circumstances.

9.2 CONVENTIONAL MEMORY TECHNOLOGIES

9.2.1 Classification of Conventional Memories

Memories used in mobile devices are summarized in Table 9.1. There are two categories: one is volatile memory, which loses data when the power is off; the other is nonvolatile memory, which retains data even without power supply. DRAM (Dynamic Random-Access Memory) and SRAM (Static Random-Access Memory) are volatile memories, but flash memories, such as NAND-, AND-, and NOR-type, are nonvolatile memories. Of course, if nonvolatility is achieved without penalty, volatile memory will disappear. However, in reality, time required for write is generally much slower in nonvolatile memories, typically ms level and is not suitable for frequent data communication with a CPU. Although both volatile and nonvolatile memories are important, we focus here on volatile memory, which acts as a part of a computer and becomes the bottleneck part in the processor system for new applications.

SRAM has an irreplaceable feature that it is built with transistors for ordinary CMOS logic circuits, and therefore, it does not require any special process steps in addition to those required for CMOS logic that are anyway necessary to build peripheral circuits like decoders, sense amplifiers, and input/output circuits. Also, an SRAM retains data with ultra-low power consumption, if it is very carefully optimized to minimize leakage at the expense of some performance, which allows us to keep data with only a small backup battery. Because of this, SRAM has become a very important component in cellular phones. However, the drawback of SRAM is its bit cost. An SRAM is expensive because it has a six-transistor configuration, whereas DRAM is formed with only one transistor and one capacitor. An SRAM

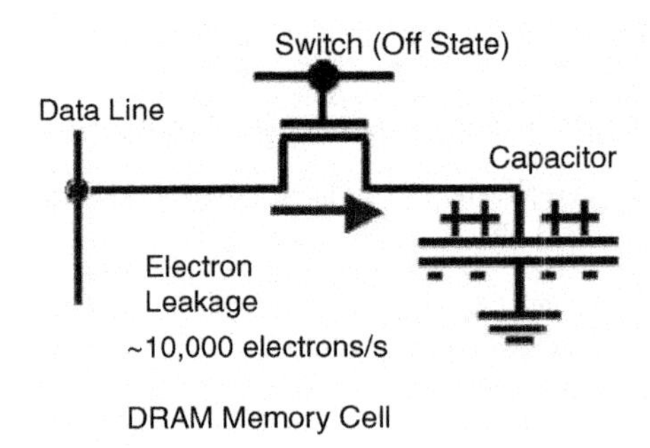

FIGURE 9.3 Origin of DRAM refreshing.

cell typically consumes silicon area of 60 to 100 F^2 per bit, where F is the minimum feature size and F^2 is commonly used as a normalized unit to compare among memory cell technologies, whereas DRAM cell area is around 10 F^2.

DRAM requires a special capacitor-formation process, and therefore, wafer process cost is higher than that of the planar CMOS logic; however, the bit cost is low, because of the one-transistor one-capacitor cell configuration. DRAM has a stack or trench capacitor process, to achieve enough storage capacitance; however, in either case, the difficulty of achieving enough capacitance value in a cell is becoming more difficult as the feature size becomes smaller. This is because the readout signal of a DRAM cell at the data line is determined by the capacitance ratio between the stored capacitance and data-line capacitance, and it has to be constant, typically 5, and therefore, the cell store capacitance cannot be reduced. Also, the power consumption is high because of the refreshing operations, as will be discussed below.

Memory devices for mobile/ubiquitous applications require capacity as large as DRAM, and power consumption as low as SRAM, which is the target and challenge for new memory candidates, such as SESO.

9.2.2 Origin of DRAM Power Consumption

If one tries to lower bit cost to a lower level than that of a six-transistor SRAM, one should use a kind of DRAM-like dynamic operation. If we look back at the history of DRAM, the cell configuration evolved toward the direction to reduce cost: starting from six-transistor SRAM to four-transistor cell which eliminated the load MOSs, then to three-transistor cell and eventually to one transistor and one capacitor. The one-transistor one-capacitor DRAM stores data as a charge on a capacitance, and the write operation is done through a switch MOS transistor. However, the off-state of the switch transistor is not perfectly off, but leakage cannot be completely suppressed as shown in Figure 9.3. The origin of the leakage is the junction leakage around the pn junction region of the switch MOS transistor. To avoid the data loss due to the leakage, DRAM is designed to be refreshed, that is, every cell is read out and rewritten periodically, and this refreshing requires power. This power is not acceptable from the battery-lifetime requirement for mobile devices, typically 100 to 1000 μA per chip.

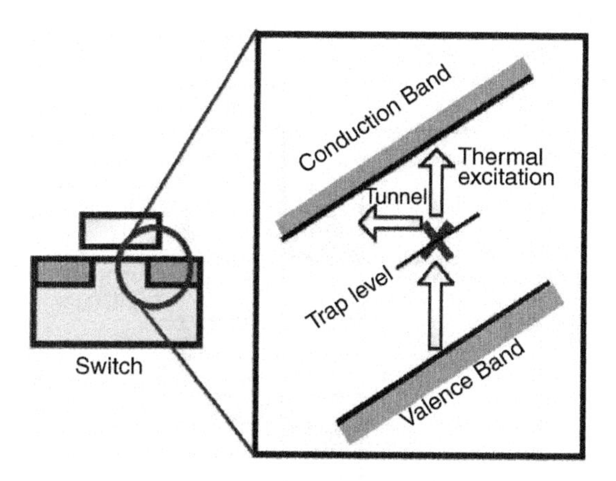

FIGURE 9.4 DRAM memory cell switch and origin of the leakage.

To reduce the refreshing power, simply reducing leakage is effective. The physical origin of the leakage is the transfer of electrons (or holes) beyond bandgap from valence band to conduction band via trap levels as shown in Figure 9.4. Sometimes the internal electric field enhances such undesirable electron transfers. One conventional approach to reduce leakage is to reduce trap levels near the pn junctions. However, no method that dramatically reduces trap levels is known.

9.3 BANDGAP ENLARGEMENT IN NANOSILICON

SESO reduces the leakage in a different manner.[1] The bandgap of the silicon is enlarged by thinning the silicon film to the nanostructure region in which a quantum phenomenon is essential, as shown in Figure 9.5.

One of the immutable laws of quantum mechanics is that a particle has inevitable uncertainty in energy and momentum when it is confined in a nanoscale region. The uncertainty principle is expressed as follows:

$$\Delta p \, \Delta x \sim h,$$

where p is the momentum, x is the position and h is the Planck constant. This uncertainty raises the lowest energy level of a particle, known as zero-point energy. Even at absolute zero temperature, without thermal excitation, this energy exists.

This zero-point energy appears even in one-dimensional confinement. In nanoscale silicon film sandwiched by silicon dioxide insulators, an electron has higher energy than the ground-level based on this principle. In this case, the lowest conduction energy is raised from the conduction band edge. The lowest energy level of the conduction band electron is raised by ΔE and expressed by the following:

$$\Delta E = p^2/(2m)$$

in which m is the effective mass of a carrier and

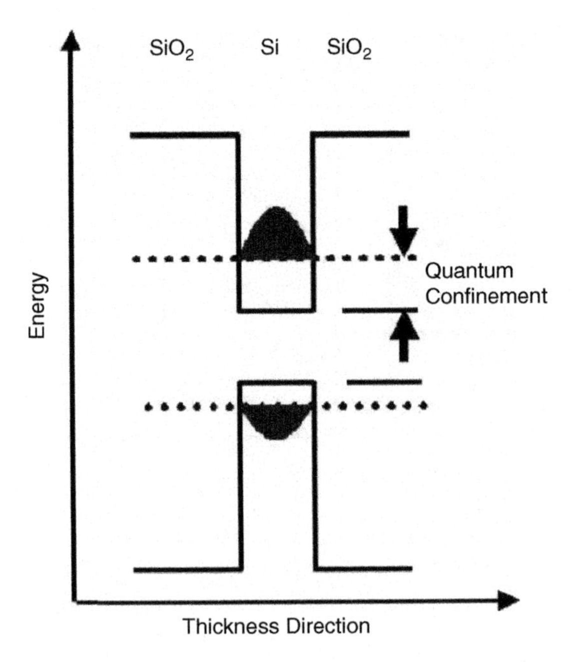

FIGURE 9.5 The origin of the quantum confinement energy and effective enlargement of the bandgap.

$$p = hk/(2\pi)$$

where k is the wave number and it is defined by

$$k = 2\pi/l$$

where l is the wavelength of an electron. In a quantum well having infinite barrier height, the lowest wave has a wave function of zero at both edges. Because of this, the wavelength l is

$$l = 2d$$

where d is the thickness of the silicon film. Combining all these, we obtain:

$$\Delta E = h^2/(8md^2)$$

Total energy on the lowest energy surface including the horizontal carrier motion is expressed as

$$E = h^2/(8m_z d^2) + p_x^2/(2m_x) + p_y^2/(2m_y)$$

in which m_z is the effective mass in the z direction, which is the perpendicular direction to the film, m_x and m_y are the effective mass in x and y direction, respectively, and p_x

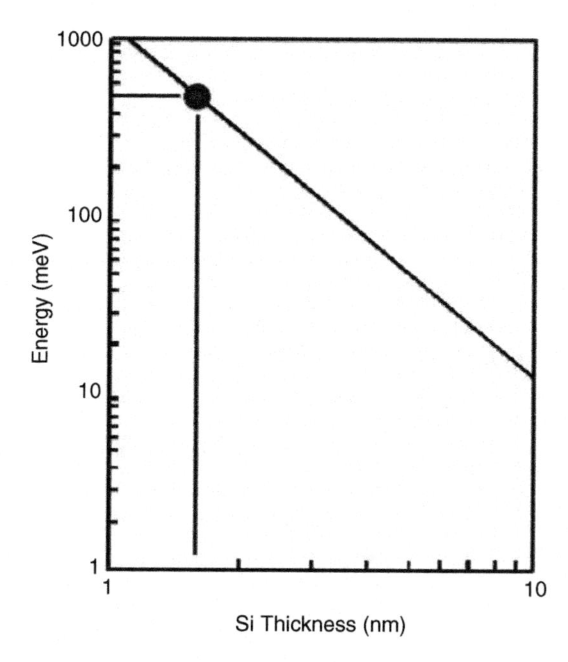

FIGURE 9.6 The quantum-confinement-induced bandgap enlargement. Conduction band contribution is shown.

and p_y are the momentum in the x and y direction, respectively. Infinitely high barrier approximation is useful because the conduction band edge of the silicon dioxide is high enough (about 3 eV higher than that of silicon).

Careful treatment is needed because the silicon conduction band bottoms are sixfold degenerate without confinement; namely, this equivalent valley has the same energy, however, with one-dimensional confinement they are split into three branches: one lower, the other two higher, and each branch has a different effective mass for horizontal direction motion to the film. Because the crystalline orientation of silicon is random in polycrystalline silicon used in the real device, here we roughly estimate the bandgap enlargement by using a conductivity effective mass of 0.26 as shown in Figure 9.6. Based on this analysis, when the silicon thickness is 2 nm, the bandgap enlargement is as large as 300 meV as shown in Figure 9.6. Also because of the hole vertical confinement, the valence band bottom level is shifted. Here the effective mass of the silicon heavy hole is 0.49. Therefore, about half of the bandgap enlargement is added by valence band hole confinement to that solely due to conduction-band electron confinement.

The thermal excitation element of the leakage is an exponential function of the bandgap over kT. The 300 meV corresponds to five orders of magnitude lower leakage than that without. Of course this is again rough estimation, because the field-effect enhanced tunneling and other effects are neglected. But it does clearly demonstrate that the nanoscale structure strongly has the potential to affect the leakage by orders of magnitude.

9.4 SESO TRANSISTOR

9.4.1 History: Single-Electron Devices to SESO

A SESO transistor is a nanoscale thin-film transistor, in which bandgap-enlarged nanosilicon film is used as the active channel region.[1] Nanosilicon thin-film transistors have been studied in the 1990s as a basic test structure to explore the possibility of single-electron devices,[2–12] that is, single-electron memories,[3–10] single-electron transistors,[11] and single-electron transfer devices.[12]

A major difference between the structures used in these single-electron devices experiments and SESO transistor is the uniformity of the nanosilicon film. In single-electron devices, nanosilicon film is intentionally formed rugged, namely, the thickness is different from one site to the other. This is because the ruggedness is intended to create very deep potential energy differences for an electron, resulting in naturally formed nanodots and nanowires.

The basic operations of these nanosilicon single-electron devices are as follows[3]: the structure is a thin-film transistor. One special feature is the channel region is ultrathin: the test device typically used 3-nm silicon film for channel. In addition, the film is intentionally formed granularly to introduce random quantum-confinement potential. This is because to operate a single-electron device at room temperature, one needs to use 10-nm dot; however, lithography to achieve sub-10-nm size is yet to come in 2030 or later. The temperature requirements for single-electron device operations or Coulomb blockade is as follows:

$$kT < q^2/(2C_{tt}),$$

where k is the Boltzmann constant, q is the electron charge, and C_{tt} is the total capacitance of a dot. By using the natural nano-Grand-Canyon-like structure, one can make nanowire and nanodots without using nanolithography. For example, by setting the gate voltage at an appropriate level near the threshold voltage of the transistor, a narrow percolation channel between source and drain is formed. This condition is used in the first room-temperature operation of a single-electron transistor.[3,4]

When one raises the gate voltage further, an electron is injected into an isolated dot, which raises the potential energy for an electron through source-to-drain. By using this dot as a storage dot, one can achieve single-electron memory. After an electron is injected into a dot, one can read out the single-electron charge by the current difference between source and drain. One interesting feature of this device is that the current change is quantized because of the number of the stored electrons is an integer.

The author and a coworker Ishii found in these single-electron device experiments that the leakage current of the test devices is unexpectedly low, despite their polycrystalline structure. This low leakage is naturally explained by the quantum confinement model. After this early-stage work, the authors built SESO to confirm the effect as shown in the following sections.

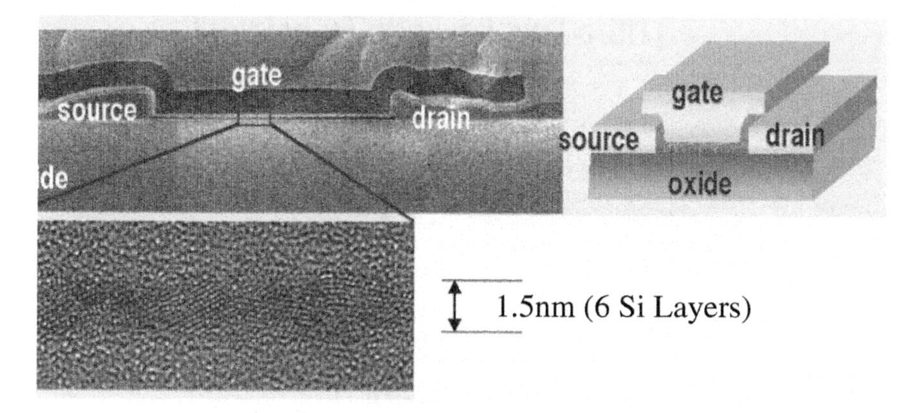

FIGURE 9.7 Cross-sectional SEM and TEM photograph of the SESO transistor and schematic SESO transistor structure.

9.4.2 Fabricated SESO Transistor

A SESO transistor is a thin-film transistor, in which bandgap-enlarged nanosilicon film is used as the active channel region. Because of the quantum-confinement-induced bandgap enlargement, the leakage current under off condition is low. This nanostructured channel is attached to thicker source and drain n+ regions to have contacts and reduce series resistance.

Experimental SESO transistors shown in Figure 9.7 are fabricated using conventional silicon fabrication processes.[1] The heart of the SESO transistor, 1.5-nm polysilicon that has only six silicon layers, is formed by low-pressure chemical vapor deposition (LPCVD). Stable formation of the nanoscale film is possible by careful preparation of the deposition surface and deposition conditions, that is, gas, pressure, temperature, and flow. Unlike the single-electron devices above, the nanosilicon film is formed very uniformly to minimize device-to-device variations. One concern is that the film is not a single crystal, but polycrystalline silicon, in which there are imperfections and trap states on the grain boundaries. The trap states in middle of the bandgap generally increase the transition probability of carriers between valence band and conduction band, which might offset the low leakage feature of the SESO transistor.

The fabricated SESO transistor has the gate length of 0.4 μm, the gate width of 0.5 μm, and the gate-oxide thickness of 25 nm. Because it was the first trial, all the parameters are set with good margin. When we talk about the 1.5-nm thickness of the fabricated device, sometimes one suspects that this successful operation is only seen in limited selected devices in a wafer or a lot. In reality, it is not. Basically all the devices fabricated work quite similarly. Based on nearly 10 years experience only nanosilicon formation for single-electron devices as is introduced above, the control of the LPCVD nanosilicon formation is quite stable and uniform in our process (this is due especially to the expertise of T. Mine).

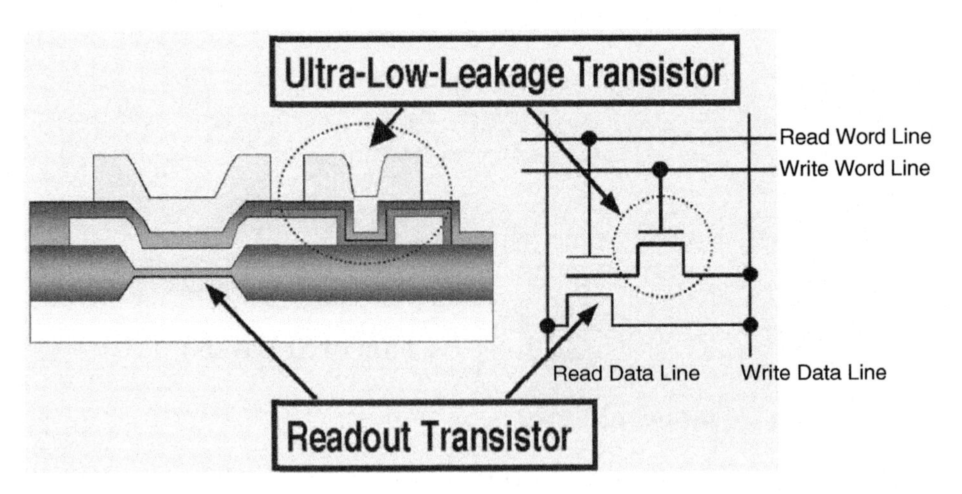

FIGURE 9.8 SESO memory cell circuit diagram and cross-sectional view.[1] Copyright 2000 IEEE.

In test devices, ultralow leakage characteristics, which have not been achieved with other devices, have been demonstrated. The leakage current under off condition is lower than the measurable level with a current meter (usually 1 fA is the low-current resolution limit). As it turned out, by data retention characteristics of the cell, the leakage current of the SESO transistor is as low as 10^{-19} A, that is, one electron leakage per second, which exhibits the near ideal switch characteristics. When one recalls that typical refreshing time of a DRAM is around 0.1 s, the electron leakage of the SESO transistor is less than one electron in a DRAM refreshing cycle. The transistor is named after this fact.

On current is about 10 nA when the gate and drain voltages are 2 V. This is, of course, much smaller than the bulk MOSFET. One reason for this low current is its low mobility. In this nanosilicon film, the grain size is very small. From Transmission Electron Microscope (TEM) observations, the grain size is comparable to the film thickness. Because of this, an electron has many scattering events in conduction. However, the required current for the memory operation is also as low as 10 to 100 nA. This is because the SESO transistor does not have to charge the heavy load capacitance like data line, but it charges a small bulk-MOS gate capacitance. And there is much to be improved in this first trial device, such as gate oxide, gate length, and channel thickness, and we are trying to achieve on-current improvement. One version of improved device characteristics is reported in Reference 13.

9.5 SESO MEMORY

The SESO transistor acts as an ideal switch for a memory cell as shown in Figure 9.8. The cell is a kind of dynamic memory cells having a two-transistor configuration. As will be discussed later, a three-transistor configuration is also possible, with better operating margin against various variations of fabrication and operating conditions, at the cost of somewhat larger cell area; however, for now, we use the two-transistor configuration for explanation. The two-transistor DRAM cell has the gain, which means that charge much larger than the stored one can be driven. This is why it is categorized as a gain cell.[14–19] Sometimes the terminology “gain cell” is misleading. Although it has gain, the stored charge cannot be much reduced as compared with that of a one-transistor one-capacitor cell. This is because the charge leakage time cannot be improved without improving the switch transistor leakage. Therefore, even with the gain cell, the large capacitance to store the charge is required.

In contrast, SESO memory stably retains the charge without special capacitance that requires complex fabrication process. The gate capacitance of a normal bulk-MOS transistor is used to store charge. This is less than 1 fF, which is two orders smaller than the typical DRAM cell capacitance, but still enables data retention as long as 10 seconds by the ultralow leakage characteristics.

The basic structure and the operation are as follows. The source-drain path of the SESO transistor is connected between the floating gate and the write data line. The gate voltage of the SESO transistor is called write word line. Write is done by charging (or discharging) the floating-gate capacitance through the SESO transistor with its gate voltage (or write word-line voltage) high. Typical write word-line voltage is 3V in this case, but in real applications this voltage will be lowered to around 2 V by thinning the gate oxide thickness of the SESO transistor. The write word-line voltage is 0 to 1 V depending on the write data value.

The source-drain path of the readout MOS transistor is connected between read data line and write data line, although there are some variations of the cell configurations. The gate of the readout transistor is the floating gate, the storage node, and this is capacitively coupled to read word line. Read is done by applying high level on the read word line. When the storage node is charged, the threshold voltage of the readout transistor is low and the data line is discharged by the readout transistor. On the other hand, when the storage node is discharged, the threshold voltage of the readout transistor is high, and the data line stays at the starting voltage, which is usually set by precharging operation of the read data line. The voltage difference at the read data line, induced by the storage node charge, is sensed by using a sense amplifier placed at the end of the read data line.

Here, let us quantitatively analyze the threshold voltage. The charge in the readout transistor channel is expressed by

$$Q_c = -C_{gco}V_g - (C_{fc}/C_{ff})(C_{gf}V_g - Q)$$

where C_{gco} is the direct fringe capacitance between the gate, that is, the read word line, and the channel, V_g is the read word line voltage, C_{fc} is the capacitance between the floating gate and readout transistor channel, C_{ff} is the total capacitance of the floating gate, C_{gf} is the capacitance between gate and the floating gate, and Q is the stored charge in the floating gate.

The threshold voltage Vth is the gate voltage that corresponds to a specific channel charge. By increasing Q by ΔQ, the threshold voltage is shifted by

$$Vth = (C_{fc}/C_{ff})(\Delta Q/C_{gc})$$

where C_{gc} is the capacitance between the gate and the channel and is given by:

$$C_{gc} = C_{gco} + (C_{fc}/C_{ff})C_{gf}.$$

The threshold voltage shift ΔVth is simplified, when the fringe capacitance C_{gco} is neglected, to:

$$\Delta Vth = \Delta Q/C_{gf}$$

As clarified by this equation, a large threshold voltage shift is easily achieved in SESO memory, because it is very easy to lower C_{gc} by simply placing the read word line far from the floating gate.

The operation of the SESO memory itself seems to be similar to that of the flash memory operation; however, the SESO memory operation is more sophisticated and stable, because it does not require the forced injection of charge through insulator by high voltage. To make this feasible, the electric field between the floating gate and the channel is stronger than the electric field between the floating gate and the word line. In the flash memory community, coupling ratio r is used to monitor the situation.

$$r = C_{gf}/C_{ff}$$

The coupling ratio should be high in flash memory for efficient charge injection. When coupling ratio is low, charge injection in flash memory is difficult. Instead, SESO memory injects charge through an independent gate-controlled switch, the SESO transistor, and very stable operation is possible.

Memory cell test device and evaluation results are shown in Figure 9.9 through Figure 9.14. The fabricated memory cell has a planar structure, because this is the first test device, however, stacked structure, in which a SESO transistor is formed on top of the read MOS transistor would be desirable to reduce cell size. Write and read operation voltages used in the experiments are shown in Figure 9.10. The current-voltage characteristics of the read transistor after write, with changes in write voltage, is shown in Figure 9.11. The threshold voltage of the read MOS is confirmed to correspond to the write voltage.

Figure 9.12 shows data retention characteristics. The normal write word-line voltage under data retention conditions is 0 V; however, positive voltages are applied to shorten the retention time intentionally. Indeed, the charge leakage is faster as higher write word-line voltages are applied; however, its retention time is long when the write word-line voltage is low as shown in Figure 9.13. Leakage current is estimated based on the retention characteristics as shown in Figure 9.14. Leakage current as low as 10^{-19} A is confirmed.

The basic array and peripheral circuit organization of the SESO memory can be designed with conventional CMOS circuitry, similar to those used in conventional memories. In this case the circuits are similar to DRAMs and flash memories.

The address signals are fed to row and column decoders, and appropriate word-line select or nonselect voltages and various control signals, such as sense amplifier select signal are generated. Differential sense amplifiers are used to sense the small signal at the read data line. The read data line signal is usually around 500 mV. The

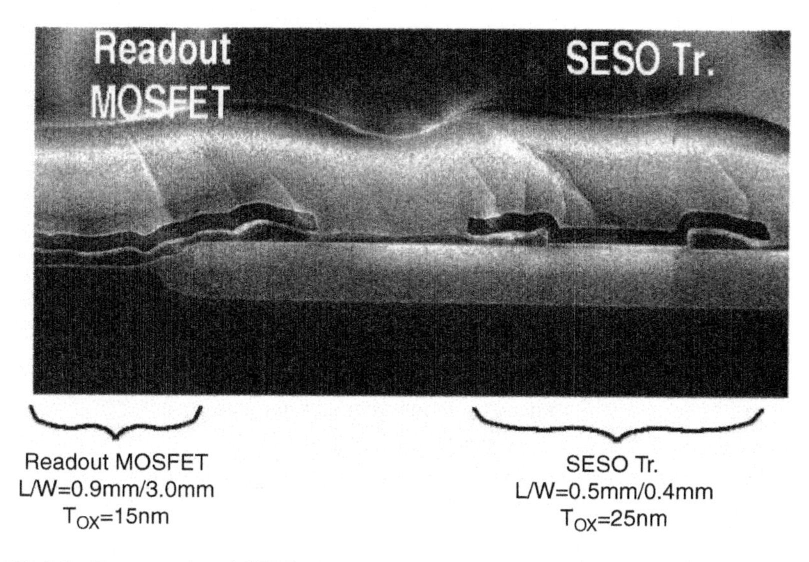

FIGURE 9.9 Cross-sectional SEM photograph of the fabricated SESO memory cell.

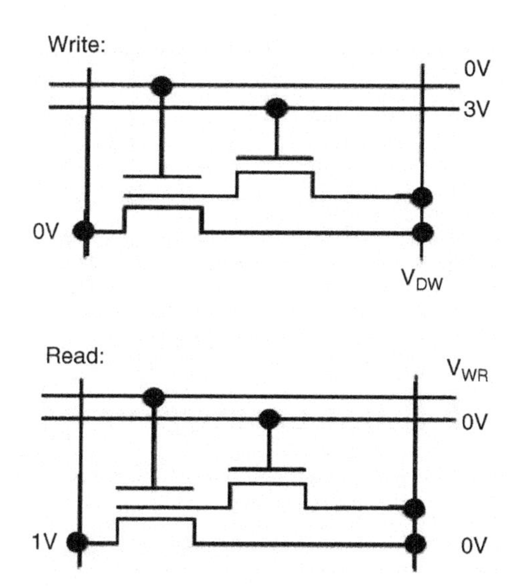

FIGURE 9.10 Write and read operation conditions of SESO memory cell.

differential sense amplifier has two inputs, in which one input is connected to the read data line, and the other input has to be set to a reference voltage. Usually the differential sense amplifier senses whether the voltage difference between two inputs are positive or negative. So making good reference voltage suitable for the cell characteristics, temperature and process variations is a very important issue in memory circuit design.

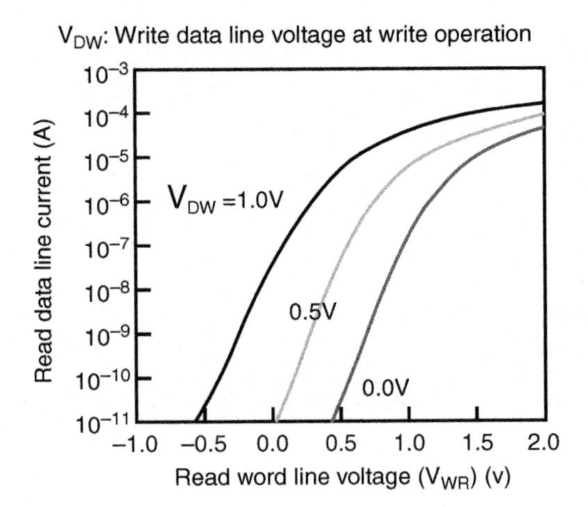

FIGURE 9.11 The threshold voltage shift of the readout transistor after write with different write-data-line voltage. Copyright 2000 IEEE.

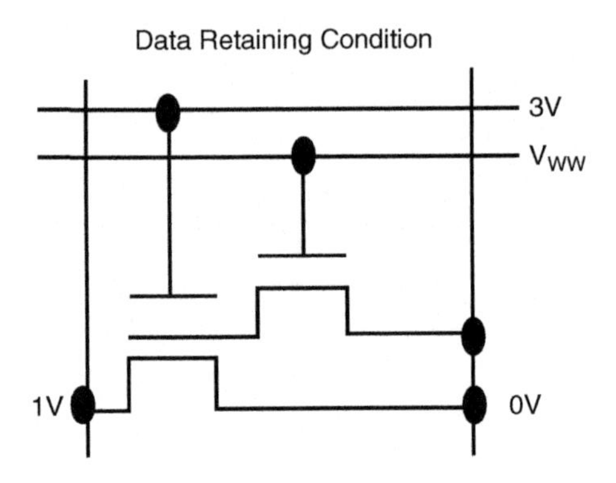

FIGURE 9.12 Data retention condition of the SESO memory cell.

9.6 MEMORY-TECHNOLOGY COMPARISON

SESO has the potential to become a memory beyond SRAM and DRAM (Table 9.2) suitable for mobile and ubiquitous applications.

First, SESO is expected to have very low standby power. Because of the long retention characteristics, the refreshing cycle time is as long as 10 sec, which allows us to enjoy 1-μA-class standby current. This is as low as those of special low-power SRAMs and lower than those of usual on-chip SRAMs.

Next, the cell area of SESO is small and SESO is suitable for high integration and low cost. SESO transistors can be formed on top of the readout MOS transistor, and therefore the effective transistor count is one. The area can be comparable to or even smaller than those of DRAMs. The fabrication process is simple and does not require the complex capacitor formation process of DRAMs. This is an advantage when SESO is integrated on the same chip with logic circuits.

TABLE 9.2
Comparison of SRAM, DRAM, and SESO Memory for Mobile Applications

Memory Type	Low Power	Large Capacity	Low Bit Cost
DRAM	No Good (~mA)	Good (64–256 Mb)	Good ($8F^2$)
SRAM	Good (~μA)	No Good (4–16 Mb)	No Good ($50–100F^2$)
SESO Memory	Good (~μA)	Good (>4 Gb possilbe)	Good ($4–20F^2$, 30% less steps)

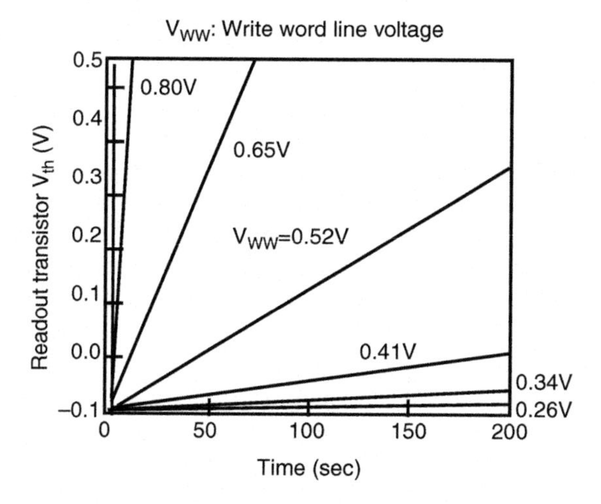

FIGURE 9.13 Dependence of threshold voltage of readout transistor on time. The parameter is write word-line voltage V_{ww} under the retention condition. Copyright 2000 IEEE.

SESO is expected to have very strong immunity against alpha-particle soft errors. It does not have a pn junction in storage node, and the very thin channel region is rarely hit by those particles.

SESO's advantage is more pronounced as we use smaller feature size to cram more bits or to lower cost. DRAM has difficulties in capacitance formation, especially in the 4-Gb-or-beyond region. SRAM has also suffered difficulty in achieving low power because of the leakage current problem. Soft errors are also very severe constraints in SRAM. SESO is free from all these problems.

9.7 SESO AS ON-CHIP RAM COMPONENT

As discussed in the introduction, on-chip RAM becomes important to achieve performance and low power expected by the market. In conventional memory architectures,

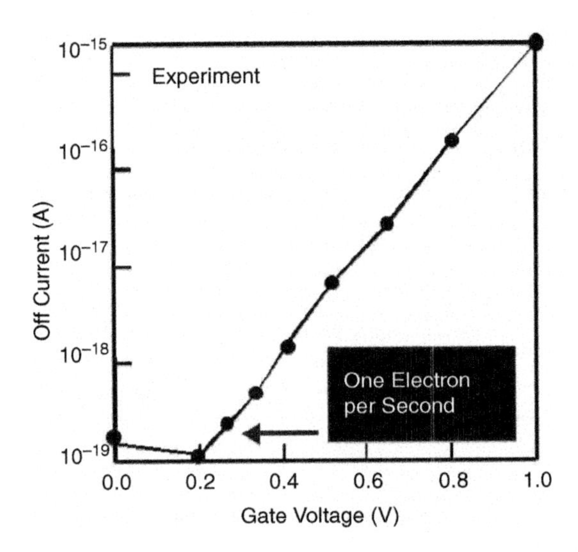

FIGURE 9.14 Off-current evaluated from date retention characteristics. Copyright 2000 IEEE.

memories are placed hierarchically. A small register file is placed close to an Arithmetic Logic Unit (ALU), then the first and second caches and main memories are put in place. As they are placed farther from the ALU, the memory capacity becomes larger. As processor architecture is enhanced to a more parallelized one, such as superscalar architecture, more data and instructions could be fed to these ALUs, which results in larger register files, larger cache memories, and faster bus architectures.

In the next phase, because the leakage problem of the transistors becomes serious and the general-purpose architecture improvement head room becomes smaller, application-specific memory and bus architectures, in which tight integration of ALU and memory configuration have to be pursued as shown in Figure 9.15. For example, processor architectures specialized for video and music media signal processing, such as MPEG and MP3, or for advanced human interface, such as face recognition, will be common. This trend drives tight integration of ALUs and memories, instead of separate CPU and memory chip organization.

SESO is very attractive when it is used for these on-chip RAM applications discussed above. The additional number of masks for SESO transistor formation is only 2 to 3. The very low power characteristics are suitable for a wide variety of mobile battery-operated volume markets.

One modification making on-chip integration easier is the use of a three-transistor cell, rather than a two-transistor cell[13] as shown in Figure 9.16. The application-specific architecture mentioned before requires a variety of memory configurations and short design and verification time. The three-transistor cell has a much larger margin against process and condition variations, because of the separate read transistor and select transistors. Large-scale memory design was done using this cell, and its effectiveness has been confirmed.

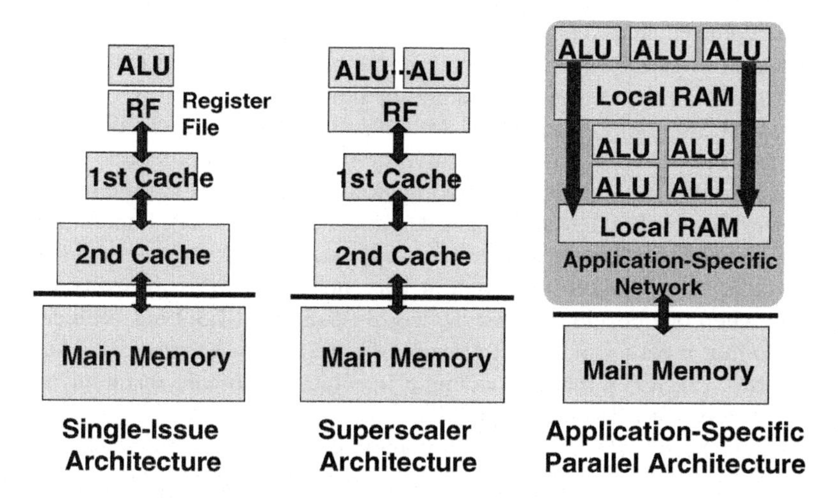

FIGURE 9.15 The change of computer architecture toward application-specific parallel processor/memory architecture.

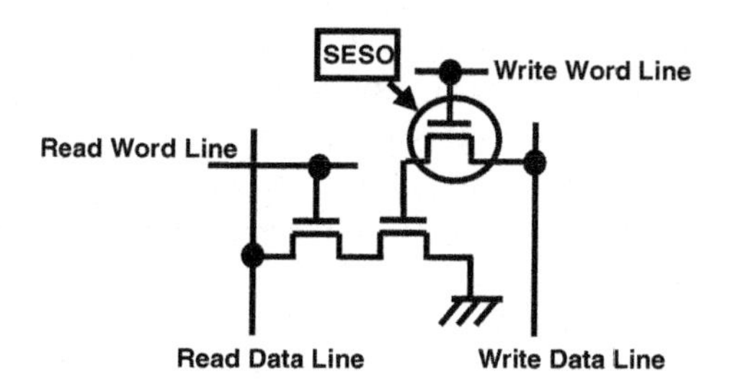

FIGURE 9.16 Three-transistor SESO memory cell.

9.8 CONCLUSIONS

SESO memory is unique in that it satisfies both low power and high density requirements, and it is being developed to target mobile and ubiquitous applications. Particularly, the μA-class standby power and the scalability are outstanding advantages.

Although a number of nanodevices for semiconductor chip innovations are proposed, most are positioned to be applied in the far future. By contrast, SESO is targeted for solving current crucial problems in the sub-100-nm regime.

The underlying strategy of SESO is to introduce a small change in the device (keep relying on conventional CMOS device process and circuits in other portions) and to enjoy great difference. We should not wait until 2030, when nanoscale lithography is to be achieved.

From now on, more concrete chip and system targets will be specified and practical use will come into sight. Remaining issues are to suppress variations when it is mass produced, and to confirm reliability under various practical environments. Not only are device improvements pursued, but also circuit innovations to support or complement the device behaviors are to be devised.

Indeed, SESO or any nanodevices are not panacea, and they should not have to be. The test they have to pass is to solve or improve the bottleneck part of the problem chain that real applications and markets require. This is difficult for most nanodevices; sometimes they are expected to improve much in one important parameter; however, if the other parameters are kept unimproved, the total market value of the product is not much improved, at least not as much as to justify the pain inevitable when one develops a new device. By contrast, SESO has been carefully checked so that it passes the test if remaining issues are to be solved. As far as the author knows, SESO is the first example in which nanoscale quantum effect is expected to solve Si-LSI's crucial problem.

Although the sub-100-nm region is commonly discussed with negative flavors and limitation arguments, I believe we live in an era in which many technology innovations will emerge.[20] Sometimes the fact that a simple scaling rule does not apply in the sub-100-nm region gives reason for the negative arguments; however, this is not a proper argument. For example, wire interconnection does not obey scaling rules as was clarified in the original Dennard's paper[21] thirty years ago. Now, 30 years later, there is no hint that any LSI or any business is severely damaged by the difficulty of wire scaling. Instead, low-k, CMP, Cu and many other wire technology innovations are evident in this field. Another example is DRAM. DRAM also is known not to obey scaling rules, but it still is the mainstream memory. Based on this thought, we should take scalability as the luxury of engineers, which alleviates the burden of making things go forward. But innovations go hand in hand with things difficult to scale. Based on this thought, SESO exhibits a good example of innovations for the sub-100-nm "Second-Golden Era of Innovations".[20]

ACKNOWLEDGMENTS

The author would like to show sincere appreciation to coworkers T. Ishii, T. Osabe, T. Mine, T. Watanabe, T. Sakata, B. Atwood, and F. Murai for their devotion to getting things done and inspiring discussions. Also the author is indebted to E. Takeda for his intriguing arguments and vision on nanodevices and the sub-100-nm region.

REFERENCES

1. Osabe, T., Ishii, T., Mine, T., Murai, F., and Yano, K., Single-Electron Shut-Off Transistor for Scalable Sub-0.1-um Memories, IEEE Intl. Electron Devices Meeting, 13.2, 301–304, 2000.
2. Likharev, K.K., Single-electron devices and their applications, *Proceeding of IEEE*, 87, 606–632, 1999.
3. Yano, K., Ishii, T., Sano, T., Mine, T., Murai, F., Hashimoto, T., Kobayashi, T., Kure, T., and Seki, K., Single-electron memory for giga-to-tera bit storage, *Proceeding of IEEE*, 87, 633–651, 1999.

4. Yano, K., Ishii, T., Hashimoto, T., Kobayashi, T., Murai, F., and Seki, K., Room-Temperature Single-Electron Memory Using Fine-Grain Polycrystalline Silicon, IEEE International Electron Devices Meeting, 541–545, 1999.
5. Yano, K., Ishii, T., Hashimoto, T., Kobayashi, T., Murai, F., and Seki, K., Room-temperature single-electron memory, *IEEE Trans. Electron. Devices*, 41, 1628–1638, 1994.
6. Yano, K., Ishii, T., Sano, T., Mine, T., Murai, F., and Seki, K., Impact of Coulomb blockade on Low-Charge Limit of Memory Device, IEEE Intl. Electron Devices Meeting, 525–528, 1995.
7. Yano, K., Ishii, T., Sano, T., Mine, T., Murai, F., and Seki, K., Single-Electron-Memory Integrated Circuit for Giga-To-Tera Bit Storage, IEEE Intl. Solid-State Circuits Conference, 266–267, 1996.
8. Ishii, T., Yano, K., Sano, T., Mine, T., Murai, F., and Seki, K., Verify: Key to the Stable Single-Electron Memory Operation, IEEE Intl. Electron Devices Meeting, 171–174, 1997.
9. Ishii, T., Yano, K., Sano, T., Mine, T., Murai, F., and Seki, K., 3-D Single-Electron Memory Cell with 2F2 Per Bit, IEEE Intl. Electron Devices Meeting, 924–926, 1997.
10. Yano, K., Ishii, T., Sano, T., Mine, T., Murai, F., Kure, T., and Seki, K., A 128-Mb Early Prototype for Gigascale Single-Electron Memories. IEEE Intl. Solid-State Circuits Conference, 344–345, 1998.
11. Yano, K., Ishii, T., Hashimoto, T., Kobayashi, T., Murai, F., and Seki, K., Transport characteristics of polycrystalline-silicon with influenced by single electron charging at room temperature, *Appl. Phys. Lett,*. 67, 828–830, 1995.
12. Yano, K., Ishii, T., Sano, T., Mine, T., Murai, F., and Seki, K., Synchronous single-electron transfer at room temperature, in Fujikawa, K., Ono, Y.A., Eds., *Quantum Coherence and Decoherence*, Elsevier Science B. V., Amsterdam, 1996.
13. Atwood, B., Ishii, T., Osabe, T., Mine, T., Murai, F., and Yano, K., A CMOS Compatible High Density Embedded Memory Technology for Mobile Applications, Symposium on VLSI Circuit, 154–155, 2002.
14. Shichijyo, H., Malhi, S.D.S., Shah, A.H., Pollack, G.P., Richardson, W.F., Elahy, M., Banerjee, S., Womack, R., and Chaterjee, P.K., TITE RAM: A New SOI DRAM Gain Cell for Mbit DRAM's. Ext. Abs. of 16th Intl. Conference on Solid-State Devices and Materials, 265, 1984.
15. Kim, W., Kih, J., Kim, G., Jung, S., and Ahn, G., An Experimental High Density DRAM Cell with a Built-in Gain Stage, *IEEE J. of Solid-State Circuit*, 29,978, 1994.
16. Shukuri, S., Kure, T., and Nishida, T., A Complementary Gain Cell Technology for Sub-1V Supply DRAMs, Intl. Electron Devices Meeting, 1006, 1992.
17. Terauchi, M., Nitayama, A., Horiguchi, F., and Masuoka, F., A Surrounding Gate Transistor (SGT) Gain Cell for Ultra High Density DRAMs, Symposium on VLSI Technology, 21, 1993.
18. Sunouchi, K., Fuse, T., Hasegawa, T., Matsubara, Y., Watanabe, S., and Horiguchi, H., A Self-Amplifying (SEA) Cell for Future High Density DRAMs, Intl. Electron Devices Meeting, 465, 1991.
19. Nakazato, K., Itoh, K., Mizuta, H., and Ahmed, H., Silicon stacked tunnel transistor for high speed and high-density random access memory gain cells, *Electronics Letters*, 35, 848, 1999.
20. Takeda, E., 21st Century CMOS Devices: Key Devices for E-Society, FUET, Oiso, Jan., 2001.
21. Dennard, R.H., Gaensslen, F.H., Yu, H., Rideout, V.L., Bassons, E., and LeBlanc, A.R., Design of ion-implanted MOSFET's with very small physical dimensions, *IEEE J. Solid State Circuits,* SC-9, 256, 1974.

10 Few Electron Devices and Memory Circuits

Kazuo Nakazato and Haroon Ahmed

10.1 INTRODUCTION

Over the last few decades, the performance of very large scale integrated (VLSI) circuits has been steadily improved by scaling down device dimensions. Today 65- to 90-nm lithography technologies have been introduced in production lines, and the size of a memory cell has been reduced to 0.02 μm^2. Two major semiconductor memories have been developed as high-density memories. One is the dynamic random access memory (DRAM), which is characterized by its unlimited number of write cycles and high speed. The other is flash memory characterized by its nonvolatility. Flash memory is rewritable read-only memory (ROM); the write cycles are limited, a block erase operation is needed before writing data, and verification is required to write data. DRAM and flash memories cannot be replaced by each other in applications. DRAMs are used as main memories of computers, and flash memories are used as an alternative to hard disks.

Memory architectures are now approaching fundamental difficulties. In this chapter we discuss the future possibilities for DRAM-type memories. For flash type memories, several new approaches have been proposed such as nanocrystal memories which use nanometer-scale islands instead of a floating gate to store charge. These approaches will be discussed in other chapters of this book. Since the approaches are very different for DRAM and flash memories, we will concentrate our attention on DRAM architecture which has several restrictions such as the need to maintain high speed and low supply voltage.

In section 2 of this chapter we discuss the difficulties of present day DRAMs and consider DRAM gain cells as one of the solutions. In Section 3 the phase-state low-electron number drive memory (PLEDM) is described as a high-density DRAM gain cell. The PLEDM has smaller cell size compared to a current DRAM cell and has scalability; that is, the stored charge can be reduced according to the reduction of the cell size. Further into the future, DRAM gain cells can be single-electron memories, where the precise number of electrons is controlled by multiple-tunnel junctions (MTJ). In the ultimate, single-electron memories can be based on just one electron representing one bit of information.

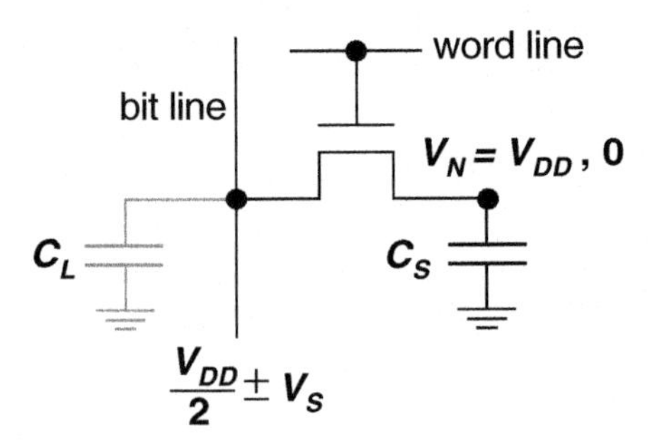

FIGURE 10.1 1-T DRAM cell.

10.2 CURRENT SEMICONDUCTOR MEMORIES

10.2.1 LIMITATIONS OF THE DRAM

The capacity of semiconductor memory chips has been continuously increased by more than six orders of magnitude (1 Kb to 4 Gb) over the last 30 years ever since the advent of DRAMs in the early 1970's.[1] DRAMs based on a one-transistor one-capacitor (1-T) cell,[2] as shown in Figure 10.1, are used as main memories in computers because of their high capacity and high speed. Since there is no gain in a 1-T cell, it requires a large cell capacitor C_S to produce a sufficiently large sense signal.

When the data are read, the bit line is precharged to a voltage $V_{DD}/2$ and then the pass gate transistor is opened. Corresponding to the voltage difference $\pm V_{DD}/2$ between the bit line and the memory node, the bit line voltage will be changed by an amount of $\pm V_S$. The signal V_S is given by $V_{DD}C_S / 2(C_L + C_S) \sim V_{DD}C_S / 2C_L$ where C_L is bit line load capacitance which is between 100 and 200 fF, nearly fixed at this value through all the DRAM generations. The signal V_S is sensed by a peripheral sense amplifier, and must be larger than the threshold voltage mismatch of the metal-oxide-semiconductor field effect transistor (MOSFET) pair in the sense amplifier, which is around 0.1 V. From this we deduce that C_S must be larger than 20 fF ($V_{DD} \sim 1V$).

Another requirement is that the cell charge $Q_S = C_S = C_S V_{DD}/2$ must be larger than the soft-error critical charge, which is the maximum charge collected at the cell storage node by an α particle or neutron hitting the memory cell. The particles generate electron-hole pairs inside the well region giving around 10 to 20 fC/μm along the particle path. The induced electrons enter the memory node and change the stored charge value. The soft-error critical charge is 10 to 20 fC when the well depth is 1 μm.

These effects limit the minimum value of cell capacitance C_S in a DRAM and the cell capacitance has been nearly constant through the DRAM generations, although the memory cell size A has been reduced continuously, as shown in Figure 10.2. To keep the cell capacitance fixed despite reductions in cell area, the capacitor structures and the fabrication processes become complicated. The cell capacitance is defined as $C_S = \varepsilon A_S / t_i$, where ε, A_S, and t_i are permittivity of capacitor insulator, surface area of capacitor electrode, and the insulator thickness, respectively. The

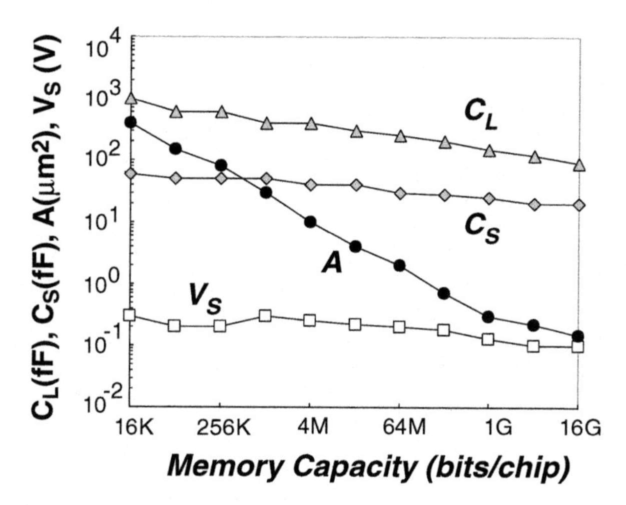

FIGURE 10.2 Trends of DRAM parameters.[3,4]

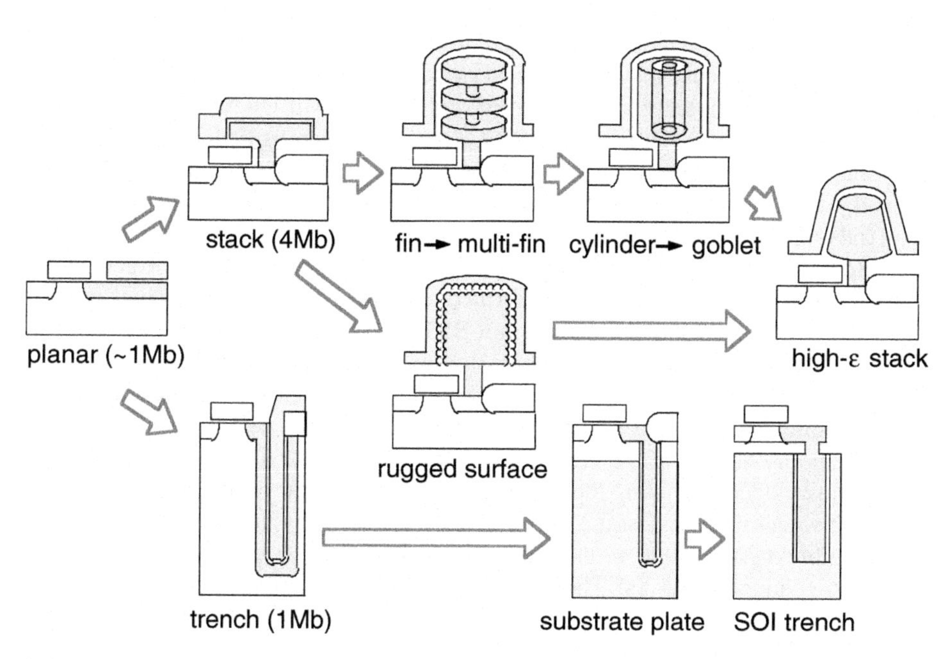

FIGURE 10.3 DRAM cell evolution.[4]

insulator thickness is limited by the electric breakdown field strength and cannot be reduced significantly. The surface area A_S can be increased by designing three-dimensional (3-D) cells such as stacked and trench capacitors. These were introduced in the Mb DRAM era, as shown in Figure 10.3. The area A_S of 3-D cells was essentially independent of the decrease in memory cell area, because capacitance was maintained by increasing the capacitor height, that is, the aspect ratio of the

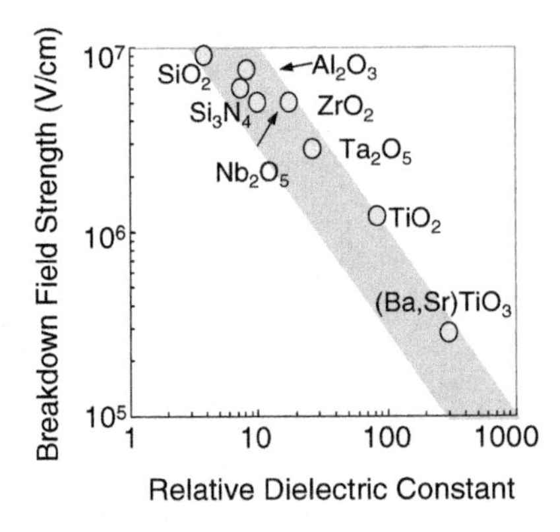

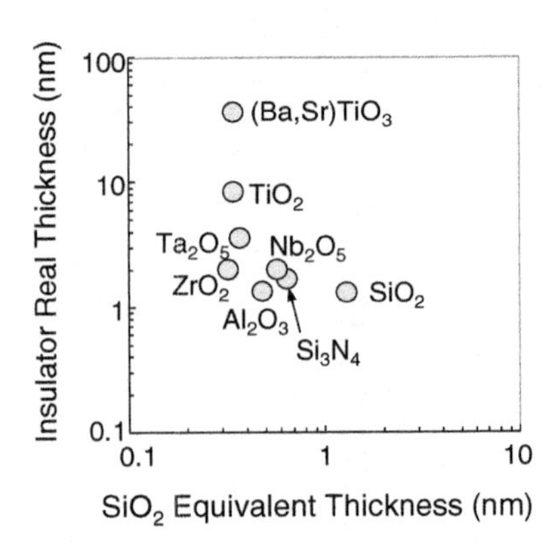

FIGURE 10.4 High-ε dielectric materials.[4]

structure. This requirement causes difficulty in fabrication. Hence A_S enhancement techniques such as hemispherical grain (HSG) or rugged polysilicon were introduced in the fabrication process which almost doubled the surface area. The employment of higher ε films was also investigated. At around 64-Kb, Si_3N_4 films were employed as the dielectric because they have almost twice the permittivity ε of SiO_2. The Ta_2O_5 dielectric again doubled ε and was employed for 256-Mb DRAMs. In the gigabit era, new materials such as BST($BaSrTiO_3$) have been investigated. We note that the maximum storage charge of the material is proportional to the product of ε and the breakdown field strength. One of the most serious problems is that high permittivity materials have generally lower breakdown field strength, as shown in Figure 10.4. Therefore, the thickness of the capacitor dielectric cannot be reduced, which means that high value capacitors cannot be formed as anticipated from the higher permittivity of the newer materials. Although a BST film has an extremely large permittivity, two orders of magnitude higher, the improvement in capacitance is only a factor of 3, and the thickness t_i increases by 40 times. Even with the new materials it is possible that the final stage of development of a 1-T cell will come when t_i reaches about one half of the feature size. At that stage the film will fill up the gap of a storage node, preventing the separation of the capacitor plates. For example, an SiO_2 equivalent thickness of 0.2 to 0.4 nm is targeted today in the high-ε films with physical thickness of 20 to 40 nm. Thus, when the feature size reaches 40 to 80 nm, a capacitor cannot be formed. The stage will be reached for DRAMs of 16- to 64-Gb.

10.2.2 DRAM Gain Cell

One of the solutions of the scaling problem is to use a gain cell, where the cell capacitor is replaced by a MOS transistor which amplifies the stored charge in each cell, as shown in Figure 10.5. The sense signal V_S can be increased from the 0.2V of the 1-T cell to $V_{DD} \sim 1$ V in a gain cell. The cell capacitance can decrease from the 20 fF of the 1-T cell to less than 1 fF, and becomes scalable, that is, the cell capacitance can be decreased following the miniaturization of the memory cell.

FIGURE 10.5 1-T to DRAM gain cell.

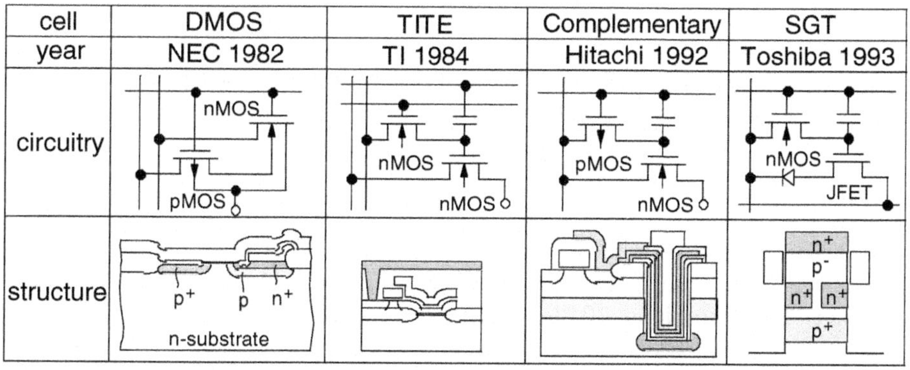

cell	DMOS	TITE	Complementary	SGT
year	NEC 1982	TI 1984	Hitachi 1992	Toshiba 1993
circuitry	nMOS, pMOS	nMOS, nMOS	pMOS, nMOS	nMOS, JFET
structure	p+, p, n+, n-substrate			n+, p-, n+, n+, p+

FIGURE 10.6 Proposed gain memory cells. DMOS,[5] TITE,[6] Complementary,[7] and SGT.[8]

It should be recalled that the early DRAMs were designed with gain cells. In 1-Kb DRAMs, initially four transistors with 3.5-line gain cell, and later three transistors with 4.5-line gain cell, were used. In 4-Kb DRAMs the memory cell changed to a three-transistor 3.5-line gain cell. However, these cell sizes were relatively large because of the large number of transistors and lines. From 16-Kb DRAM generations the present 1-T cell has been used because it consists of just one transistor and one capacitor with 2.5 lines.

After 16-Kb DRAMs several gain cells were proposed, as shown in Figure 10.6, but these were not used as commercial products. The main reason was that the cell size became larger, or the structure became too complicated to make a small cell. Small and simple gain cells are required to replace the 1-T cell for future generations in the gigabit era.

10.3 A NEW DRAM GAIN CELL — THE PLEDM

A phase-state low electron-number drive memory (PLEDM) was proposed using a stacked tunnel transistor (PLEDTR).[9–11] The memory cell is vertically structured and effectively occupies the area of just one transistor. The cell size is $5F^2$ where F is the minimum feature size, smaller than 6 to $8F^2$ of the 1-T cell. Its read and write times are simulated as 20 and 5 nsec, respectively. In principle, it is possible with PLEDM to have a retention time longer than 10 years,[12,13] enabling a nonvolatile memory to be realized.

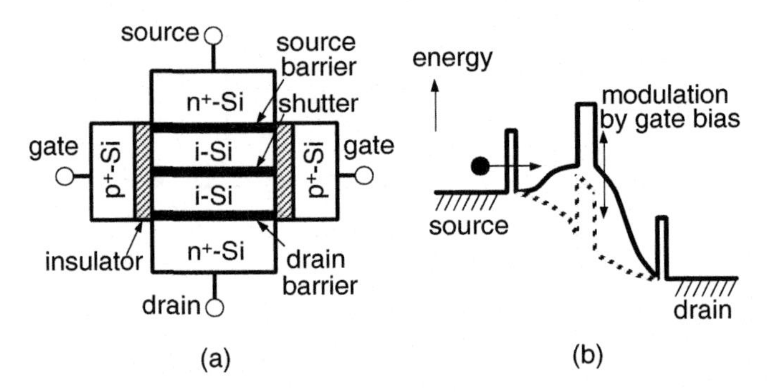

FIGURE 10.7 (a) Schematical cross section of PLEDTR. (b) Conduction band energy diagram showing device operating principle.

10.3.1 PLEDTR

PLEDTR is a vertical, fully depleted, double-gate SOI-MOSFET (silicon-on-insulator metal-oxide-semiconductor-field-effect-transistor) with barriers in the channel region, as shown in Figure 10.7(a). Gate voltage modulates the internal potential in the intrinsic silicon region and the central shutter barrier or barriers (CSB) also move up and down energetically following the internal potential, as shown in Figure 10.7(b). The CSBs reduce the OFF current substantially, while keeping a high ON current in the device. The device may be regarded as a three-terminal version of the heterostructure hot-electron diode (H^2ED) based on the transition from a tunneling current to a thermionic emission current at a semiconductor hetero-junction.[14] The role of source and drain barriers is (1) to adjust the source impedance to the CSB, (2) to act as diffusion barriers keeping a low impurity level within the channel, and (3) to reduce leakage current such as GIDL (gate induced drain leakage current) at the drain side.

The transistors with the double-tunnel barriers were fabricated on silicon dioxide using standard 0.2 μm silicon technology as shown in Figure 10.8.[15] All the transistor regions, source, drain, channel, and gate, are polycrystalline silicon films. The thin tunnel junctions were formed by thermal nitridation of silicon; after deposition of the silicon layer, the surface of the silicon is directly converted to silicon nitride by heating at 900°C for 3 min in an NH_3 ambient environment. The thickness of the nitride layer is self-limited to around 2 nm with a barrier height around 2 eV.[16] The source and drain regions were heavily phosphorous doped to 3×10^{20} cm^{-3}. The gate was boron doped to 5×10^{19} cm^{-3} to have a high threshold voltage. The gate separation width was 0.2 μm, and the gate width; a dimension perpendicular to the plane of Figure 10.7(a) was 0.4 μm. The gate insulator was formed by using 10-nm-thick silicon dioxide.

In order to modulate the internal potential by the gate, the relation between channel length L and gate separation width D is important. The strength of gate modulation can be represented by the subthreshold voltage slope s, which is approximately given by the scaling theory for a double-gate SOI MOSFET,[17]

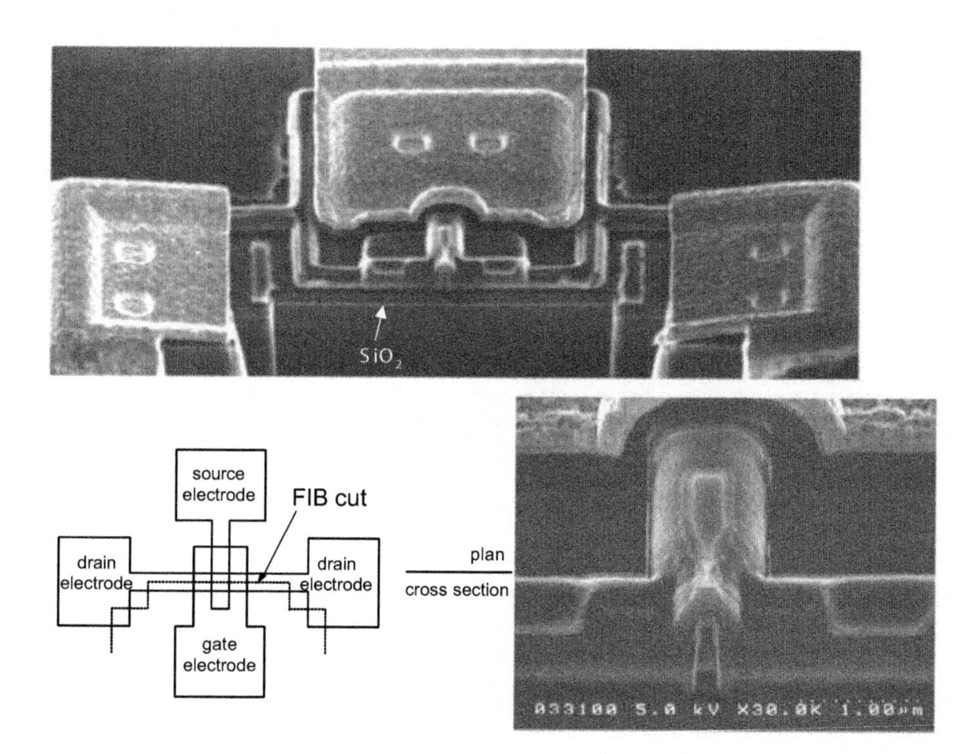

FIGURE 10.8 Scanning electron micrograph of fabricated PLEDTR.

$$s = \ln(10)\frac{k_B T}{e}\frac{1}{1-\cosh^{-1}(L/2\lambda)}, \tag{10.1}$$

where k_B is the Boltzmann constant, T is the absolute temperature, e is the absolute value of electronic charge, and λ is the scaling length given by

$$\lambda = \sqrt{D^2/8 + Dt_{ox}\varepsilon_{Si}/2\varepsilon_{SiO_2}}, \tag{10.2}$$

ε_{Si} and ε_{SiO_2} are permittivities of silicon and silicon dioxide, respectively. In order to obtain $s < 80$ mV/decade at room temperature, $L/2\lambda$ must be greater than two. With the gate separation width $D = 0.2\,\mu$m and gate insulator thickness $t_{ox} = 10$ nm, $\lambda \sim 90$ nm, the channel length L was designed to be 350 nm to obtain good current-voltage characteristics.

Drain current was measured as a function of gate voltage for several drain voltages as shown in Figure 10.9. The threshold voltage was 0.5 to 1 V, and subthreshold voltage swing s of 100 mV/decade was obtained. The ON current has a strong dependence on V_{DS} and weak dependence on V_{GS} since the drain current is determined by tunnel current through the source barrier. The OFF current is rather

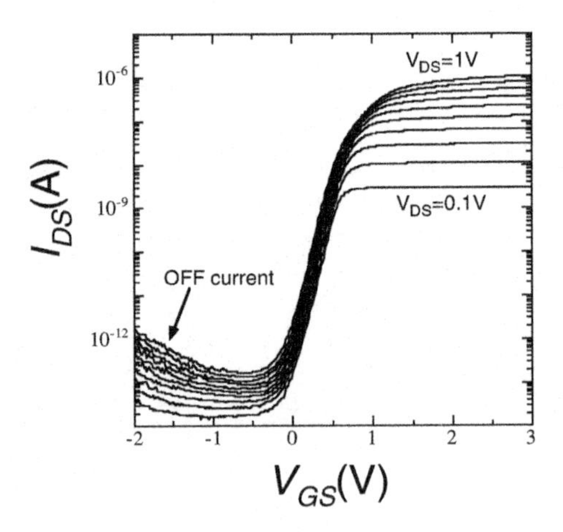

FIGURE 10.9 Measured drain currents. V_{DS}: 0.1 V step.[15]

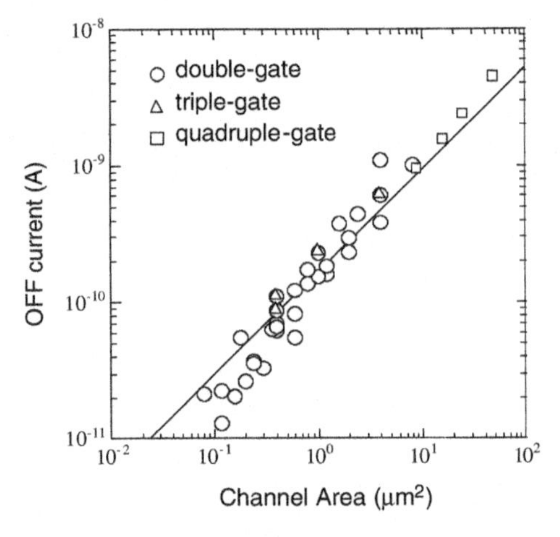

FIGURE 10.10 Channel area dependence on OFF current (V_{GS} = −2 V, V_{DS} = 1.5 V).[15]

high, about 1 pA. The channel area dependence of the OFF current was measured for several transistors as shown in Figure 10.10, which indicated that the OFF current is dominated by the current through the bulk region in the channel.

The channel region was maintained at a low impurity level, lower than 4×10^{17} cm^{-3} as confirmed by SIMS analysis (Figure 10.11). The silicon nitride films act as very effective impurity diffusion barriers between the source and drain regions and the channel region. However, this channel impurity concentration is still higher than the level required to achieve full depletion in the channel region in the OFF state. The maximum depletion length is around 50 nm at a channel concentration of 4×10^{17} cm^{-3}. To pinch off the channel region completely, the gate separation width D must be less than the maximum depletion length, and the channel concentration must be lower than 3×10^{16} cm^{-3} for D = 0.2 μm.

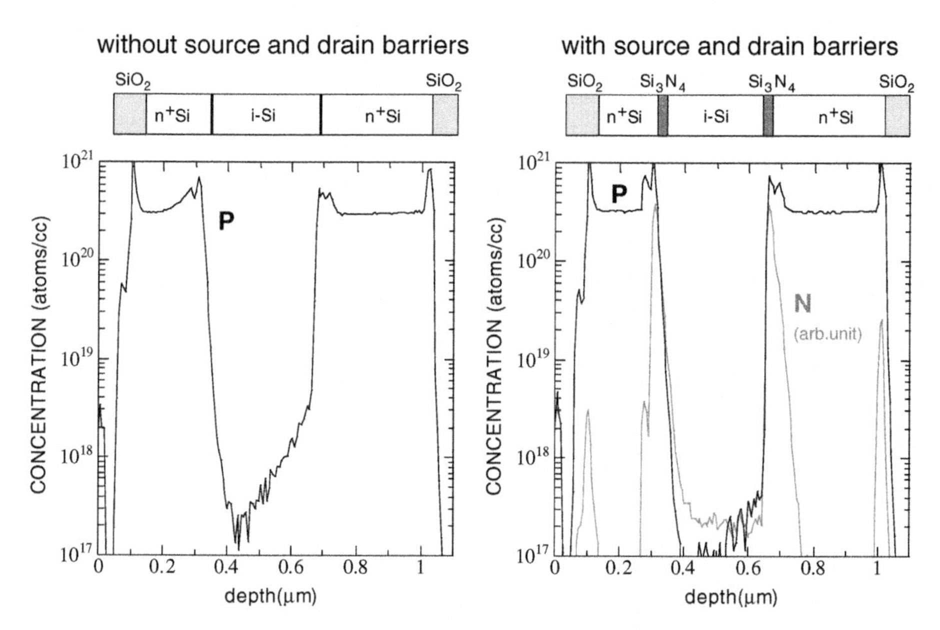

FIGURE 10.11 Distributions of phosphorous atoms in PLEDTR measured by SIMS without (left) and with (right) source and drain barriers formed by rapid thermal nitridation at 900°C.

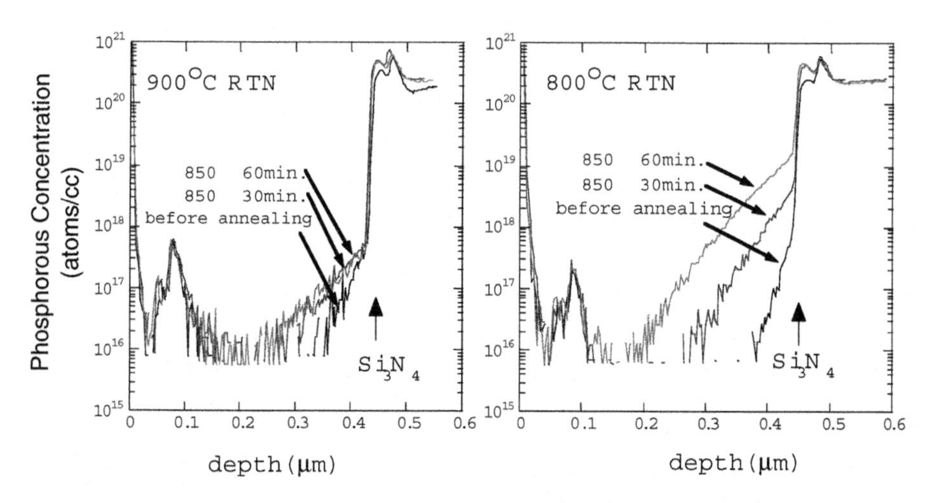

FIGURE 10.12 Diffusion of phosphorous by thermal annealing. The Si_3N_4 barriers are formed by rapid thermal nitridation at 900°C (left) and 800°C (right).

The change of phosphorous distribution by thermal annealing was investigated for a wafer before forming a heavily doped polysilicon layer for the source region. As shown in Figure 10.12, auto-doping of phosphorous atoms was detected at boundaries of the intrinsic polysilicon layers, and the amount of phosphorous atoms was not changed by thermal annealing. These results indicated that the origin of the

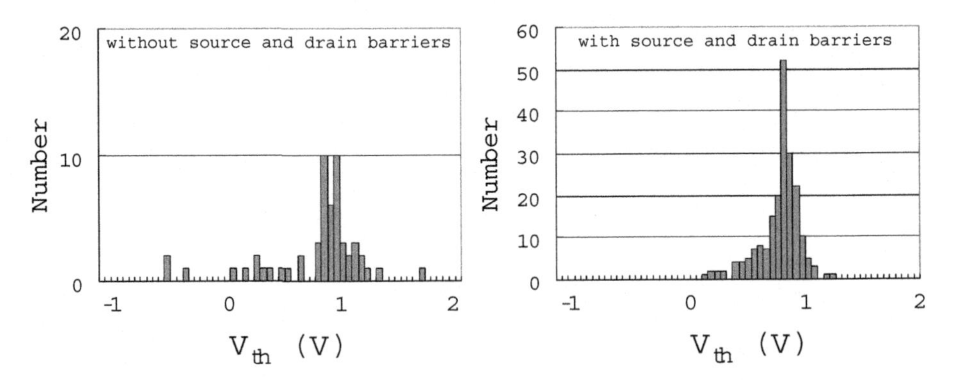

FIGURE 10.13 Distribution of threshold voltage of transistors without and with tunnel barriers.[15]

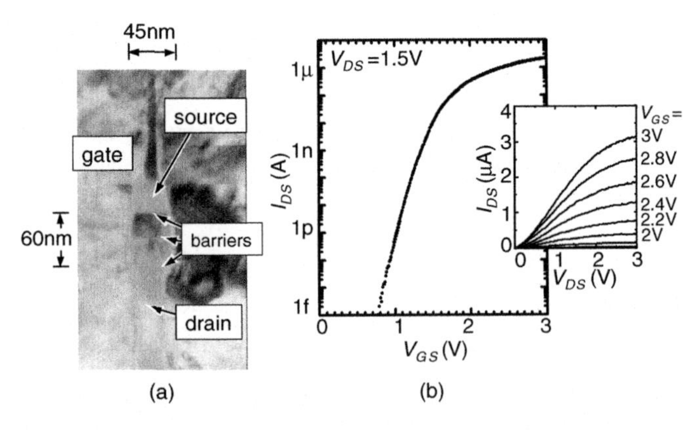

FIGURE 10.14 (a) Transmission electron micrograph of 60 nm channel PLEDTR. (b) Measured drain currents.[10]

residual impurities in the channel region is not from diffusion from the source and drain regions, but contaminants of Si Chemical Vapor Deposition (CVD) equipment.

The threshold voltage distribution was compared for transistors with and without the silicon nitride tunnel barriers in Figure 10.13. In the transistors without the tunnel barriers the channel impurity concentration cannot be controlled, and so the variation of threshold voltage is very large. On the other hand, the transistors with the source and drain tunnel barriers were found to have well-controlled threshold voltages.

PLEDTRs with 60-nm channel and triple tunnel barriers were also fabricated, as shown in Figure 10.14(a). The channel length L was 60 nm and the gate insulator was formed by 6 nm of silicon dioxide. The gate separation width D was 45 nm determined by TEM observation, and the gate width was 0.4 μm. Drain current was measured as a function of gate and drain voltages as shown in Figure 10.14(b). The leakage current was less than 1 fA, which is the limit of the sensitivity of our measurement system. The low leakage current is obtained because the gate separation width is shorter than the maximum depletion length, even with rather high channel

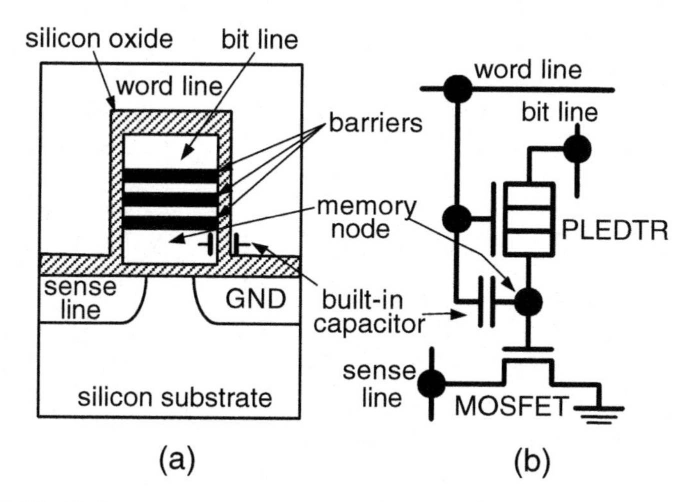

FIGURE 10.15 (a) Cross section of PLEM gain cell and (b) equivalent circuit diagram.[11]

concentration. The subthreshold voltage slope of 96 mV/decade observed is explained well by scaling theory Equation (1), giving 130 mV/decade, with $L/2\lambda \sim 1.2$.

10.3.2 PLEDM Cell

PLEDTR enables the construction of a high-density memory because each memory cell occupies the area of just one transistor, as shown in Figure 10.15. A PLEDTR is stacked onto the gate of a conventional MOSFET with a built-in coupling capacitor to realize a memory cell. High-speed write is possible by transferring electrons from the top electrode (bit line) onto the memory node through the ON-state PLEDTR. Since the OFF-state PLEDTR can confine electrons very effectively, the stored information can be kept for a long time without a refresh operation. Since the information is read via the current in a MOSFET, this cell has gain and a large S/N ratio.

Standby, read, and write cycles are all controlled by voltage V_W on the word line, $V_W^{(S)}$ (–2V), $V_W^{(R)}$ (0.5V), and $V_W^{(W)}$ (3V), respectively (Figure 10.16). The generation of negative word line voltage is described in Reference 18. In the standby cycle the built-in coupling capacitor C_C causes the memory node voltage V_N to be lower than the threshold voltage V_{th} of the sense MOSFET. In the read cycle V_N becomes higher than V_{th} when the memory state is high, and lower than V_{th} when the memory state is low. In the write cycle the PLEDTR is opened, and V_N becomes the bit line voltage, 1.5 V for the high memory state and 0 V for the low memory state. Figure 10.16(b) and (c) show the results of a mixed-level device and circuit simulation of two memory cells designed using 0.13 μm design rule. A sequence of writing high state (W_H), standby (S), read (R), refresh (r), S, R, writing low state (W_L), S, R, r, S, R, with 10 nsec/20 ns/5 nsec standby (pre-charge) /read/write time is simulated. V_{DD} = 1.5 V. V_{W2} is kept at –2 V (unselected) after writing high or low state. The refresh inverts the memory state. The inverting cell concept is described in Reference 19. Although the drain-source current in the ON state of a PLEDTR

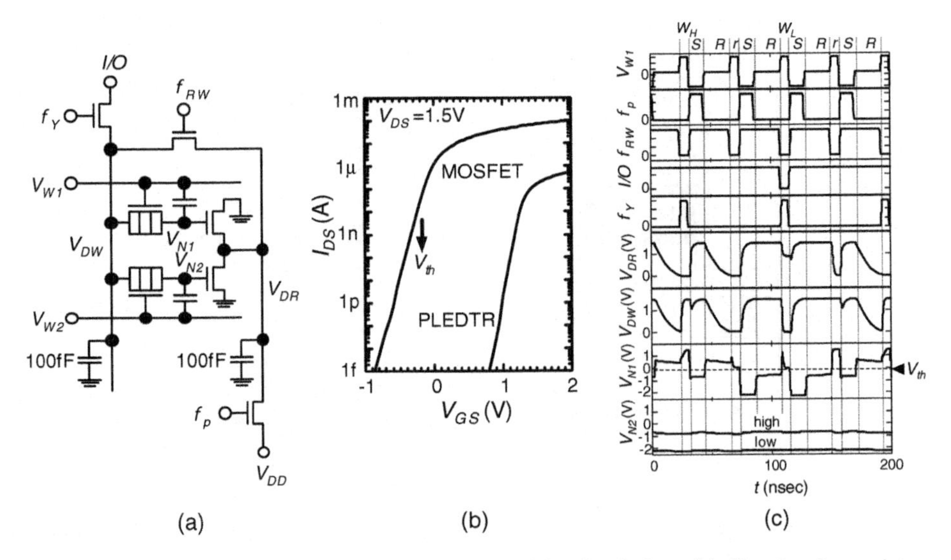

FIGURE 10.16 (a) Schematic circuit diagram used in simulation. (b) Simulated transistor characteristics in memory cell using 0.13 μm design rule. (c) Simulated waveforms.[10]

is small, around 1 μA, high-speed write can be realized because of the reduced stored charge, which is determined by the gate capacitance of the sense MOSFET and estimated as 0.2 to 0.3 fC. On the other hand a high ON current is available from the sense MOSFET to drive the bit line capacitance, 200 fF in this simulation.

The memory node voltage V_N in the standby and read cycles was calculated as a function of the coupling capacitance C_C in Figure 10.17. $V_H^{(R)}$ and $V_L^{(R)}$ are, in the read cycle, in high and low memory states, respectively. $V_H^{(S)}$ and $V_L^{(S)}$ are in the standby cycle. The voltage difference on the memory node between high and low memory states, $V_H - V_L$, can be larger than the writing voltage difference, 1.5 V in this case, because of the change of memory node capacitance between inversion and depletion states of the sense MOSFET. Random read access in a cell array is possible when V_{th} is set inside the hatched area, for example, between 0.5 V and –0.5 V at a coupling capacitance of 0.04 fF. This coupling capacitance can be realized for a 50-nm-thick memory node (t_N), without needing to form an additional capacitor.

The schematic circuit diagram of a memory device and the layout of the memory cell with $5F^2$ cell size are shown in Figure 10.18. The refresh circuit consists of just one transistor per column with the same pitch as the memory cell, without sense amplifiers, so the cell area occupancy ratio increases substantially. The fabricated PLEDM cell array is shown in Figure 10.19.

10.4 SINGLE-ELECTRON MEMORY

Reduction of the cell size of semiconductor memory enhances the storage capacity. However, size reduction results in a decrease in the number of electrons that determine the memory cell state, and thus fluctuations in this number become relatively large so that the electrons cannot be controlled by conventional methods. Such a situation

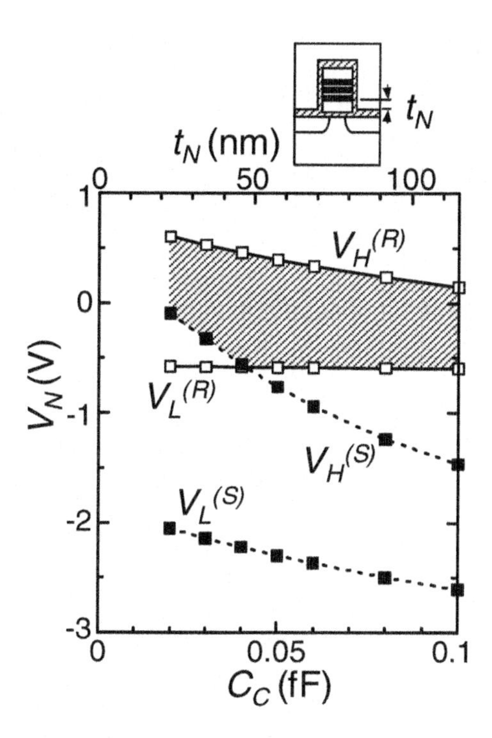

FIGURE 10.17 Simulated memory node voltages.

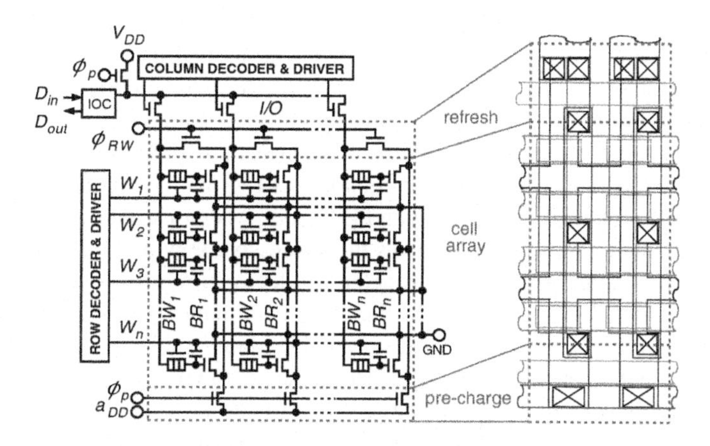

FIGURE 10.18 Schematic circuit diagram of memory device and the layout.

will arise when the number of electrons for operation becomes less than 100. A new memory principle was proposed in which individual electrons are fully controlled by the Coulomb blockade effect.[20,21]

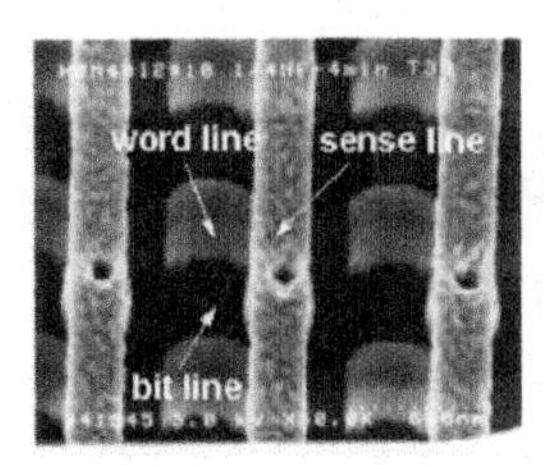

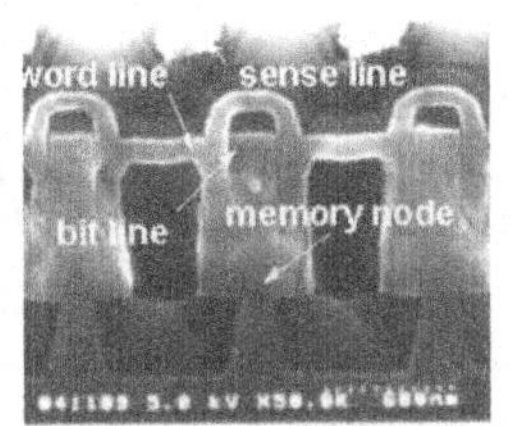

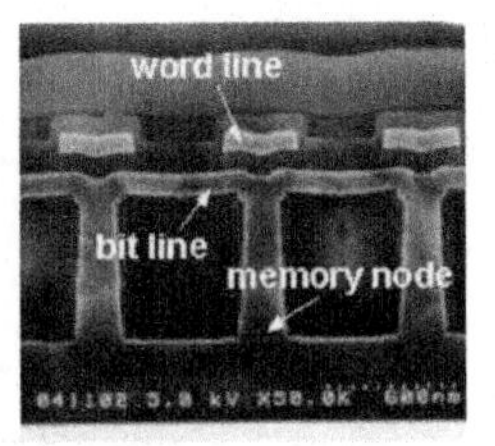

plan view cross sectional view

FIGURE 10.19 Scanning electron micrographs of fabricated PLEDM cell array.

10.4.1 Single-Electron Devices

Single-electron charging effects have attracted attention since Likharev pointed out the possibility of single-electronics[22–25] and because modern nanotechnology enables us to fabricate very small structures.[26,27] Single-electronics is based on the so-called Coulomb blockade. If an electron tries to enter a small isolated region the electrostatic energy of the region would increase; thus the electron cannot enter if the charging energy, $e^2/2C$ is larger than the thermal energy k_BT, where C is the island capacitance. Since the capacitance of the island is roughly proportional to its linear dimensions, this effect can be observed only in very small structures and at very low temperatures for reasonable magnitudes of capacitance. For example, the capacitance of an island of 10 μm circumference is of the order of 1 fF so that the temperature must be less than 1 K for any single-electron effects to be observed. In nanoscale structures of around 10 nm in circumference, the capacitance can be reduced to the order of 1 aF and single-electron effects may be observed at room temperature.

The basic electrical characteristic caused by single-electron charging effects is the Coulomb gap. When a voltage difference V is applied to a suitable system, an electron can obtain an energy eV by movement. But, only when this energy is greater than the charging energy, that is, only if $e|V| > e^2/2C$, can the electron pass through the system. Thus, the system resistance is high in the Coulomb blockade regime, $-e/2C < V < e/2C$, and becomes low outside this regime. Using this effect, Fulton and Fulton and Dolan,[28] and Kuzmin and Likharev,[29] demonstrated a single-electron transistor in which gate voltage controls the successive flow of single electrons, and Geerligs et al.[30] demonstrated the transfer of electrons one by one, synchronized with an external AC gate voltage. For the static confinement of discrete numbers of electrons, a single-electron box with one stable state[31] and a single-electron trap with several stable states[32,33] have also been demonstrated.

One of the most important elements for utilizing such charging effects is the Multiple-Tunnel Junction (MTJ) in which a series of small islands is formed.[34] The charging energy of the island creates an energy barrier which blocks the entrance of electrons into the MTJ so that multistable states of different numbers of electrons can be formed. The MTJ is also important in suppressing co-tunneling effects; that is, electron tunneling simultaneously across more than one junction.[35–38] A direct

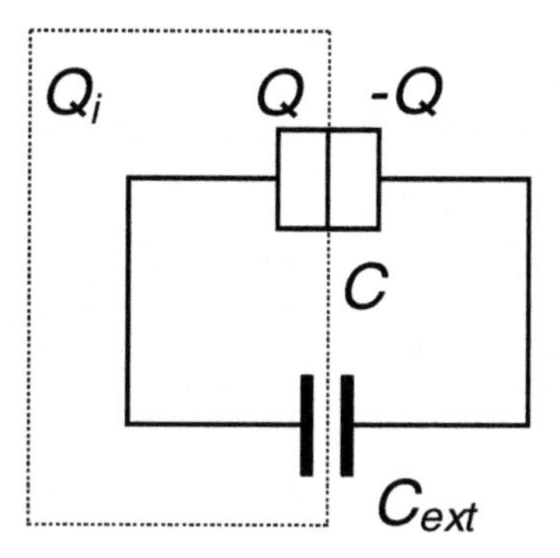

FIGURE 10.20 Equivalent circuit for the calculation of tunneling rate through a junction.

application of such MTJs is the single-electron memory in which one bit of information can be represented by the excess or shortfall of a precise number of electrons.

10.4.2 Operation Principle of Single-Electron Memory

10.4.2.1 Local Stability

The single-electron effect is based on two principles: tunneling occurs in units of single-electrons, and secondly, that the tunneling rate is determined by the amount of change of charging energy before and after the tunneling. It is assumed that the tunnel resistance is larger than the resistance quantum $R_K = h/e^2 \approx 26$ kΩ, where h is Planck's constant, to avoid smearing by quantum fluctuations and to ensure electron localization in islands. It is also assumed that the electromagnetic environment of leads is of low impedance. In the first approximation, tunneling occurs through one tunnel junction at a time, that is, the simultaneous tunneling of electrons across several junctions is neglected in this approximation. Therefore, when the tunneling through a junction is considered, the other junctions behave simply as capacitors and the tunneling rate is determined by the equivalent circuit as shown in Figure 10.20, in which Q is the charge on one side of the tunnel junction, C is its capacitance, and C_{ext} is the equivalent capacitance in parallel with the tunnel junction. The charging energy of the island with charge Q_i is given by $E = Q_i^2/2(C + C_{ext})$. When an electron enters the island through the tunnel junction, the charge of the island changes from Q_i to $Q_i - e$, and the charging energy increases by $\Delta E = \{(Q_i - e)^2 - Q_i^2\}/2(C + C_{ext})$. Using a relation $Q = CQ_i/(C + C_{ext})$ the difference of charging energy is given by

$$\Delta E = e\left(Q_c - Q\right)/C \tag{10.3}$$

where the critical charge Q_c of the tunnel junction is given by

$$Q_c = \frac{e}{2\left(1 + C_{ext}/C\right)}. \tag{10.4}$$

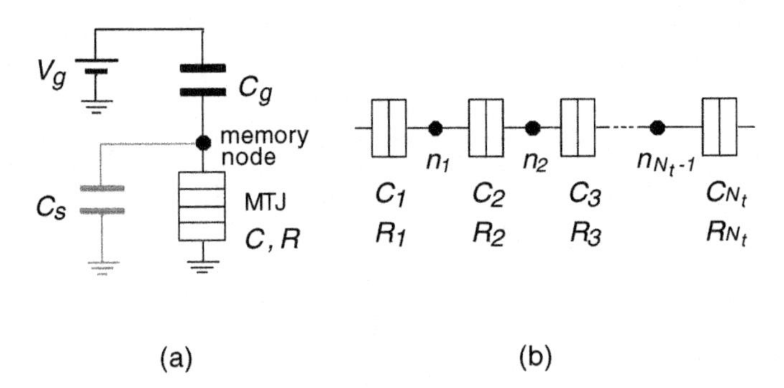

FIGURE 10.21 Circuit diagram of (a) principal parts of a single-electron memory and (b) the multiple-tunnel junction(MTJ).

By the same argument, when an electron exits the island, the charging energy increases by

$$\Delta E = e\left(Q_c + Q\right)/C \tag{10.5}$$

In the Coulomb blockade regime, $-Q_c < Q < Q_c$, the energy increases both by the entrance and the exit of an electron and thus tunneling through the junction is suppressed.

Equations (10.3) through (10.5) are known as the global rule,[30] which is written in terms of the parameters of a tunnel junction, but the critical charge Q_c is determined by the properties of the whole circuit. Since the capacitance C_{ext} is obtained by a linear circuit calculation, the global rule gives a systematic method for calculating the characteristics of single-electron devices.

We now consider a single-electron memory circuit, the principal parts of which consist of a gate-capacitor and an MTJ as shown in Figure 10.21(a). Electron transfer to or from the node is possible only through the MTJ, which is formed by N_t tunnel junctions connected in series as in Figure 10.21(b). In this section we assume for simplicity that all tunnel junctions in the MTJ have the same capacitance. A more general case will be discussed in the next section. The passage of single-electrons through the MTJ is blocked when the modulus of the charge Q on one side of this device is less than the critical charge given by

$$Q_c = e\frac{C}{C_\Sigma}\frac{1+\Delta}{2}. \tag{10.6}$$

Here C_Σ is the total capacitance $C + C_g + C_s$, where C is the capacitance of the MTJ, C_g is the gate capacitance, and C_s is the stray capacitance; Δ determines a multistate condition given by

$$\Delta = \left(1 - \frac{1}{N_t}\right)\left(\frac{C_\Sigma}{C} - 1\right) \tag{10.7}$$

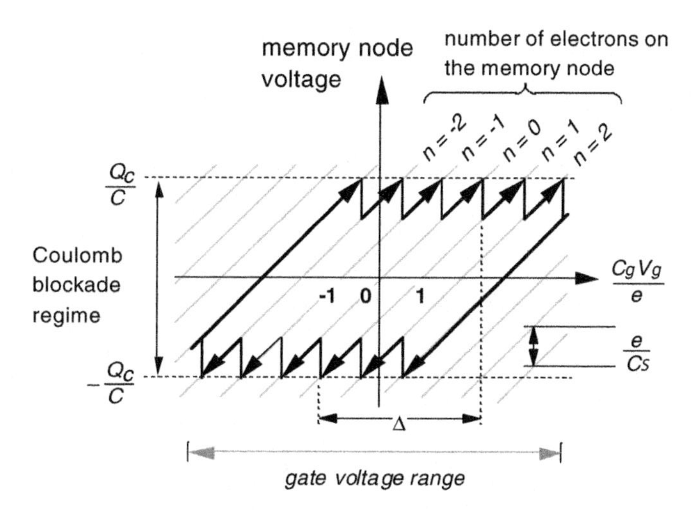

FIGURE 10.22 Principle of operation. The memory node voltage is plotted as a function of gate voltage for cyclic operation.

The memory node voltage V depends both on the voltage V_g applied to the gate and on the charge stored on the node,

$$V = \frac{e}{C_\Sigma}\left(\frac{C_g V_g}{e} - n\right) \tag{10.8}$$

where n is the number of excess electrons on the node. Equation (10.8) is plotted in Figure 10.22 where cyclic operation of gate voltage is assumed. Within a Coulomb blockade regime, $-Q_c/C < V < Q_c/C$, electrons cannot enter or exit the memory node. When V reaches the boundary of this Coulomb blockade regime, one electron enters or leaves to keep the electron state inside the Coulomb blockade regime. By applying a gate-voltage pulse V_g with magnitude larger than $e\Delta/C_g$, the number of electrons on the memory node can be changed. In Figure 10.22 the lower branch corresponds to a stable state with an excess of two electrons, and the upper branch to a stable state with a shortfall of two electrons from the charge neutrality state. In general, one bit of information can be represented by $+N$ and $-N$ electron number states, where N is given by the integer part of $(\Delta + 1)/2 \sim C_\Sigma/2C$. If the capacitances are chosen to satisfy the condition $\Delta < 1$, a binary code can be represented by the presence or absence of one electron.[25] The operating scheme has a large operational range since it is free from the offset charge on the memory node.

The master equation for a configuration of electrons $\{n\} = (n_1, n_2, \ldots, n_{Nt})$ is given by

$$\frac{\partial}{\partial t}P(\{n\}) = \sum_{\{n'\}}\Gamma(\{n'\}\to\{n\})P(\{n'\}) - \sum_{\{n'\}}\Gamma(\{n\}\to\{n'\})P(\{n\}). \tag{10.9}$$

In orthodox theory[24] the transition probability Γ is given by a single-electron tunnel process through one tunnel junction i and is determined by its tunnel resistance R_i and the difference of energy ΔE before and after tunneling,

$$\Gamma(\{n\} \to \{n'\}) = \frac{1}{e^2 R_i} \frac{\Delta E(\{n\} \to \{n'\})}{\exp\left(\Delta E(\{n\} \to \{n'\}/k_B T) - 1\right)}. \tag{10.10}$$

One of the parameters which characterize the single-electron memory cell is the area of the hysteresis loop defined by $A_h = \oint V dV_g$. From Equation (10.10) the maximum memory node voltage is given by ~ $e/2C + (k_B T/e)\ln(CR_i\Gamma)$. Thus the following expression gives a good approximation for the area of the hysteresis loop,

$$A_h = A_{h0}\left\{1 + (a/\gamma)\ln(v_s/v_{s0})\right\} \tag{10.11}$$

where A_{h0} is the area of the hysteresis loop at absolute zero temperature, v_s is the sweep rate of gate voltage and v_{s0} is given by

$$v_{s0} = \frac{e}{C_g \tau}. \tag{10.12}$$

In Equation (10.11) and Equation (10.12) the parameter $\gamma = e^2/(2Ck_B T)$ and τ is a time constant of the MTJ, $\tau = N_t CR$, where R is the total resistance of the MTJ. Calculation by the Monte Carlo method showed that Equation (10.11) determines the dependence of sweep rate and temperature over a wide region for $v_s < v_{s0}$ with parameter a around 4.5.[39]

10.4.2.2 Global Stability

We now consider the global stability of the states in a single-electron memory cell. For this purpose the charging energy is more suitable rather than the energy difference,

$$\Delta E(\{n\} \to \{n'\}) = E(\{n'\}) - E(\{n\}) \tag{10.13}$$

where $\{n\}$ and $\{n'\}$ are electron states before and after tunneling, respectively. In the single-electron memory case, the charging energy is obtained as

$$E = \frac{e^2}{2C_\Sigma}\left(\frac{C_g V_g}{e} - \sum_{i=1}^{N_t} x_i n_i\right)^2 + \frac{e^2}{2C}\sum_{i=1}^{N_t-1}\sum_{j=1}^{N_t-1} x_{\min(i,j)}\left(1 - x_{\max(i,j)}\right) n_i n_j \tag{10.14}$$

where $x_i(0 < x_1 < x_2 < \ldots < x_{N_t} = 1)$ is the position of the i-th island in the MTJ weighted by the capacitance of each tunnel junction defined in Figure 10.21(b),

$$x_i = C\sum_{j=1}^{i}\frac{1}{C_j} \tag{10.15}$$

and C is the total capacitance of the MTJ,

$$\frac{1}{C} = \sum_{j=1}^{N_t}\frac{1}{C_j} \tag{10.16}$$

The memory node voltage V is given by

$$V = \frac{e}{C_\Sigma}\left(\frac{C_g V_g}{e} - \sum_{i=1}^{N_t} x_i n_i\right) \tag{10.17}$$

where n_{N_t} is the number of electrons on the memory node n. The first term of Equation (10.14) is the charging energy of the memory node. The electrons in the MTJ induce charge $ex_i n_i$ on the memory node through the MTJ's capacitance. The charging energy of the memory node takes a minimum value when the electronic charge on the memory node compensates for the induced charges by the gate voltage and electrons in the MTJ, that is, when the charge neutrality condition is satisfied as far as possible. The second term of Equation (10.14) is the charging energy of the MTJ,[40] which is determined by only the MTJ's capacitance. The charging energy of the MTJ is decoupled from the memory node and takes a smaller value, the fewer the electrons in the MTJ. We now consider single-electron capture and escape processes on the memory node. From Equation (10.14) the minimum energy process is that just one electron passes through the MTJ. When one electron enters into an island on the site i in the MTJ, the charging energy is given by[41]

$$E = \frac{e^2}{2C_\Sigma}\left(\frac{C_g V_g}{e} - x_i - n\right)^2 + \frac{e^2}{2C}x_i\left(1 - x_i\right). \tag{10.18}$$

The second term of this equation has a simple meaning. In Figure 10.23(a) the capacitance between the site i and ground is given by $C_1 = C/x_i$, and the capacitance between the site i and the memory node is given by $C_2 = C/(1 - x_i)$. These capacitors give the charging energy of the MTJ, $e^2/2(C_1 + C_2) = e^2 x_i(1 - x_i)/2C$, which creates an energy barrier to the electron since this charging energy takes a maximum value, $e^2/8C$, when the electron comes to the center of the MTJ. Equation (10.18) is plotted

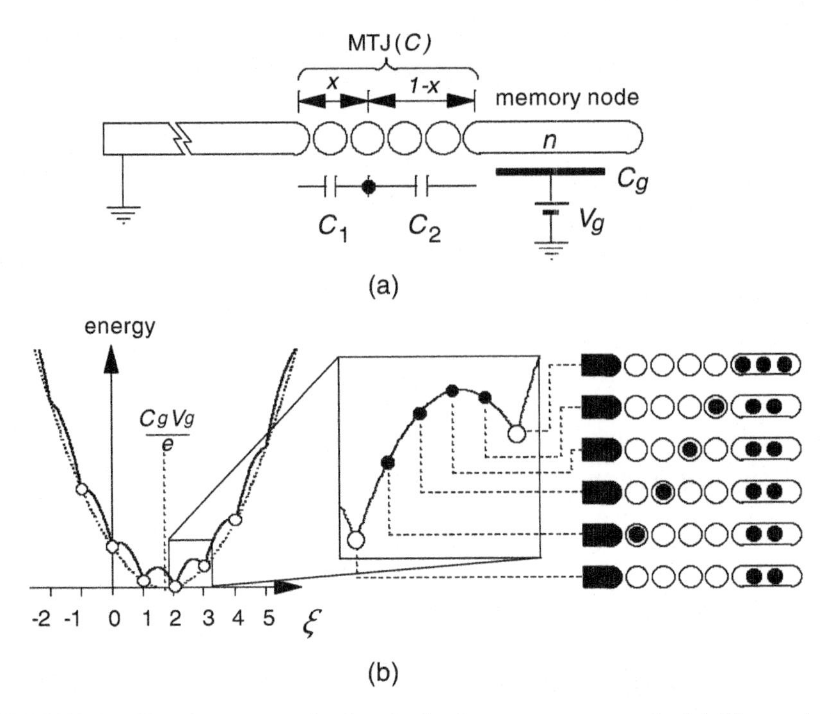

FIGURE 10.23 Charging energy in the single-electron memory cell. (a) The equivalent capacitors for the calculation of charging energy of the MTJ. (b) The electrostatic energy of the system plotted for a generalized electron number ξ defined by Equation (10.19). Open circles show metastable states. The noninteger ξ represents a one-electron transport process through the MTJ.

for energy in Figure 10.23(b) by introducing the generalized electron number on the memory node,

$$\xi = n + x_i \left(n: \text{integer}, \ 0 < x_i \le 1 \right). \tag{10.19}$$

Referring to the box inset of Figure 10.23(b), the electron capture process at the memory node is described by ξ increasing from two to three, and the electron escape process is described by ξ decreasing from three to two. The axis ξ gives a saddle point path of the single-electron capture and escape processes since a one-electron process gives a minimum energy process. It should be noted that the capacitance C of the MTJ creates the energy barrier between two successive integer numbers, whereas the capacitance of the memory node, C_Σ, lowers the energy barrier by displacing it from C_gV_g/e. Thus, around a most stable state which ensures charge neutrality, a finite number of stable states are created,

$$-N < n - C_gV_g/e < N, \tag{10.20}$$

determined by the ratio of the memory node capacitance to the MTJ capacitance, $N = C_\Sigma/2C$, which gives the same result as Figure 10.22. As shown in Figure 10.23

the islands in the MTJ are represented by the discreteness of the variable ξ. From Equation (10.15) the capacitance of each tunnel junction determines the separation of ξ between successive lattice points. If all tunnel junctions have the same capacitance, the values of the variable ξ form a regular lattice. In general the lattice points distribute irregularly, but if the irregularity is not strong, the overall characteristics may be described by a universal curve, $E(\xi) = e^2 (C_g V_g/e - \xi)^2/2C_\Sigma + e^2 \mathit{frac}(\xi)\{1 - \mathit{frac}(\xi)\}/2C$, when the number of tunnel junctions in the MTJ is large. Such a MTJ may be regarded as a continuous MTJ. Assuming that the tunnel resistance R_i is same in all junctions, the master Equation (10.9) can be replaced by the following Smoluchowski equation,

$$\frac{\partial}{\partial t}P(\xi)=\frac{1}{2\gamma\tau}\frac{\partial}{\partial\xi}\left[\left\{\frac{\partial}{\partial\xi}\left(\frac{E(\xi)}{k_BT}\right)\right\}+\frac{\partial}{\partial\xi}\right]P(\xi). \tag{10.21}$$

One of the important characteristics of the single-electron memory is the retention time of the information which can be represented by a mean first-passage time (MFPT). The MFPT is defined as the average time elapsed until the process starting out at point ξ_0 leaves a prescribed domain of a state space for the first time.[42] The MFPT for a process described by Equation (10.21) can be obtained as follows,[43]

$$t(\xi)=2\gamma\tau\int_{\xi_0}^{\xi}dx\exp\left[\frac{E(x)}{k_BT}\right]\int_{-\infty}^{x}dy\exp\left[-\frac{E(y)}{k_BT}\right], \tag{10.22}$$

where ξ is the first passage point ($\xi_0 < \xi$, C_gV_g/e). The MFPT is plotted in Figure 10.24 when $\xi_0 = -N$ and $C_gV_g/e = -0.5$. The information can be recognized by the positive or negative charge on the memory node, $\xi < C_gV_g/e$ or $\xi > C_gV_g/e$. The time in which the information is completely destroyed is given by the MFPT at $\xi = C_gV_g/e$. As shown in Figure 10.24 the MFPT increases exponentially as $\exp(\gamma/4)$ when γ is greater than 50 and the step nature of the MFPT appears. The MFPT increases as $\exp\left(-28/\sqrt{N}\right)$ when the number of electrons used to represent information, N, increases.

In this section we have analyzed a single-electron memory based on a semiclassical model. One of the important effects neglected in this section is the co-tunneling effect, or the so-called macroscopic quantum tunneling of charge. For example, when an electron co-tunnels through five tunnel junctions simultaneously from $\xi = 3$ to $\xi = 2$ in Figure 10.23(b), the electron will not feel the energy barrier and the co-tunneling probability is large because the energy decreases after the co-tunneling. The co-tunneling time through N_t junctions is given by

$$t^{(N_t)}/\tau \cong 2(2\pi)^{2N_t-2} N_t^{-4N_t}(2N_t-1)!\{(N_t-1)!\}^2 (R_i/R_K)^{N_t-1}(V_c/V)^{2N_t-1}, \tag{10.23}$$

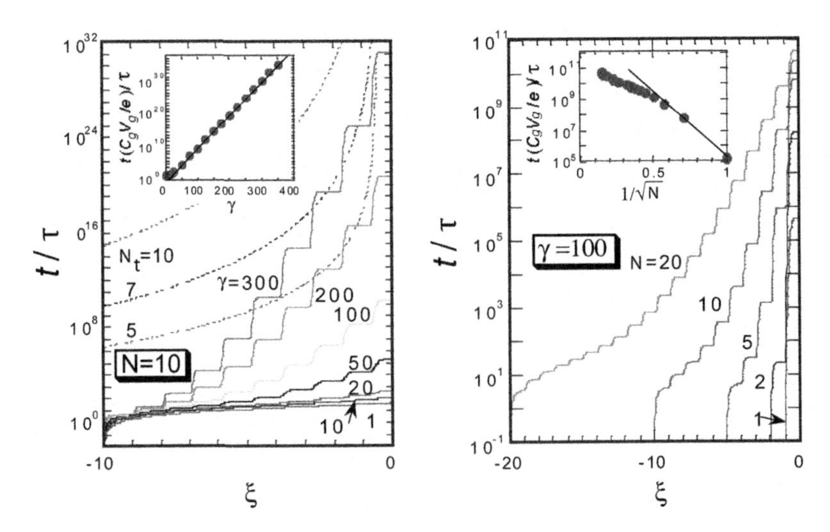

FIGURE 10.24 Calculated mean first-passage time starting at a point $-N$ to reach a point ξ for the first time, with two parameters γ and N. The time to reach a most stable point $\xi = C_g V_g / e$ is plotted in the insets of the figures. The dotted lines correspond to the co-tunneling time, Equation (10.23), where N_t is the number of tunnel junctions in the MTJ.[39]

where R_i is the resistance of one junction, R_K is the resistance quantum, $V_c = Q_c / C \approx e/2C$, and V is the memory node voltage.[35] Therefore, in order to reduce the co-tunneling rate, the number of tunnel junctions in the MTJ and the tunnel resistance of each junction must be increased. In Figure 10.24 the co-tunneling time is plotted when R_i = 500 kΩ and N = 10. In order to obtain a co-tunneling time longer than the intrinsic retention time, the number of tunnel junctions in the MTJ must be larger than five for $\gamma = 200$.

For parameters $N_t \sim 7$, $R_i \sim 1$ GΩ, $C_i \sim 0.1$ aF, the time constant $\tau = N_t R_i C_i$ is given by 0.7 nsec. At a temperature of 110°C, $\gamma = e^2 N_t / 2 C_i k_B T \sim 170$, and the retention time is around 10 sec, comparable with the present DRAM cell retention time. The realization of capacitance of the order of 0.1 aF, however, requires minimum feature size less than 1 nm.

10.4.3 Experimental Single-Electron Memory

10.4.3.1 First Experimental Single-Electron Memory

As discussed earlier, to utilize the Coulomb blockade effect the structures must be reduced in size so that the charging energy is larger than the thermal energy. A number of different techniques have been employed to realize ultrasmall tunnel junctions, including double-angle evaporation of Al[26] and Schottky gate confinement of the two-dimensional electron gas (2-DEG) formed at the GaAs/AlGaAs hetero-interfaces.[44,45] These methods provide only a single tunnel junction and the overall size becomes large when multiple-tunnel junctions are constructed by connecting these elements. To realize very small MTJ structures, side-gated structures have been investigated in δ-doped GaAs.[34] The δ-doped layer is less than 10 nm thick, with

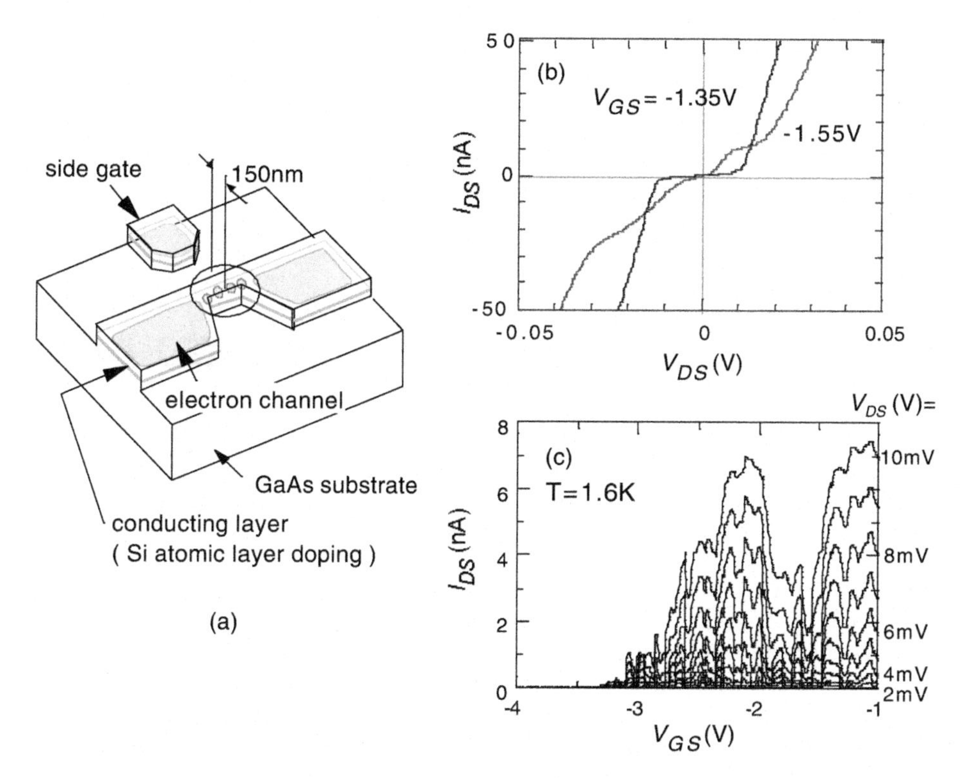

FIGURE 10.25 (a) MTJ using a side-gated constriction in δ-doped layer. (b) Measured drain current as a function of drain voltage of a constriction at a temperature of 0.3 K. (c) Drain currents as a function of side-gate voltage.[34]

an electron density as high as 10^{13} cm^{-2}. The thin layer reduces the capacitance. In addition, δ-doped layers are suitable for miniaturization because of the small depletion lengths resulting from the high carrier concentration and the relatively shallow depth of the 2-DEG.[46] Furthermore, disorder in the δ-doped layer creates several small islands in a constriction without the need for lithography to define the individual islands.

The δ-doped GaAs wafer was grown by Metalorganic Chemical Vapor Deposition (MOCVD). The electron channel formed in the δ-doped layer is situated 30 nm below the GaAs surface and is a few atomic layers in thickness which is doped with Si to a concentration of 5×10^{12} cm^{-2}. The carrier concentration and electron mobility were estimated to be 4×10^{12} cm^{-2} and 2×10^{3} cm^2/Vsec, respectively, from magneto-resistance measurements made at liquid-helium temperature. An MTJ was formed by an etched constriction with a width of 150 nm and length of 400 nm as shown in Figure 10.25(a). The pattern was defined by electron beam (EB) lithography and wet etching. The depth of the trench was controlled to 120 nm by adjusting the etch time.

DC current-voltage characteristics were measured using standard equipment. Within a certain side-gate voltage range Coulomb blockade effects were observed.

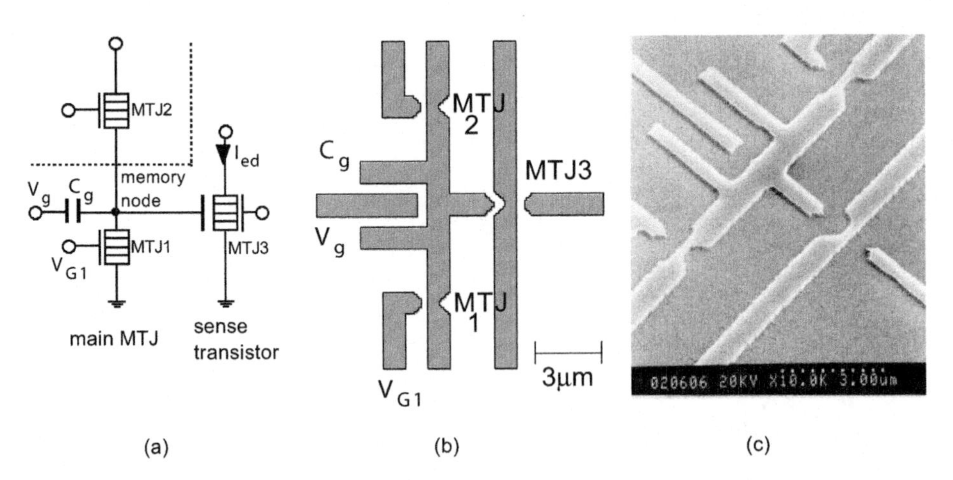

FIGURE 10.26 (a) Schematic circuit diagram for an experimental single-electron memory device. The Coulomb blockade sense transistor detects the electron state on the memory node. (b) Mask layout and (c) scanning electron micrograph.[39]

The width of the Coulomb gap and the slope at zero voltage change periodically as a function of side-gate voltage as shown in Figure 10.25(b). A Coulomb staircase appears at several gate voltages. The total capacitance C of the MTJ is estimated to be 5 aF from the Coulomb gap $e/2C \approx 15\ mV$. The current as a function of side-gate voltage is plotted in Figure 10.25(c). Clear Coulomb blockade oscillations are observed, which are explained by postulating a series of microsegments split by potentials of donor atoms and forming a series of single-electron transistors. In a single-electron transistor[28,29] the current through two ultrasmall tunnel junctions connected in series is controlled by the gate voltage, V_{GS}, applied to a gate capacitor, C_G which couples the island between two junctions. The current-voltage characteristic is periodic in the gate voltage with a period of $\Delta V_{GS} = e/C_G$. The gate capacitance C_G is estimated to be 1 aF from $e/C_G \approx 0.2V$ by taking the broadest period in Figure 10.25(c). The observed period in the side-gate voltage, however, is not constant being composed of several components, which cannot be explained by only one single-electron transistor. Several narrow peaks appear inside a broad peak and disappear at high drain voltages. These features can be explained by the formation of a series of single-electron transistors.[34] The formation of several islands is supported by another measurement in the highly pinched-off region of operation in that negative resistances due to resonant tunneling among microsegments have been observed, and by the calculation of conductance modeled using standard single-particle recursive Green's function techniques.[47] There are large (≈30 meV) potential fluctuations within the channel, and the size of islands is estimated around 10 nm.

The experimental single-electron memory device consists of three MTJs and two capacitors as shown in Figure 10.26. MTJ1 controls single-electron transfer to the memory node. MTJ2 is used to characterize MTJ1; the side-gate voltage dependence on the conductance of MTJ1 is measured by opening MTJ2. When memory characteristics are measured, MTJ2 is pinched off by applying a large negative

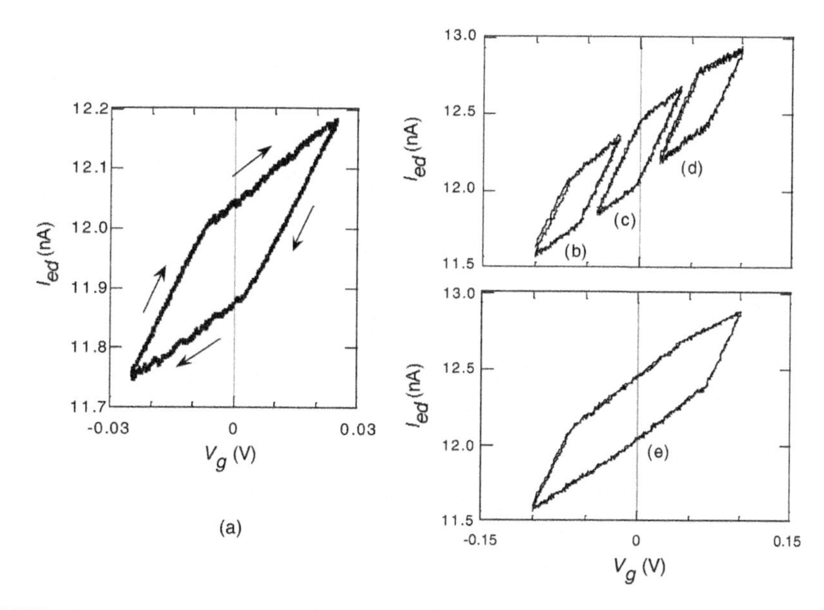

FIGURE 10.27 (a) Memory operation characteristics, showing hysteresis at $T = 0.03$ K. The sweep rate of gate voltage is 70 μV/s. (b)-(e) Universal characteristics of hysteresis. The sweep rate is 2 mV/s. $T = 0.03$ K.[39]

voltage to the side gate. MTJ3 is used as a sense transistor to detect the memory node voltage. The sensitivity to the memory node voltage is estimated to be around 17 nA/V.

The characteristics of the single-electron memory were measured in a dilution refrigerator with a base operating temperature of ~30 m K. The characteristic is shown in Figure 10.27(a), for one cycle between 0.025 V and –0.025 V. Clear and reproducible hysteresis is observed. From a rough estimation of capacitance, $C = 5$ aF by the characterization of one constriction and $C_g = 200$ aF and $C_s = 200$ aF by the numerical calculations for coplanar electrodes; the upper and the lower branches correspond to ±40 electrons. The sense current in the upper and lower branches of the hysteresis is not constant because there is a direct coupling between the gate voltage and MTJ3. The hysteresis shown in Figure 10.22 is quite universal, independent of not only the absolute value of the gate voltage but also the gate voltage range. These universal characteristics are clearly observed in Figure 10.27(b)~(e) for several gate voltage ranges.

The hysteresis can be characterized by the area of the loop which has the merit that the component from the direct coupling between gate voltage and sense transistor is subtracted. Figure 10.28 is a contour plot of the area as a function of side-gate voltage of MTJ1 and temperature. Within a selected side-gate voltage range hysteresis is observed above 4.2 K.

The area of the hysteresis loop is plotted as a function of gate voltage range, sweep rate, and temperature in Figure 10.29. The measured data are well explained by orthodox theory (Equation [10.11]), with $C_\Sigma/C_g \sim 1.7$ and $C \sim 5$ aF.

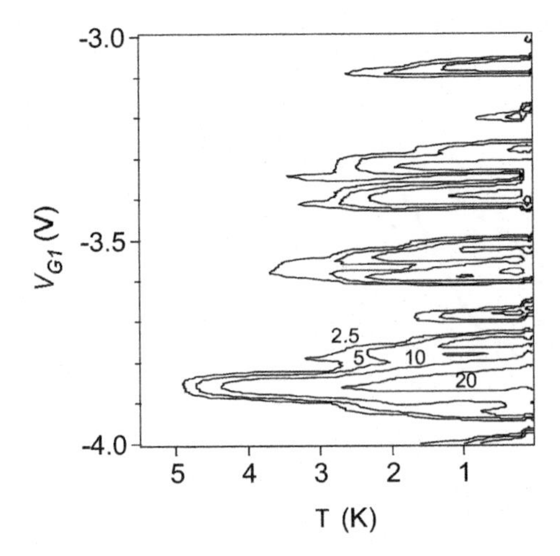

FIGURE 10.28 The area (nA mV) of hysteresis loop plotted as a function of side-gate voltage of MTJ1 and temperature. Sweep rate: 2 mV/sec. Gate voltage range: –0.05 ~ 0.05 V.[39]

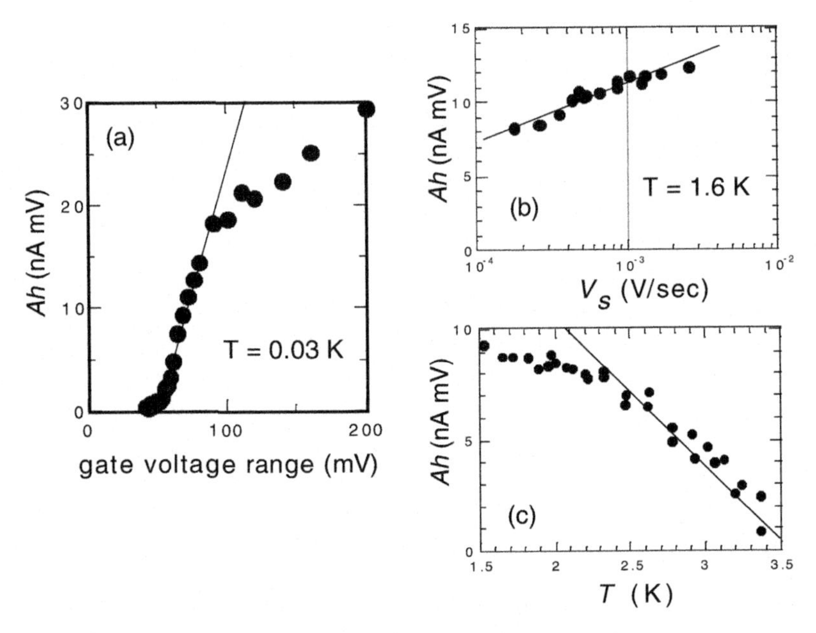

FIGURE 10.29 The area of hysteresis loop plotted as a function of (a) gate voltage range, (b) sweep rate, and (c) temperature. The center parameters are sweep rate: 2 mV/sec, gate voltage range: –0.05 ~ 0.05 V. Straight lines are predicted by orthodox theory (Equation (10.11).[39]

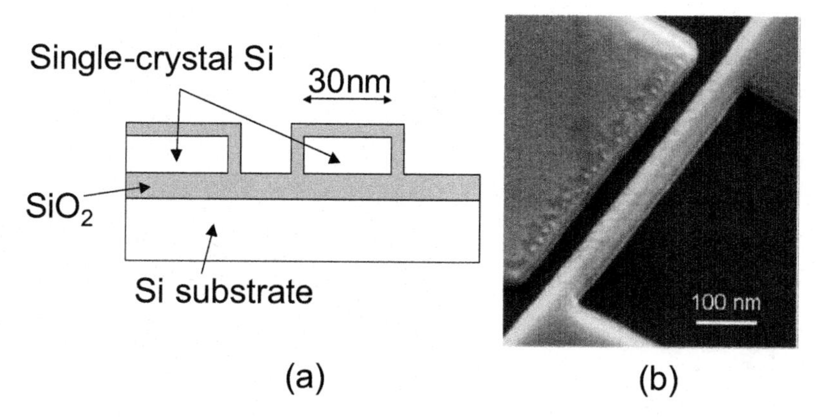

FIGURE 10.30 (a) Schematic cross-section of SOI MTJ. (b) The scanning electron micrograph.

10.4.3.2 Silicon Single-Electron Memory

Although many different materials systems have been used to demonstrate Coulomb blockade, the demonstration of the effect in silicon has been of particular importance because the material is widely used and fabrication processes have been developed for MOS transistors that can be readily adapted for single-electronics. A device structure showing clear Coulomb blockade effect in SOI was first demonstrated by Ali and Ahmed.[48]

The structure is a silicon nanowire as shown in Figure 10.30. A 40-nm-thick silicon layer of a SOI substrate was formed with 350 nm of buried oxide by separation by implantation of oxygen (SIMOX). The silicon was uniformly doped to 1×10^{19} cm^{-3}, above the metal-insulator transition, by implanting phosphorus and was annealed at 950°C for 30 minutes, initially in an oxidizing environment to grow a 25-nm-thick oxide cap and minimize loss of dopant. The oxide cap was then removed in buffered HF. The structure, including interconnects and bonding regions, was patterned in resist on an SOI wafer using a combination of optical and e-beam lithography. Metals were evaporated and lifted off, making a mask for reactive ion etching which was used to transfer the pattern into the silicon with buried oxide acting as an insulating substrate. Once the masking metals had been removed, the chip was oxidized in dry oxygen for 15 minutes at 1000°C which passivated the charge traps and decreased the lithographic size of the nanowire. Side gates were also patterned to control the transport in the nanowire. Clear Coulomb blockade effects were observed in the current-voltage curve as shown in Figure 10.31.[49]

The formation of small islands inside the nanowire is simulated by assuming that ionized dopant atoms are randomly placed over the entire volume of the nanowire.[50] The electron density is evaluated for a range of Fermi energies by self-consistently solving the Thomas-Fermi approximation in two dimensions at $T = 0$ K. The result is shown in Figure 10.32. When negative gate voltage increases, the Fermi energy is lowered, and the electron channel is split into small islands due to the potential fluctuations by ionized dopant atoms.

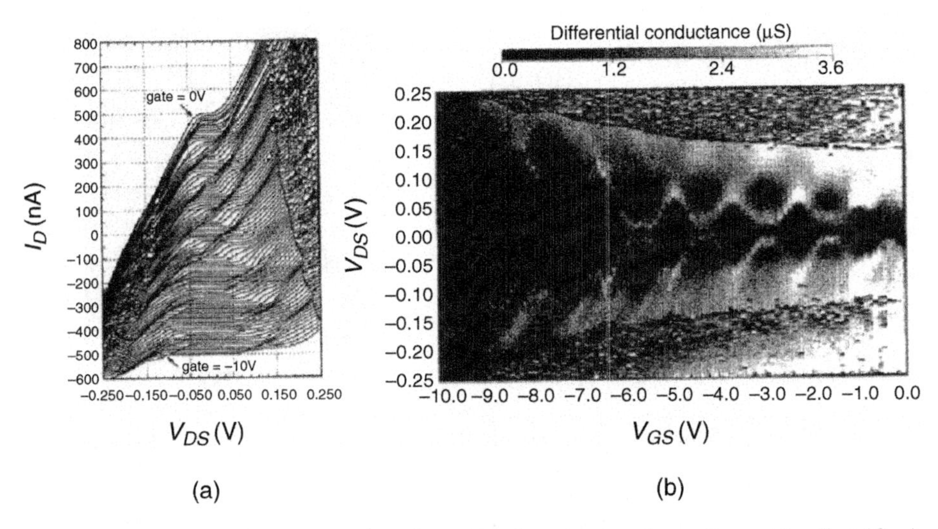

FIGURE 10.31 (a) Drain currents plotted as a function of gate voltage. Current is offset 10 nA per 0.1 V gate step. (b) Differential conductance plotted as a function of drain and gate voltages.[49]

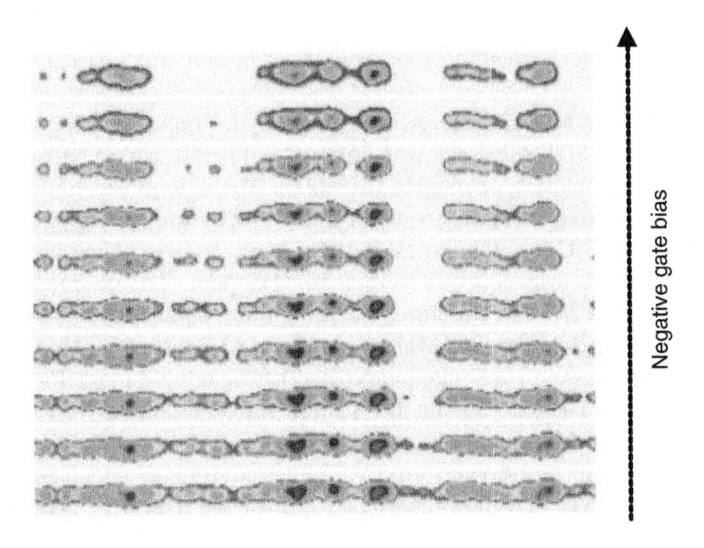

FIGURE 10.32 Calculated electron-density distribution, showing the formation of small islands inside a nanowire. The sample is 20 nm wide and 400 nm long with a doping concentration of 10^{20} cm^{-3}.[50]

A silicon single-electron memory was fabricated as shown in Figure 10.33.[51] The MTJ's silicon nanowire is 300 nm long and approximately 35 nm wide. The fabricated radius of the storage node is measured to be 30 nm but the effective area might be less than defined by this radius. The single-electron transistor's nanowire is 120 nm long and approximately 35 nm wide. Electrical characteristics for the

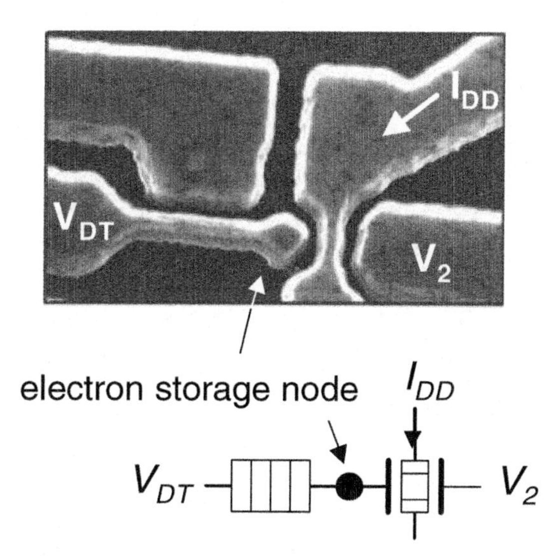

FIGURE 10.33 Scanning electron micrograph of the silicon single-electron memory.[51]

very short nanowires used for the sense transistors suggest that they contain just one dominant island with two tunnel barriers rather than the multiple-tunnel junctions in long wires. On the other hand, electrical characteristics for MTJ devices show multiple periodicity in the oscillation of conductance as a function of gate voltage.

The single-electron transistor is very sensitive to any change of voltage on either its trimming gate or the storage node. At a temperature where $e^2/2C_{SET} \ll k_BT$ (where C_{SET} is the total capacitance of the dominant island in the single-electron transistor) the device acts as a very stable single-electron transistor. When either the trimming-gate or the storage node voltage is swept, the current I_{DD} through the single-electron transistor oscillates with a well-defined period.

In the experiment designed to detect the arrival and departure of just one electron, the single-electron transistor is biased with a constant voltage on its drain and the trimming gate, V_2, is used to set the operating point on a steep part of the slope of an oscillation in I_{DD}; either a negative or a positive slope may be chosen. Any change in charge on the storage node is reflected as a change in the current flowing through the sense transistor. The MTJ supplying the electrons to the storage node is biased with its trimming gate, so that it has the widest possible Coulomb blockade region; thus no charge can flow through the device until the Coulomb gap voltage (V_C) has been overcome. As the drain voltage of the MTJ (V_{DT}) is decreased no transfer takes place until the voltage across it is less than $-V_C$. Once V_{DT} with respect to the node voltage V is less than $-V_C$ then one electron can flow onto the node. This single electron causes the voltage of the node to fall by e/C_Σ (C_Σ is the total capacitance of the storage node).

The current through the sense single-electron transistor as a function of its trimming gate voltage is shown in Figure 10.34(a). Single-electron charging effects cause well-defined conductance oscillations observable over several periods. However, when the drain voltage, V_{DT}, of the MTJ is swept, the conductance oscillations

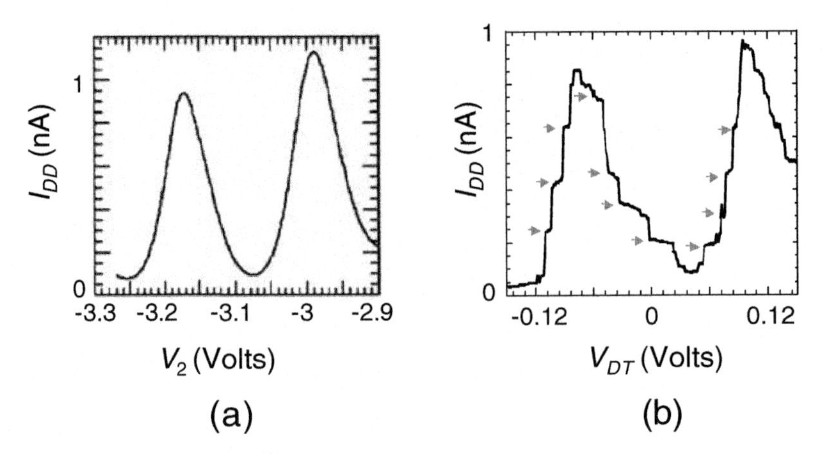

FIGURE 10.34 (a) Sense transistor characteristics showing current oscillation as a function of trimming gate voltage. (b) Sense transistor characteristics as a function of the change of electron numbers on the node as the MTJ drain voltage V_{DT} is swept. The temperature is 4.2 K.[51]

of the single-electron transistor are again measured, quantized effects are clearly observed, demonstrating that the transfer of discrete charge onto the storage node is being detected.

Using the single-electron transistor characteristic in Figure 10.34(a), the operating point was set on the negative slope of the characteristic at a point where the sensitivity is about 14 nA/V. We can assume that the coupling between the storage node and the sense transistor is nearly the same as the coupling of the trimming gate to the sense transistor since the distances between these and the nanowire are very closely the same. With this setting it is possible to use a single-electron transistor to detect just one electron on the storage node.

This system can be operated to count a precise number of electrons. If the voltage V_{DT} is swept over a slightly larger range then, as shown in Figure 10.35(a), three steps of equal height are seen in the sense current for the forward sweep, and also exactly three in the reverse sweep, returning to the original current level, which signifies the counting of three electrons from and to the storage node. Larger voltage sweeps result in more steps and hence more electrons being transferred to and from the storage node.

In Figure 10.35(b) the MTJ is now operated by sweeping the drain V_{DT} over a small range so that only one stepped decrease in sense current I_{DD} is seen as the voltage is increased and only a single stepped increase back to the original level is seen as the voltage is decreased. The first step is interpreted as the transfer of just one electron *from* the storage node and the second step as a transfer *to* the storage node returning it to its initial state. This transfer to and from the storage node can be repeated many times by cycling V_{DT} and is completely reproducible over many cycles.

The interpretation of the results can be related to the physical dimensions of the devices. The step height seen in Figure 10.35 is about 0.11 nA while the measured sensitivity of the sense transistor is 14 nA/V, giving a voltage change in the storage node of around 8 mV. If the storage node is approximated by a disk of radius 30

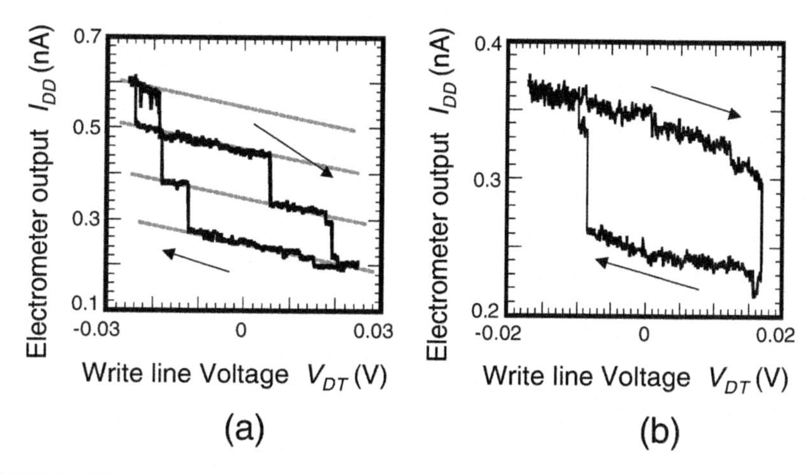

FIGURE 10.35 (a) The steps in I_{DD} show that three electrons are counted on and off the storage node. (b) The steps in I_{DD} demonstrate the transfer of one electron on and off the storage node. The temperature is 4.2 K.[51]

nm, its self-capacitance is approximately 10 aF so that the addition or removal of just one electron would change its potential by approximately 16 mV. This compares reasonably well with the measured sensitivity if we take into account that stray capacitances that would increase the effective capacitance of the node significantly have not been included.

10.4.4 Single-Electron Memory Array

In the previous sections, one single-electron memory cell was described. In this section, the operation of a memory cell array is described.

The method of selective write is described in Figure 10.36. Writing to the memory by using a combination of bit line voltage V_B and word line voltage V_W pulses allows selection of the memory cell. If the magnitude of the write pulse function V_B is less than $C_\Sigma V_C / C_g$, then no electrons are written to the memory when $V_W = 0$ (unselected word line). If at the same time, however, an enable function is applied at V_W on the selected word line, the voltage across the MTJ is greater than V_C and charge is written to the node. As shown in Figure 10.36(b), the enable function takes the form of a positive pulse followed immediately by a negative pulse, both of magnitude less than V_C. During a positive write pulse, the negative swing of the enable function causes the voltage across the MTJ to be greater than V_C and the memory state changes while the positive swing of the enable function in combination with the write pulse causes no change to the state of the memory. During a negative write pulse coincident with an enable pulse, the stored electrons are removed from the memory node via the MTJ; the memory node then becomes positive.

Figure 10.37(a) shows the enable pulse train, the write pulses and the sense transistor output and demonstrates the selectivity of the fabricated silicon single-electron memory.[52] The pulses used in this measurement were 1 msec in duration. The test sequences consisted of coincident write and enable pulses along with

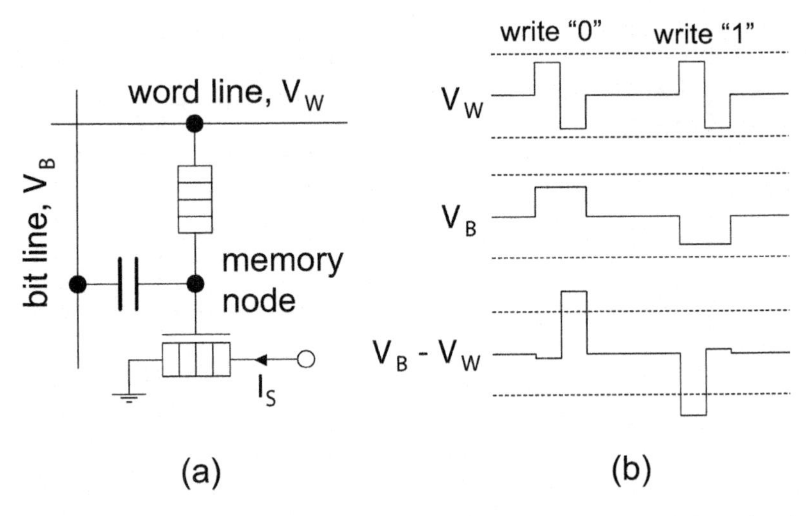

FIGURE 10.36 (a) Circuit equivalent of the memory cell and sensing transistor showing the word and bit lines of the memory. (b) The enable pulses on word line and write pulses on the bit line. Coulomb blockade regions are between dotted lines.

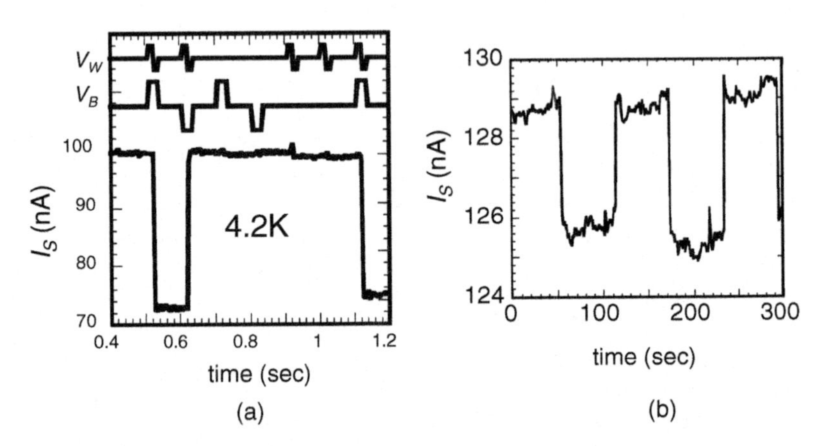

FIGURE 10.37 (a) Nondestructive selectivity of memory cell is demonstrated here. The enable pulses, write pulses, and sense transistor output are shown. (b) The response of the memory to 5 nsec write pulses, retention time is clearly greater than 1 min at 4.2 K.[52]

individual enable and write pulses. The Coulomb gap of the MTJ was estimated to be ~ 0.36 V, the write pulses (applied to V_B) were 1.5V in magnitude and the enable pulses were 0.36 V peak to peak. It can be seen from Figure 10.37(a) that when either the enable function or the write pulses are applied separately, the memory does not change state; however, when the two coincide the memory changes state and two memory levels are distinctly observed.

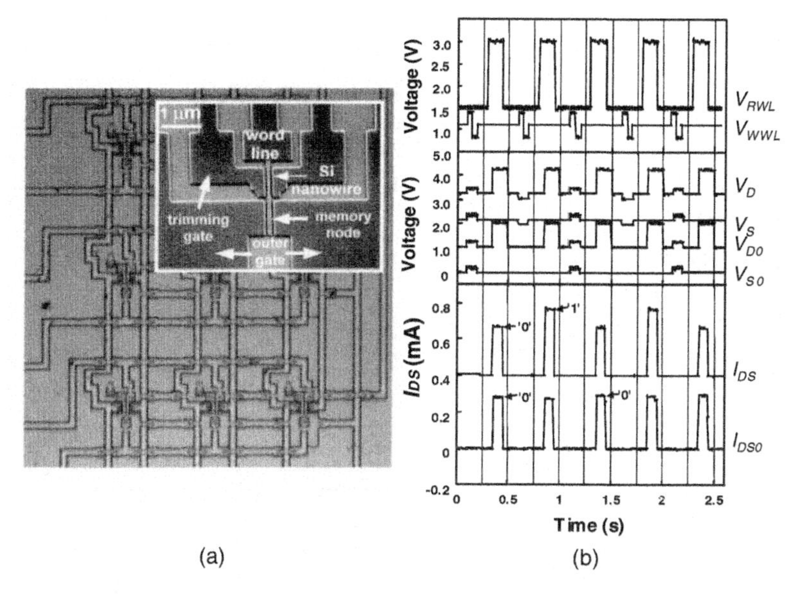

FIGURE 10.38 (a) Optical micrograph of a 3 × 3 array of split-gate SET/MOSFET memory cells in SOI material. The inset shows a scanning electron micrograph of a single memory cell, where a nanowire single-electron MTJ is integrated with a split-gate MOSFET. The memory-node area is 1 μm × 70 nm. (b) Cell selection in a 3 × 3 array at 20 K.[54]

In order to assess the write speed, positive and negative pulses of magnitude greater than V_C and separated by one minute intervals were applied to the word line. As shown in Figure 10.37(b) the circuit responded for pulses of widths down to 5 nsec applied to the device. During the 1-minute intervals between pulses, the memory level is not significantly reduced which demonstrates the long retention time of the memory circuit. It should be noted that in order to maintain the memory levels as pulse width was decreased, the amplitudes of the write pulses were increased in order to allow the required number of electrons representing a memory state to leave or enter the storage node. The pulses used for the experiment were +2 V and -0.5 V at the bit line.

The read time is determined by $C_L V_{DD}/I_S$ where I_S is the current through the sense transistor and C_L is the bit line load capacitance. To obtain a read time less than 10 ns, the sense current must be higher than 10 μA for C_L = 100 fF. Since the drivability of a single-electron transistor is low, it is difficult to reduce the read time. A hybrid single-electron tunneling (SET)-MOSFET memory was proposed to overcome the problem, where a conventional MOSFET is used as the sense transistor.[53] The SET MTJ is used for controlling the precise number of electrons stored on the memory node. The structure is called the L-SEM (lateral single-electron memory) shown in Figure 10.38, in which the MTJ is integrated with a memory node in the top silicon layer of SOI material. The gate of the MOSFET is a split gate consisting of a memory node and an outer gate. The outer gate is connected to the read word line. The drain, source and channel regions of an n-channel MOSFET lie in the substrate silicon. A 3 × 3 array was fabricated and the cell selection was demonstrated.[54]

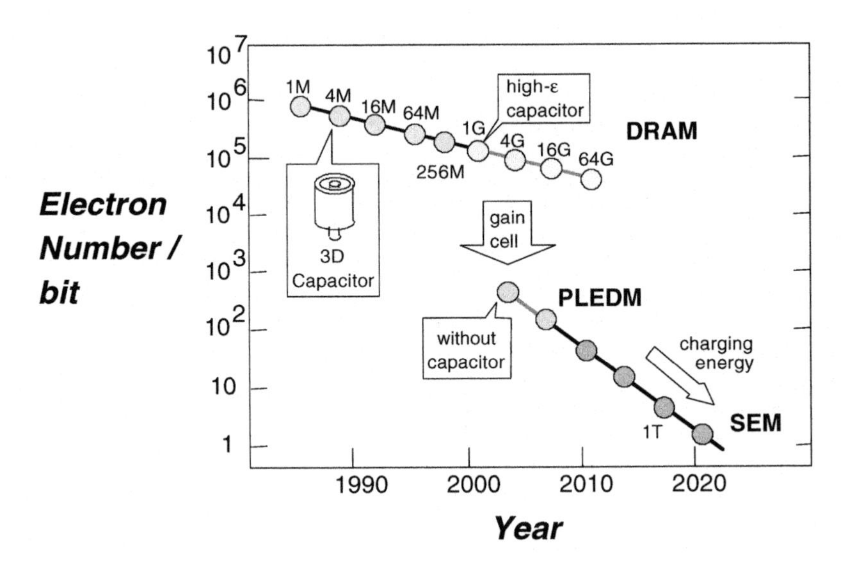

FIGURE 10.39 Electron number per bit is plotted for DRAM 1-T cell. The prediction of few electron number DRAMs, PLEDM and SEM (single-electron memory) is also shown.

Figure 10.38(b) shows the write and read waveforms observed for the 3 × 3 array. To write "0"("1") to the L-SEM cell, a common positive(negative) bias voltage $V_D = V_S$ is applied both to the source and drain of the sense MOSFET, and a cyclic voltage V_{WWL} is applied to the write word line. To read the data, positive voltages are applied on the outer gate and the drain of the sense MOSFET. In Figure 10.38(b), V_{D0} and V_{S0} are the drain and source voltages on the MOSFET on the neighboring column and on the same row. It was shown that the access of a cell does not perturb the states on the other cells. High-speed write and read operation of the L-SEM has been demonstrated theoretically.[55,56] The write time is determined by $\tau = N_t CR$ where N_t is the total number of junctions in MTJ, and C, R are total capacitance and resistance of the MTJ, respectively. The write time can be lower than 10 nsec, comparable to the present day DRAM.

10.5 CONCLUSION

In the present day DRAM, the electron number cannot be reduced, and the maintenance of a large storage capacitor in the circuit is approaching the physical limitations in design and fabrication. One of the solutions to this problem is to use a gain cell, where the stored node is separated from the large interconnect load. The PLEDM has been proposed as a small DRAM gain cell in which a stacked silicon tunnel transistor is integrated inside the gate of a conventional MOSFET. Beyond this stage, if the feature size is reduced, single-electron charging effects or quantum confinement effects can be utilized to reduce the leakage current of the write transistor, resulting in few or ultimately single-electron DRAMs. The electron-number-per-bit decrease in the next few decades is plotted in Figure 10.39.

REFERENCES

1. Regitz, W.W. and Karp, J.A., A 3-transistor cell, 1024 bit 500 ns MOSRAM, *ISSCC Digest of Technological Papers*, p. 42, 1970.
2. Dennard, R. H., U. S. Patent 3,387,286, 1968.
3. Itoh, K., *VLSI Memory Chip Design*, Springer-Verlag, Heidelberg, 2001.
4. Itoh, K., Sunami, H., Nakazato, K., and Horiguchi, M., Pathways to DRAM design and technology for the 21st century, *Silicon Material Science and Technology VIII,* San Diego, 1998.
5. Terada, K., Takada, M., Kurosawa, S., and Suziku, S., A new VLSI memory cell using DMOS technology (DMOS Cell), *IEEE Trans. Electrons Devices,* 29, 1301, 1982.
6. Shichijo, H., Malhi, S.D.S., Shah, A.H., Pollack, G.P., Richardson, W.F., Elahy, M., Banerjee, S., Womack, R., and Chatterjee, P.K., TITE RAM:A new SOI DRAM gain cell for Mbit DRAM's, *Ext. Abstract of 16th SSDM*, 265, 1984.
7. Shukuri, S., Kure, T., and Nishida, T., A Complementary Gain Cell Technology for Sub-1V Supply DRAMs, *IEDM Tech. Dig.*, 1006, 1992.
8. Terauchi, M., Nitayama, A., Horiguchi, F., and Masuoka, F., A Surrounding Gate Transistor (SGT) Gain Cell for Ultra High Density DRAMs, *VLSI Tech. Dig.,* 21, 1993.
9. Nakazato, K., Piotrowicz, P.J.A., Hasko, D.G., Ahmed, H., and Itoh, K., PLED - Planar Localised Electron Devices, *IEDM Tech. Dig.*, 179, 1997.
10. Nakazato, K., Itoh, K., Mizuta, H., and Ahmed, H., Silicon stacked tunnel transistor for high-speed and high-density random access memory gain cells, *Electronics Letters*, 35, 848, 1999.
11. Nakazato, K., Itoh, K., Ahmed, H., Mizuta, H., Kisu, T., Kato, M., and Sakata, T., Phase-state Low Electron-number Drive Random Access Memory (PLEDM), *ISSCC Digest of Technical Papers*, 132, 2000.
12. Mizuta, H., Nakazato, K., Piotrowicz, P.J.A., Itoh, K., Teshima, T., Yamaguchi, K., Shimada, T., Normally-off PLED (Planar Localised Electron Device) for non-volatile memory, *VLSI Tech. Dig.*, 128, 1998.
13. Mizuta, H., Wagner, M., and Nakazato, K., The role of tunnel barriers in phase-state low electron-number drive transistors (PLEDTRs), *IEEE Trans. Electron Devices,* 48, 1103, 2001.
14. Hess, K., Higman, T.K., Emanuel, M.A., and Coleman, J.J., New ultrafast switching mechanism in semiconductor heterostructures, *J. Appl. Phys.*, 60, 3775, 1986.
15. Kisu, T. and Nakazato, K., Silicon Stacked Transistor with Source and Drain Tunnel Barriers, *ESSDERC,* 532, 1999.
16. Moslehi, M.M., and Saraswat, K.C., Thermal nitridation of Si and SiO2 for VLSI, *IEEE Trans. Electron Devices*, 32, 106, 1985.
17. Suzuki, K., Tanaka, T., Tosaka, Y., Horie, H., and Arimoto, Y., Scaling theory for double-gate SOI MOSFET's, *IEEE Trans. Electron Devices*, 40, 2326, 1993.
18. Tanaka, H., Aoki, M., Sakata, T., Kimura, S., Sakashita, N., Hidaka, H., Tachibana, T., Kimura, K., A precise on-chip voltage generator for a gigascale DRAM with a negative word-line scheme, *IEEE J. Solid-State Circuits*, 34, 1084, 1999.
19. Martino, W., and Croxon, B.F., The Inverting Cell Concept for MOS Dynamic RAMS, *ISSCC Digest of Technological Papers*, 12, 1972.
20. Nakazato, K., Ahmed, H., and White, J.D., U. S. Patent 5,677,637, 1992.
21. Nakazato, K., Blaikie, R.J., Cleaver, J.R.A., and Ahmed, H., Single-electron memory, *Electronics Letters,* 29, 384 1993.

22. Likharev, K.K., On the possibility of fabricating analog and digital integrated circuits based on discrete single-electron tunneling, *Mikroelektronikz*, 16, 195, 1987.
23. Likharev, K.K., Correlated discrete transfer of single electrons in ultrasmall tunnel junctions, *IBM J. Res. Devel.*, 32, 144, 1988.
24. Averin, D.V. and Likharev, K.K., Single electronics: A correlated transfer of single electrons and cooper pairs in systems of small tunnel junctions, in Altshuler, B.L., Lee, P.A., and Webb, R.A., Eds., *Mesoscopic Phenomena in Solids*, 173, Elsevier, Amsterdam, 1991.
25. Averin, D.V. and Likharev, K.K., Possible Applications of Single Charge Tunneling, in Grabert, H. and Devoret, M.H.D., Eds., *Single Charge Tunneling,* 311, Plenum Press, New York, 1992.
26. Dolan, G.J., Offset masks for lift-off photoprocessing, *Appl. Phys. Lett.,* 337, 1977.
27. Dolan, G.J. and Dunsmuir, J.H., Very small (~20 nm) lithographic wires, dots, rings, and tunnel junctions, *Physica: Rev. Condensed Matter* 152, 7, 1988.
28. Fulton, T.A. and Dolan, G.J., Observation of single-electron charging effects in small tunnel junctions, *Phys. Rev. Lett.*, 59, 109, 1987.
29. Kuzmin, L.S. and Likharev, K.K., Direct experimental observation of discrete correlated single-electron tunneling, *JETP Lett.*, 45, 495, 1987.
30. Geerligs, L.J., Anderegg, V.F., Holweg, P.A.M., Mooij, J.E., Pothier, H., Esteve, D., Urbina, C., and Devoret, M.H., Frequency-locked turnstile device for single electrons, *Phys. Rev. Lett.*, 64, 2691, 1990.
31. Lafarge, P., Pothier, H., Williams, E.R., Esteve, D., Urbina, C., and Devoret, M.H., Direct observation of macroscopic charge quantization, *Z. Phys. B — Condensed Matter,* 85, 327, 1991.
32. Fulton, T.A., Gammel, P.L., and Dunkleberger, L.N., Determination of Coulomb-blockade resistances and observation of the tunneling of single electrons in small-tunnel-junction circuits, *Phys. Rev. Lett.,* 67, 3148, 1991.
33. Lafarge, P., Joyez, P., Pothier, H., Cleland, A., Holst, T., Esteve, D., Urbina, C., and Devoret, M.H., Direct observation of macroscopic charge quantization: A Millikan experiment in a submicron solid state device, *C. R. Acad. Sci. Paris,* t.314, Serie II, 883, 1992.
34. Nakazato, K., Thornton, T.J., White, J., and Ahmed, H., Single-electron effects in a point contact using side-gating in delta-doped layers, *Appl. Phys. Lett.*, 61, 3145, 1992.
35. Averin, D.V. and Odintsov, A.A., Macroscopic quantum tunneling of the electric charge in small tunnel junctions, *Phys. Lett.*, A140, 251, 1989.
36. Averin, D.V. and Nazarov, Yu V., Virtual electron diffusion during quantum tunneling of the electric charge, *Phys. Rev. Lett.,* 65, 2446, 1990.
37. Geerligs, L.J., Averin, D.V., and Mooij, J.E., Observation of macroscopic quantum tunneling through the Coulomb energy barrier, *Phys. Rev. Lett.*, 65, 3037 1990.
38. Hanna, A.E., Tuominen, M.T., and Tinkham, M., Observation of elastic macroscopic quantum tunneling of the charge variable, *Phys. Rev. Lett.,* 68, 3228, 1992.
39. Nakazato, K., Blaikie, R.J., and Ahmed, H., Single-electron memory, *J. Appl. Phys.*, 75, 5123, 1994.
40. Ingold, G.L. and Nazarov, Yu V., in Grabert, H., and Devoret, M.H., Eds., *Single Charging Tunneling*, 88, Plenum Press, New York, 1992.
41. Nakazato, K. and Ahmed, H., The multiple tunnel junction and its application to single-electron memories, *Adv. Mater.,* 5, 668, 1993.
42. Hanggi, P., Talkner, P., and Borkovec, M., Reaction-rate theory: Fifty years after Kramers, *Rev. Mod. Phys.*, 62, 251, 1990.

43. Pontryagin, L.A., Andronov, A., and Vitt, A., On the statistical treatment of dynamical systems, *Zh. Eksp. Teor. Fiz.*, 3, 165, 1933 [translated by Barbour, J.B. and reproduced in *Noise in Nonlinear Dynamics*, ed. by Moss, F. and McClintock, P.V.E, Cambridge University Press, Cambridge, Vol. 1, p. 329, 1989].
44. Meirav, U., Kastner, M.A., and Wind, S.J., Single-electron charging and periodic conductance resonances in GaAs nanostructures, *Phys. Rev. Lett.*, 65, 771, 1990.
45. Kouwenhoven, L.P., Johnson, A.T., van der Vaart, N.C., Harmans, C.J.P.M., and Foxon, C.T., Quantized current in a quantum-dot turnstile using oscillating tunnel barriers, *Phys. Rev. Lett.*, 67, 1626, 1991.
46. Feng, Y., Thornton, T.J., Harris, J.J., and Williams, D.A., Side gating in δ-doped quantum wires, *Appl. Phys. Lett.*, 60, 94, 1992.
47. Blaikie, R.J., Nakazato, K., Oakeshott, R.B.S., Cleaver, J.R.A., and Ahmed, H., Lateral resonant tunneling through constrictions in a δ-doped GaAs layer, *Appl. Phys. Lett.*, 64, 118, 1994.
48. Ali, D. and Ahmed, H., Coulomb blockade in a silicon tunnel junction device, *Appl. Phys. Lett.*, 64, 2119, 1994.
49. Smith, R.A. and Ahmed, H., Gate controlled Coulomb blockade effects in the conduction of a silicon quantum wire, *J. Appl. Phys.*, 81, 2699, 1997.
50. Evans, G.J., Mizuta, H., and Ahmed, H., Modelling of structural and threshold voltage characteristics of randomly doped silicon nanowires in the Coulomb-blockade regime, *Jpn. J. Appl. Phys.*, 40, 5837, 2001.
51. Stone, N.J. and Ahmed, H., Single-electron detector and counter, *Appl. Phys. Lett.*, 77, 744, 2000.
52. Stone, N.J., Ahmed, H., and Nakazato, K., A high-speed silicon single-electron random access memory, *IEEE Electron Device Lett.*, 20, 583, 1999.
53. Durrani, Z.A.K., Irvine, A.C., Ahmed, H., and Nakazato, K., A memory cell with single-electron and metal-oxide-semiconductor transistor integration, *Appl. Phys. Lett.,* 74, 1293, 1999.
54. Durrani, Z.A.K., Irvine, A.C., and Ahmed, H., Coulomb blockade memory using integrated single-electron transistor/metal-oxide-semiconductor transistor gain cells, *IEEE Trans. Electron Devices,* 47, 2334, 2000.
55. Katayama, K., Mizuta, H., Müller, H.-O., Williams, D., and Nakazato, K., Design and analysis of high-speed random access memory with coulomb blockade charge confinement, *IEEE Trans. Electron Devices, 46,* 2210, 1999.
56. Müller, H.-O. and Mizuta, H., Memory Cell Simulation on the Nanometer Scale, *IEEE Trans. Electron Devices,* 47, 1826, 2000.

11 Single-Electron Logic Devices

Yasuo Takahashi, Yukinori Ono, Akira Fujiwara, and Hiroshi Inokawa

11.1 INTRODUCTION

Silicon metal-oxide semiconductor field-effect transistors (MOSFETs) act as simple electrical switches that allow us to build large-scale integrated circuits (LSIs) with very simple architectures. Owing to progress in fabrication technologies, we can now build LSIs that include tremendous numbers of transistors. However, this increase of transistors causes a large power dissipation in a small Si chip, which will limit the further growth of functionality of the LSIs in the not-too-distant future. To overcome the limitation, we need new functional devices that operate under different principles and dissipate as little power as possible.

The progress in fabrication technology has also made it possible to make small structures on the order of nanometer scale. In fact, MOSFETs with the gate lengths of less than 10 nm have already been achieved.[1,2] This length is small enough to observe quantum size effects even in silicon though the effective masses of carriers are relatively large. We can use the quantum size effects to reduce power consumption and build new functional devices. One simple way to reduce the power consumption is to use single-electron devices (SEDs). SEDs have to have nanometer-scale islands because their operation principle uses the Coulomb blockade effect, where the charge repulsion among the electrons confined in a small island plays an important role. The power dissipation of an SED circuit can be reduced because of the controllability of the flow of electrons by means of one-by-one tunneling. One of the great advantages of SEDs is that the operation principle is simple and does not require coherency of carriers. We only need small structures to cause charge repulsion. In addition, the operation principle guarantees that the performance of SEDs will improve as their size is reduced. In contrast, MOSFETs require complicated structures and dopant profiles to ensure their operation as a simple electrical switch as device size is reduced. These two features of SEDs, low power consumption and stable operation in small structures, are the primary requirements for future LSI devices.

Another way to reduce the power consumption is to use the coherency to suppress the voltage loss. If we can transfer electrons or holes coherently, there will be no voltage loss. Some quantum-effect devices using the interference of electron waves have been proposed and tested their fundamental operation at low temperature.[3,4]

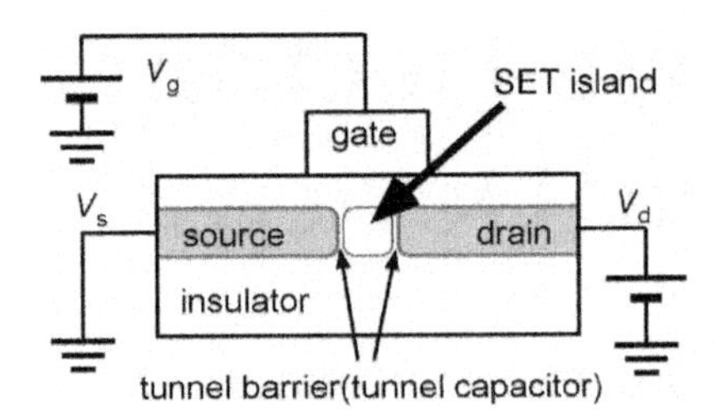

FIGURE 11.1 Schematic structure of an SET.

However, the output signal is quite small because of the difficulties in keeping the coherent length much longer than the sample length after attaching complicated electrodes. As a result, it is difficult to use coherency for achieving high performance with low power consumption at present. It is well known that the carrier coherency and quantum size effect in a small semiconductor island, which is the main part of SEDs, is also important for the transport of the carriers. This chapter focuses on device applications for future LSIs.

It is very important for device applications that the operation characteristics of SEDs are quite different from that of conventional MOS or bipolar transistors. This means we have to develop suitable circuits for SEDs. Many kinds of basic single-electron logic circuits have been proposed and experimentally demonstrated. Some of them perform a simple switching operation like MOS transistors, which is advantageous because we can employ the highly advanced technologies developed for CMOS circuit designs. Although we can use a SED as a simple switch, the Coulomb blockade and one-by-one electron tunneling dominate the current flow in the device. This causes a low drivability, which is the biggest drawback of SEDs. However, SEDs have special features not found in conventional MOSFETs. These features can be exploited to achieve high functionality that will make circuits simple, efficient, and fast.

This chapter discusses the logic circuit applications of SEDs. Section 2 outlines the operation principles and operation characteristics of SEDs and discusses the stability of their operation, which is the most important feature for their practical application. The operation characteristics strongly depend on the base materials and fabrication methods. Section 3 describes some fabrication methods developed for Si SEDs. Section 4 introduces the applications of SEDs to logic circuits. Though their small size is beneficial for high-density memory LSIs, the low-power operation nature and functionality of SEDs are more suitable for logic circuit applications. Some of the applications covered in this section have actually been experimentally demonstrated.

11.2 SINGLE-ELECTRON TRANSISTOR (SET)

The most primitive SED is a single-electron transistor (SET), which has the simple three terminal structure as shown in Fig. 1.[5] The device contain a small island together with a gate electrode coupled capacitively to the island. Source and drain electrodes are attached to the island via a tunnel barrier.

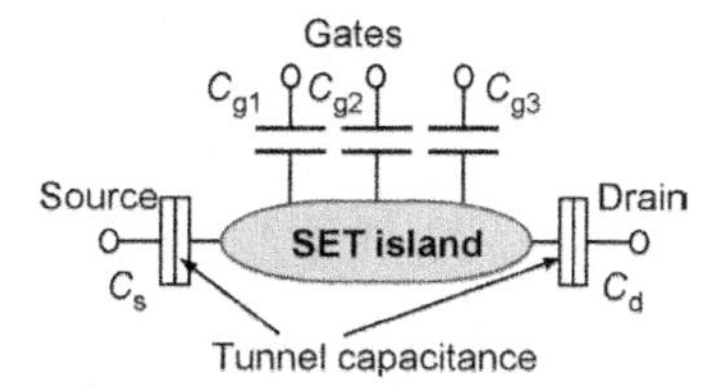

FIGURE 11.2 Equivalent circuit of a multigate SET.

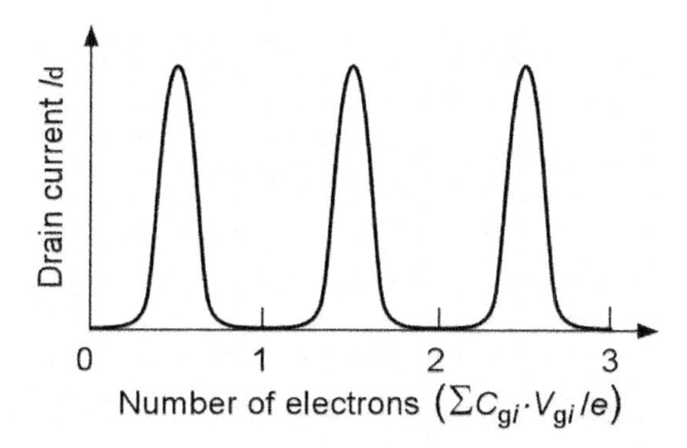

FIGURE 11.3 Gate-voltage vs. drain current characteristics of a multigate SET.

The operation principle of an SET is charge balance between the gate electrode and island. So, the SETs inherently have multiple gates as shown in the equivalent circuit in Figure 11.2. Each gate electrode is directly coupled to the SET island by a capacitor C_{gi}. To grasp the operation principle, the source and drain terminals are grounded for simplicity. When a voltage V_{gi} is applied to the i-th gate, charge $C_{gi}V_{gi}$ accumulates in the gate electrode. Consequently, the total charge accumulated in gates is $\Sigma C_{gi}V_{gi}$. However, since tunnel barriers isolate the island from the source and drain, the number of electrons in the island should be a fixed integer. The total charge in the island $-Ne$, where N is the number of electrons and e the elementary charge, is balanced with that in the gate when $\Sigma C_{gi}V_{gi} = Ne$ is satisfied. In this condition, since the number of electrons N is stable, N cannot change. That is, current does not flow through the SET island. This is the Coulomb blockade condition. As the gate voltages increase the total charge increases in the gates and the charges become unbalanced. Then, when $\Sigma C_{gi}V_{gi} = (N+1/2)e$, the electrostatic potential of the two states, one for N electrons and the other for N+1 electrons becomes equal, which means that the island can contain N or N+1 electrons. Therefore, electrons flow one at a time when a small voltage is applied between the source and drain electrodes. The number of electrons in the island is N+1 after a single electron tunnels from the source to the island. The number returns to N after an electron tunnels from the island to the drain. By repeating this sequence, a current due to single-electron tunneling flows. As the gate voltage is increased further, the number of electrons N+1 becomes stable. As a result, the source-drain conductance exhibits oscillatory characteristics as a function of gate voltages as shown in Figure 11.3.

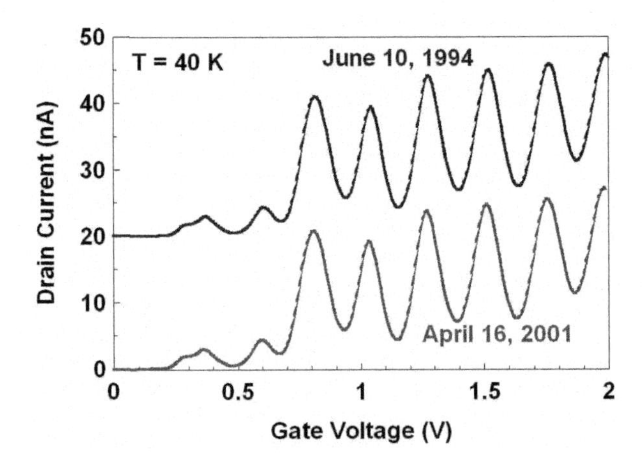

FIGURE 11.4 Current oscillations of an SET fabricated by PADOX as a function of the gate voltage measured at 40 K at a drain voltage of 1 mV. The upper data were obtained in June 1994, and lower data in April 2001.

As shown in Figure 11.3, the source-drain current is determined by the sum of $C_{gi}V_{gi}$. The drain current for an N-th input SET, $I_d\left(V_{g1}, V_{g2} \cdot\cdot V_{gi} \cdot\cdot V_{gN}\right)$, is written as,

$$I_d\left(V_{g1}, V_{g2} \cdot\cdot V_{gi} \cdot\cdot V_{gN}\right) = f\left(\sum_i^N C_{gi}V_{gi} / e\right) \tag{11.1}$$

The drain current takes a minimum when the sum $\Sigma C_{gi}V_{gi}/e$ is an integer because the Coulomb blockade sets in. Conversely, when the sum is a half integer, the current flows because the Coulomb blockade is lifted. This feature, in which the characteristics are determined by the sum of the products of gate voltage and gate capacitance, is very similar to the neuron MOSFET,[6] where multiple gates are coupled with the channel of the MOSFET through a floating gate. The biggest difference is that the gates in an SET can couple directly to the SET island, whereas the neuron MOSFET requires a floating gate for attaching coupled capacitances, which is not suitable for stable operation of circuits. Once the floating gate is charged up, discharging it is difficult. In addition, we incur an area penalty in attaching coupled capacitances. The device structure of the multigate SET is not only effective in reducing device area, but also in eliminating the need for a floating gate. The characteristics of SETs are inherently capable of supporting multiple gates and gate-level voltage summation.

Another special feature of the SET is its oscillatory characteristics, which are useful in achieving periodical functions, such as adder or parity check circuits, which are widely used in current logic circuits. This kind of periodical function is not possible in conventional devices. This feature will be discussed in detail in Section 11.4.

From the operation principle of SETs described above, it is clear that any conductive materials can be used as a base material. SETs have been fabricated

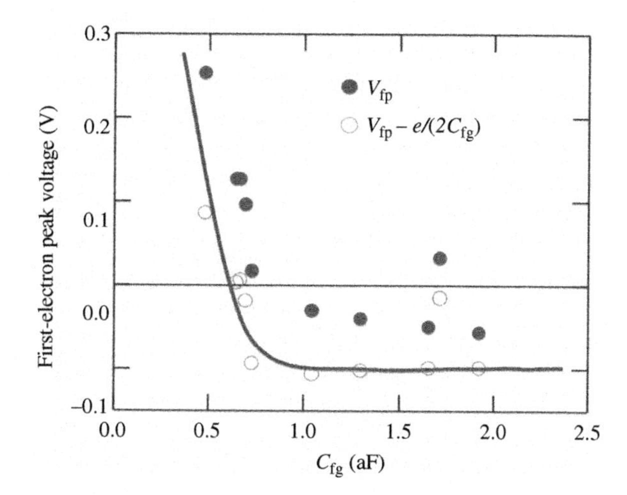

FIGURE 11.5 Relationship between gate voltage for the first electron peak and gate capacitance measured for many SETs. V_{fp} is the gate voltage for the first electron peak, and V_{fp}-$e/2C_g$ corresponds to the threshold voltage when the charging effect is subtracted.

using C60[7] or carbon nanotubes.[8] The first experimental demonstration of an SET was achieved in a metal-insulator system in 1987.[9,10] Clear conductance oscillations due to the Coulomb blockade in a semiconductor island were first observed in a double-gated Si MOSFET by Scott-Thomas et al.,[11] though these characteristics originated in small islands that had been unintentionally formed in a narrow one-dimensional wire. This result led to the investigation of small semiconductor dots formed especially on the two-dimensional electron gas of III-V compound semiconductor heterointerfaces.[12] Many investigations from the physics viewpoint have been carried out by using metals and III-V compound semiconductors. One of the greatest drawbacks of SETs fabricated by using these materials is that the operating characteristics sometimes change due to the offset charge effect. Many researchers have believed that this instability is inherent to SETs because the charge sensitivity of the SETs is quite high. However, this belief is somewhat strange because, as described above, the operation principle of an SET is a charge balance between the gate and SET island. Since this is quite similar to the situation in MOSFETs, in which a charge balance between the gate and channel is attained, the operation stability of SETs should be almost the same as that of MOSFETs.

In fact, SETs fabricated by using MOS processes are definitely stable against long-term drift, which is indispensable for the actual usage of SETs.[13] Figure 11.4 shows the conductance vs. gate-voltage characteristics of an SET measured once and then again 7 years later. During the 7-year interval, the SET was kept at room temperature, although the device was cooled down to 40 K for the two measurements. The two curves are the same within the accuracy of the measurement system. This indicates that there is no effective movable offset charge in the Si SET, probably due to the very high stability of the Si/SiO_2 system. Eliminating the movable offset charge is indispensable for practical application of the electron device to ensure the reliability of the function of the circuits. The SET used in this measurement was fabricated by using the pattern-dependent oxidation (PADOX) method,[14–16] which will be described in the next section.

The next question is whether the device includes fixed offset charges. In conventional MOSFETs, fixed charges sometimes cause threshold voltage shifts, which make the switching condition of each MOSFET irregular. To control the characteristics of SETs, since the elimination of fixed offset charges is also important, we have to examine the threshold voltage of SETs. For SETs, there is no definition of threshold voltage because of the oscillatory I-V_g characteristics. So, we have to detect the first electron peak in return for measuring the threshold voltage. The first electron peak of SETs fabricated by the PADOX method can be measured because the doping concentration in the island is quite low.[17] We can easily distinguish the first peak from other peaks. The relationship between the first-electron-peak voltage V_{fp} and gate capacitance C_g measured for several SETs is shown in Figure 11.5. All SETs were fabricated by the PADOX method. The C_g was estimated from the averaged peak interval for the electron number between five and eight, where periods of the oscillations are almost constant. In a few-electron regime, it is well known that the period sometimes shows irregularity. In discussing the threshold voltage of SETs, $V_{fp} - e/2C_g$ is used to subtract the charging effect according to Coulomb blockade theory. It should be noted that the gate capacitance of SETs fabricated by the PADOX method shows almost a linear relationship with the island size.[18] As shown in Figure 11.5, $V_{fp} - e/2C_g$ is determined by island size, and is almost constant when C_g is larger than 0.8 aF. The rapid increase in V_{fp} when $C_g < 0.8$ aF is attributed to a grand-state energy increase due to the quantum size effect. The result clearly shows that the threshold voltages of Si SETs are governed by island size, which indicates that there is little effect of fixed offset charges. This means that we can control the electrical characteristics of SETs if we can fabricate them with high size accuracy.

11.3 FABRICATION OF SI SETS

One of the great difficulties is fabrication of a small SET island. The operating temperature T is limited to low value according to $kT<e^2/2C_\Sigma$, where C_Σ is the total capacitance of the SET island; $C_\Sigma = \Sigma C_{gi} + C_s + C_d$ for the SET in Figure 11.2. For room-temperature operation, the island must have capacitance of the order of 1 aF at least. Although any conductive materials can be used as a base material, silicon is one of the best candidates because it allows us to use the sophisticated fabrication processes developed for CMOS LSIs.[19] In addition, as discussed above, the Si/SiO_2 system is stable with movable and fixed offset charges.

Another difficult issue in fabricating SETs has been how to attach the source and drain electrodes to either side of the small island via the very small tunnel barrier. One way to make such structures is to deplete the two-dimensional electron gas layer by electric fields from the gate electrodes attached over the layer, which has been demonstrated by using heterointerfaces of III-V compound semiconductors by Meirav et al.[12] Although this method is very simple and allows us to control the tunnel-barrier formation, island sizes unavoidably exceed the minimum feature size of lithography. In fact, the island capacitance of SETs reported in the early 1990s[20–23] was generally about 100 aF or more.

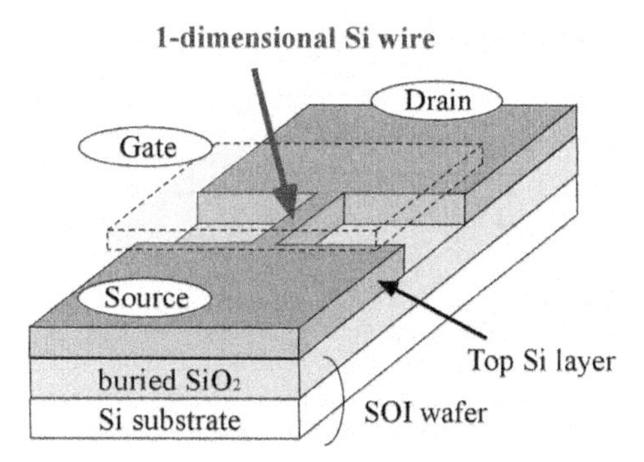

FIGURE 11.6 Initial device structure of the SET before PADOX.

One way to make small SETs is to use naturally nucleated small Si dots, or so-called Si nanocrystals.[24–27] Relatively small total island capacitance of a few aF is obtained when the nanocrystals are sandwiched between source and drain electrodes by very thin tunnel SiO_2.[27] Although the method provides a relatively small island capacitance, it is difficult to control the position of the nanocrystal dots because the dots are formed through random nucleation.

The use of randomness or fluctuations is another simple way to make small islands with tunnel barriers. In the pioneering work, Yano et al. achieved room-temperature operation of single-electron memory by using the thickness fluctuation of a very thin polycrystalline silicon film. A similar approach for a high-temperature-operated Si SET was used by Uchida et al. [28] who introduced a silicon-on-insulator (SOI) layer where undulation of the silicon layer thickness was achieved by chemical treatment. These methods also have difficulty in accurate control of island position, which is indispensable for using SETs to build a logic circuit.

A special method for making small SETs in a controlled way is pattern-dependent oxidation (PADOX). PADOX converts a small Si wire pattern on an SOI wafer into a single-electron island with a tunnel capacitor at each end.[14,16] The initial structure is a narrow and short one-dimensional Si wire oriented to the [110] direction on a thin (001) SOI wafer as shown in Figure 11.6. Typical wire width is about 30 nm. Wire height, which is the thickness of the SOI layer, is also about 20 nm. Wire length is varied to control island size. The SET fabrication mechanism is as follows. When a narrow Si wire is thermally oxidized in dry oxygen ambient atmosphere, the oxidation is suppressed due to the huge amount of compressive stress accumulated in the newly grown SiO_2, which surrounds the wire completely. It has been reported that compressive stress larger than 20, 000 atm accumulates in the SiO_2 when a 10-nm-diameter Si wire is formed by oxidizing a 30-nm-diameter Si wire.[29] When the Si wire structure shown in Figure 11.6 is subjected to thermal oxidation, the compressive stress is exerted on the middle part of the wire. Oxidation proceeds relatively faster in the wire regions around the end than at the middle because of the shear stress caused by the oxidation from the back in the wide two-dimensional layer. The first-principles calculations have shown that a band-gap reduction of about 150 meV occurs when the compressive stress on the middle of the Si wire is 20, 000 atm. This reduction cancels out the effective band-gap increase of about 50 meV

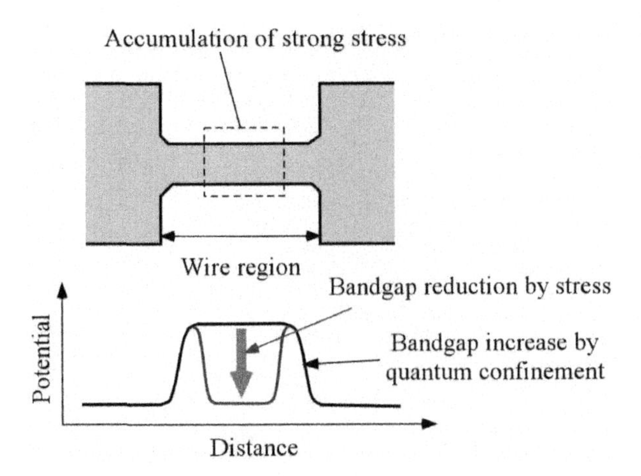

FIGURE 11.7 Schematic plane view of the device and potential diagram along the Si wire.

due to the quantum size effect in a 5- to 10-nm Si wire.[30,31] As a result, a potential profile is formed as schematically shown in Figure 11.7. The two potential hills serve as tunnel barriers:; a tunnel capacitor is formed between the Si island and the wider Si layer. This mechanism enables SETs to be made in a self-aligned manner.

11.4 LOGIC CIRCUIT APPLICATIONS OF SETS

The conductance characteristics modulated by the gate voltage allow us to make a switching device for logic circuits. As discussed above, the operation principle based on the Coulomb blockade allows the device to operate more stably as it becomes smaller. This feature is the opposite of the situation in MOSFETs, where smaller sizes cause undesirable characteristics, such as punch-through or gate leaks. For future extremely large-scale integrated circuits, it is commonly understood that devices must not only be small to achieve high density of integration but also inherently low-power-consuming. SETs are good candidates. Many applications have already been proposed and some have been demonstrated experimentally. The simplest way to apply SETs in logic circuits is to use a peak and a nearby valley as an ON and OFF state, which enables us to use SETs instead of MOSFETs. The main advantage is that we can employ advanced CMOS circuit technologies. Noteworthy is that a voltage gain larger than unity can be obtained by designing SET structures having appropriate circuit parameters, which guarantees signal propagation to any following gates. Another way of logic circuit application is to use the special features inherent to SETs. The multiple-gate capability and multipeak characteristics allow complicated functions, such as gate-level summation and multiple-valued logic. These applications are quite advantageous to enhance the low power consumption nature of SETs since they can reduce the number of devices in circuits. Finally, the most unique application of SETs is single-electron transfer logic, which sends just a single electron as a single bit. This may achieve ultimately-low-power integrated circuits.

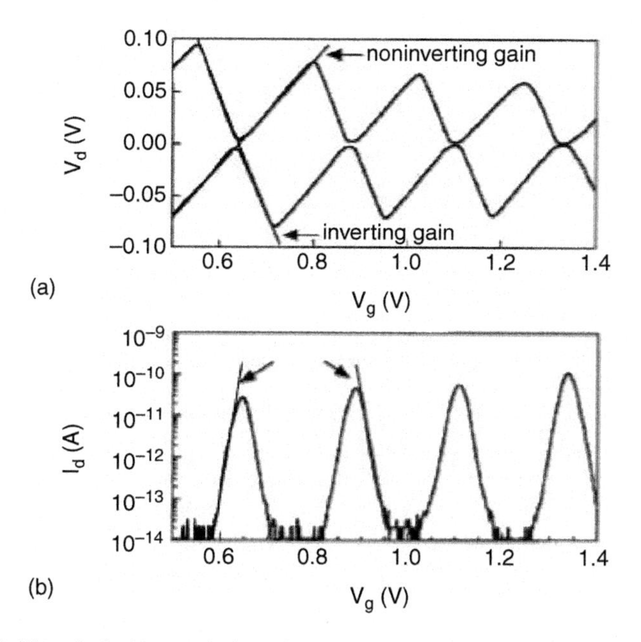

FIGURE 11.8 Electrical characteristics of an SET made by the PADOX process.[35] (a) The output drain voltage *Vd* for a fixed drain current *Id* of ±10 pA and (b) the output drain current for a fixed drain voltage *Vd* of 10 mV. The measurement temperature was 27 K,[19] Copyright 2002 Institute of Physics or IOP Publishing Ltd.

11.4.1 Fundamentals of SET Logic

When a load resistance is connected in series to an SET, logic circuits can be built similarly to the conventional MOSFET logic circuits. The voltage gain of this type of circuit is defined when the load resistance is infinite, namely, the constant current load. Figure 11.8(a) shows the output drain voltage V_d as a function of gate voltage V_g for a fixed drain current $\pm I_d$ measured for an SET fabricated by PADOX. The characteristics show so-called Coulomb diamonds. From the slope of the graph, we can define maximum inverting voltage gain G_I and noninverting gain G_{NI}. These values are determined by SET parameters as

$$G_I = C_g / C_d \tag{11.2}$$

$$G_{NI} = C_g / (C_g + C_s) \tag{11.3}$$

Although G_{NI} is always smaller than unity, G_I can exceed unity if $C_g > C_d$. Consequently, we can make logic circuits based on inverters, as is done for CMOS-type logic. As easily understood from Equation (11.2), to obtain higher inverting gain, an SET has to have larger C_g. This means that the total capacitance of the SET island tends to increase. This may make the operation temperature low. In general, it is not easy to make an SET with high gain and high operating temperature.[32–36]

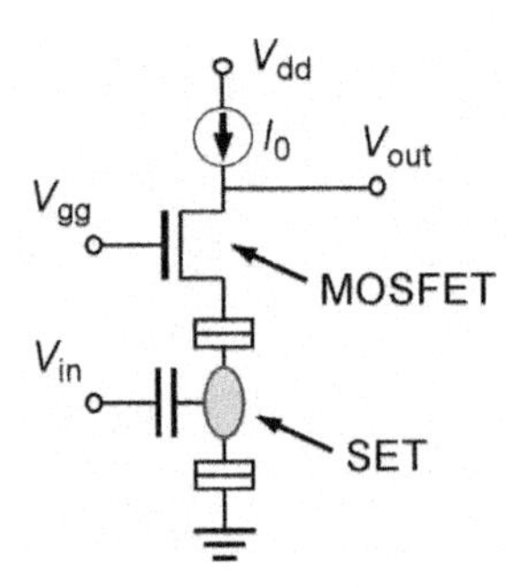

FIGURE 11.9 Equivalent circuit of the merged SET-MOSFET inverter.

Here, for comparison of SET and MOSFET characteristics, Figure 11.8(b) shows subthreshold slopes S of I_d-V_g characteristics of SET, where the I_d was measured with fixed V_d and plotted in a logarithmic scale. S is given by

$$S = \left[d\left(\ln I_d\right)/dV_g\right]^{-1} = \left(C_\Sigma/C_g\right)kT/e \tag{11.4}$$

This equation is quite similar to that for MOSFET. It also indicates that we need high inverting voltage gain G_I to obtain steep subthreshold slope. If C_s is equal to C_d, $G_I = 4$, which is possible by using a PADOX SET,[36] comes to C_Σ/C_g of 1.5, or $S = 90$ mV/dec, at room temperature. One different point in SETs is that the low-current region is limited by the existence of the next peak, which may cause the relatively large OFF current. However, if we use the first-electron peak, infinitely low current can be achieved just as in a MOSFET.

11.4.2 Merged SET and MOSFET Logic

Another important drawback of SETs is that the drain current is limited by the tunnel resistance, which has to be larger than the quantum resistance h/e^2 (25.8 kΩ) to maintain the Coulomb blockade condition. This makes the signal response quite slow when the outputs are connected to a heavy load such as long wirings. The applicable drain voltage is also limited to one smaller than e/C_Σ. This is an obstacle to driving a series of SETs with small voltage gain or external circuits that require high input voltage. SET-MOSFET logic circuits have been proposed as a way to overcome this drawback.[37,38] Figure 11.9 shows the equivalent circuit of the merged SET-MOSFET inverter with a constant current load I_0. In the circuit, the source of a MOSFET is connected to the drain of an SET. The MOSFET with fixed gate bias V_{gg} is used in order to keep the SET drain voltage nearly constant at $V_{gg} - V_{th}$, where V_{th} is the threshold voltage of the MOSFET. We can set the SET drain voltage ($V_{gg} - V_{th}$) sufficiently low to maintain the Coulomb blockade condition. Since the drain voltage is almost independent of output voltage V_{out}, large output voltage and voltage gain can be obtained.

The device operation was actually verified by using an SET and MOSFET fabricated on the same SOI wafer. The SET was fabricated by PADOX. Figure

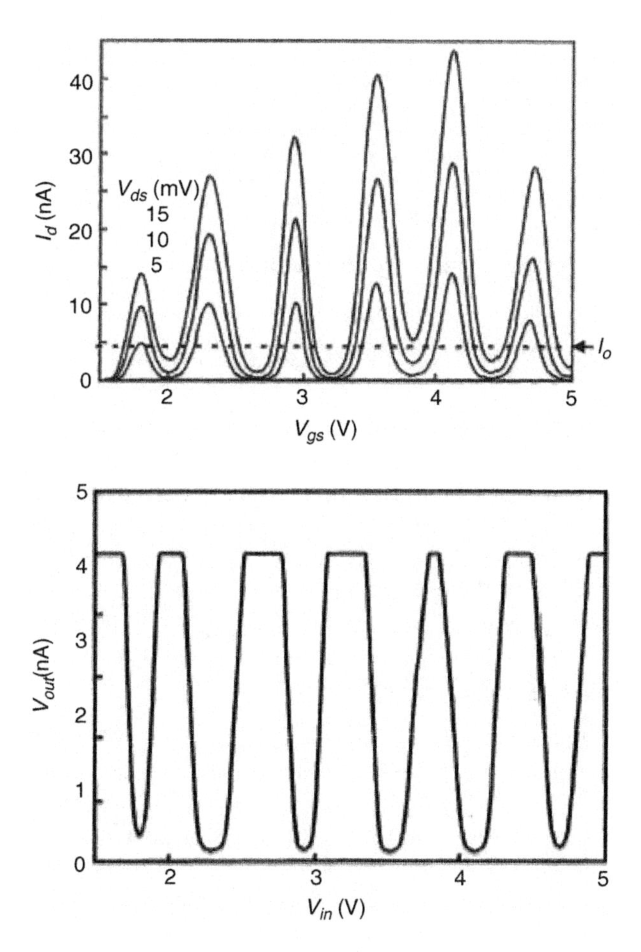

FIGURE 11.10 Id-Vg characteristics of SET fabricated by PADOX (a) and input-output characteristics of the SET-MOSFET inverter with a current load of 4.5 nA (b) measured at 27 K.[38] Copyright 2002 Japan Society of Applied Physics.

11.10(a) shows the measured drain current I_d vs. gate voltage V_{gs} characteristics of an SET for various source-drain voltages V_{ds}. The I_d-V_{gs} characteristics have a large drain voltage dependence. The voltage gain of the device is only about 0.1. Figure 11.10(b) shows the input-output characteristics of the SET-MOSFET inverter with the constant current load of 4.5 nA and V_{gg} of 1.08 V. The MOSFET is an n-type with effective channel width of 12 μm, channel width of 14 μm and gate oxide thickness of 90 nm. The threshold voltage V_{th} of the MOSFET at I_d = 4.5 nA and drain voltage of 3 V is 1.07 V, and the transconductance $G_{m(MOS)}$ of the MOSFET is 151 μS. The output voltage V_{out} and output resistance of the merged SET-MOSFET inverter is given by[38]

$$V_{out} = -G_{m(SET)} R_{d(SET)} \left(1 + G_{m(MOS)} R_{d(MOS)}\right) V_{in} \tag{11.5}$$

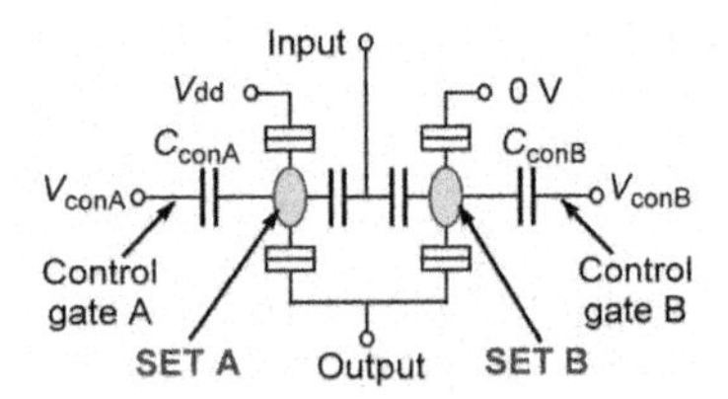

FIGURE 11.11 Equivalent circuit of the quasi-CMOS inverter.

$$R_{\text{out}} = R_{\text{d(MOS)}} + \left(1 + G_{\text{m(MOS)}} R_{\text{d(MOS)}}\right) R_{\text{d(SET)}} \tag{11.6}$$

where $G_{\text{m(SET)}}$ is the transconductance of the SET, and $R_{\text{d(SET)}}$ and $R_{\text{d(MOS)}}$ are the drain resistances of the SET and MOSFET, respectively. The voltage gain of the SET is multiplied by that of the MOSFET, which means that voltage gain of the SET-MOSFET inverter becomes quite large due to the large voltage gain of the MOSFET. In fact, the measured voltage gain of the SET-MOSFET inverter is about 40, as seen in Figure 11.10(b). The important point is that the large voltage gain can be easily obtained while preserving the oscillatory I-V_g characteristics.

11.4.3 CMOS-Type Logic Circuit

As described in Section 4.1, we can construct single-electron logic circuits in which SETs operate analogously to MOSFETs. For a higher voltage gain, a CMOS-type circuit is advantageous. If the oscillation phase is controllable, we can use an SET as both an n-switch and p-switch. Here, an n-switch means that the switch is OFF for low input gate voltage V_{in}, but ON for high V_{in}. The p-switch shows opposite operation. One possible method for controlling the oscillation phase is to use dual-gate SETs, in which one of the gates acts as an input gate and the other gate a control gate. This means that we can use the same SET as a p-switch or an n-switch. As discussed in Section 2, the oscillation phase is shifted by $\pi/2$ when voltage of $e/4C_{\text{con}}$ is applied to the control gate, where C_{con} is the gate capacitance of the control gate. Figure 11.11 shows an equivalent circuit of the CMOS-type SET inverter.[39–42] Here, SET-A is used as a p-switch, and SET-B as n-switch.

This quasi-CMOS logic circuit will allow us to utilize the sophisticated circuit design technology of the current generation of CMOS LSIs. A complementary single-electron inverter constructed by using two similar SETs, a fundamental circuit element for single-electron CMOS-type logic, has actually been fabricated on an SOI substrate by using the PADOX process. Figure 11.12(a) is an atomic force microscopy image of the fabricated single-electron inverter.[40] The structure is shown schematically in Figure 11.12(b). The key part of the circuit where the two SETs are formed is 100 nm × 200 nm. The side gates (A and B) act as control gates. The input gate, which is not shown in Figure 11.12(a), was attached over the two SETs. Figure 11.13(a) shows the source-drain conductance of each SET as a function of input gate voltage. The operation of the two SETs is complementary. Figure 11.11(b) shows the input-output transfer characteristics of the inverter for a power supply

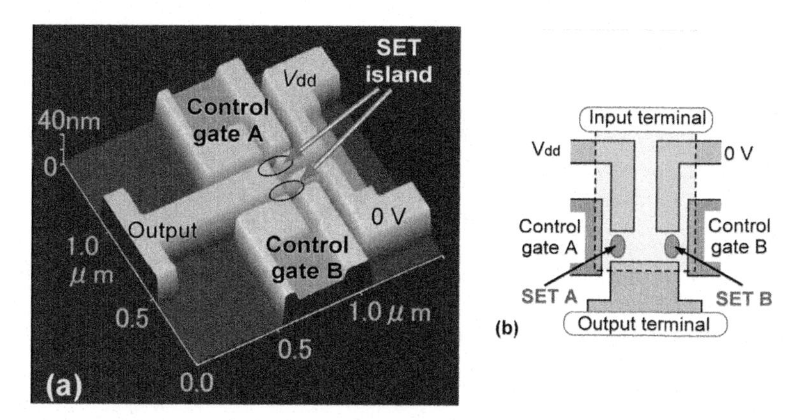

FIGURE 11.12 AFM image of the fabricated quasi-CMOS inverter where the top gate for the input is not shown (a), and the schematic top view (b).

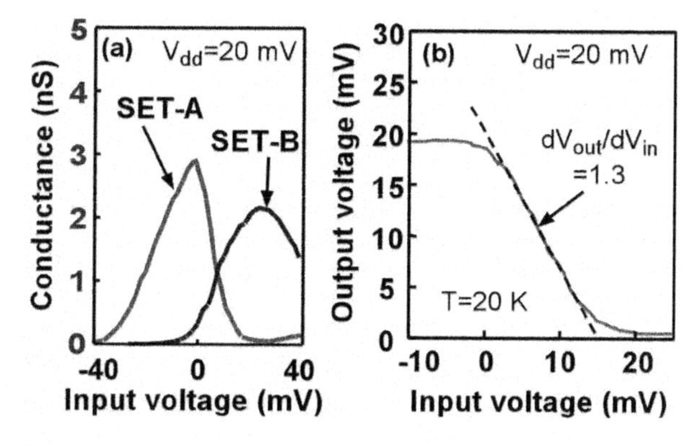

FIGURE 11.13 Input-output transfer characteristics of the inverter.[19] Copyright 2002 Institute of Physics or IOP Publishing Ltd.

voltage V_{dd} of 20 mV measured at 27 K. For this operation, we adjusted the side-gate voltages so that SET-A and SET-B work as p-type and n-type transistors, respectively. The voltage gain of the circuit is larger than unity, which relies on the high-gain SETs actually formed and guarantees signal transfer to the following gates. This also indicates that we can construct logic circuits with SETs based on CMOS-type logic.

The inverter shown in Figure 11.11 needs control gate terminals, which sometimes complicate circuit design. One proposed solution is to use a floating gate instead of the control gate as shown in Figure 11.14.[28,42] The floating gate, which is sometimes a small island or a dot, is attached to the SET island so as to couple capacitively. Charges are injected to the floating gate from the island[42] or other electrodes[28] and shift the oscillation phase of the SET. Consequently, we have to control the amount of injected charges in order to make the two SETs operate in a complementary manner.

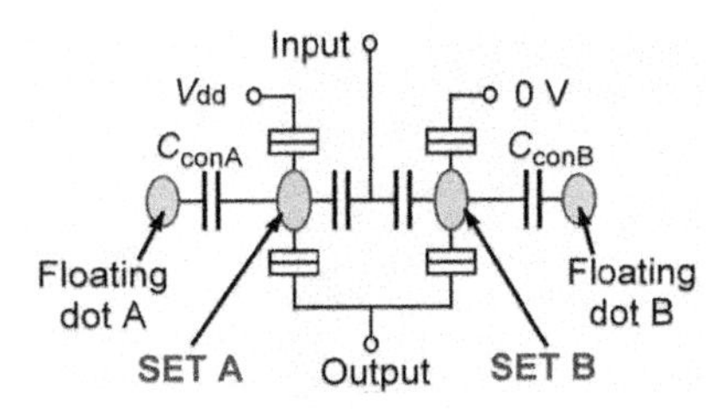

FIGURE 11.14 Equivalent circuit of the quasi-CMOS inverter where a floating gate (dot) is attached to each SET.

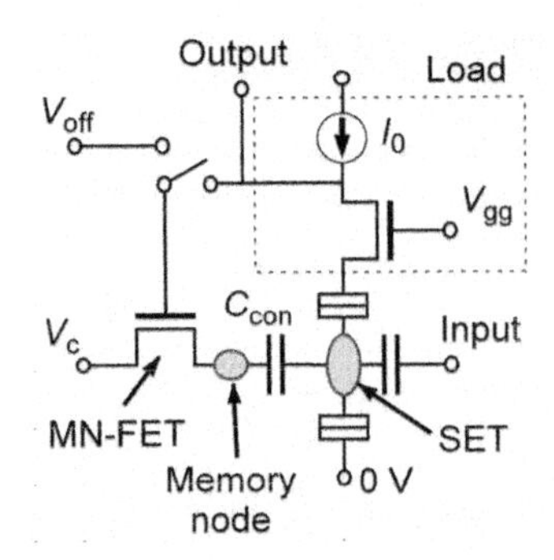

FIGURE 11.15 Equivalent circuit of the SET logic with a feedback loop, by which the oscillation phase is controlled automatically. The gate of MNFET is connected to output terminal only when the feedback is carried on, and it is connected to low voltage source V_{off} so as to shut off the MNFET.

Another technique for controlling the oscillation phase of an SET uses feedback of the output voltage to the control gate, which allows automatic phase control.[43] The basic circuit is shown in Figure 11.15. A small memory node coupled capacitively to the SET is connected to the MOSFET (MNFET), which acts as a switch for charging or discharging the memory node. In the feedback cycle, the output terminal is connected to the gate of the MOSFET, and, at the same time, an appropriate voltage is applied to the input gate of the SET. When the output voltage is high, the MNFET becomes ON, and charges are injected into the memory node. This changes the potential of the node, and the oscillation phase is shifted so as to make the output OFF. When the output signal becomes OFF, the MNFET automatically turns off. This process controls the charges in the memory node so as to make the output signal OFF. After that, the gate of the MNFET is connected to low voltage source V_{off}, whichs keep the MNFET OFF and also keeps the charges in the MNFET. The precise phase of oscillation is controlled by changing the input voltage applied to the gate of the SET.

11.4.4 Pass-Transistor Logic

As understood from Equation (11.2), it is not easy to obtain high voltage gain in high-temperature-operating SETs. In addition, the drain voltage of SETs has to be

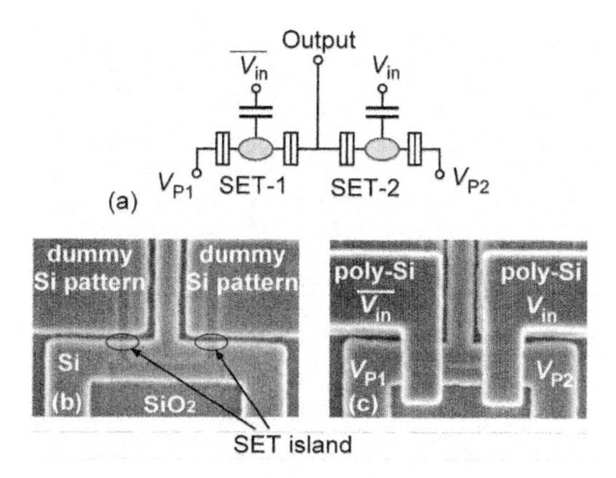

FIGURE 11.16 Equivalent circuit of a unit element of a pass-transistor logic gate (c), and SEM images before (b) and after (c) input-gate formation. The dummy Si patterns, on either side of the SET islands were formed to control the oxidation and were grounded throughout the electrical measurement. Copyright 2003. Reproduced with permission from Reference 45. The Electrochemical Society.

lower than e/C_Σ to maintain the Coulomb blockade condition. These problems can be overcome by using pass-transistor logic gates for SET logic circuits.[44] Pass-transistor logic enables us to use a drain voltage lower than the gate voltage, which keeps the SETs ON throughout both the pull-up and pull-down action. The equivalent circuit is shown in Figure 11.16(a). We used the PADOX method to make a unit element of pass-transistor logic. Figures 11.16(b) and (c) show SEM (scanning electron microscope) images of a fabricated device before and after the formation of individual small polySi gates, respectively. For the operation, pass signals, V_{p1} and $Vp2$, are input to the source side of the transistors, and one of the two routes is selected by the input of complementary signals to the gates. Almost all basic logic functions, such as NAND, NOR, and exclusive OR (XOR), can be achieved by changing the input and pass signal voltages.

One of the most important logic circuits is an adder circuit; it is the most frequently used element in arithmetic logic units (ALUs) in microprocessors and is also a base element of multipliers. In our trial, we constructed the simplest adder: the half adder, which calculates the higher (half-sum) and lower (carry-out) order bits from the addition of two one-bit operands based on the pass-transistor-logic scheme.[44] Half sum (XOR) and carry out (AND) operations were measured at 25 K using the device in Figure 11.16. The output signal exhibits the correct function of half-sum and carry-out functions as shown in Figure 11.17. Here, we set V_{p1} and V_{p2} as V_B and $\overline{V}_B$ for the half-sum circuit. For the carry-out circuit, we set V_{p1} and V_{p2} as V_B and 0 V. It should be noted that the rather slow switching is not related to the circuit itself; the cause is the huge capacitance of the measurement system. The results shown in Figure 11.17 actually guarantee that any arithmetic operations can be performed by SET circuits.

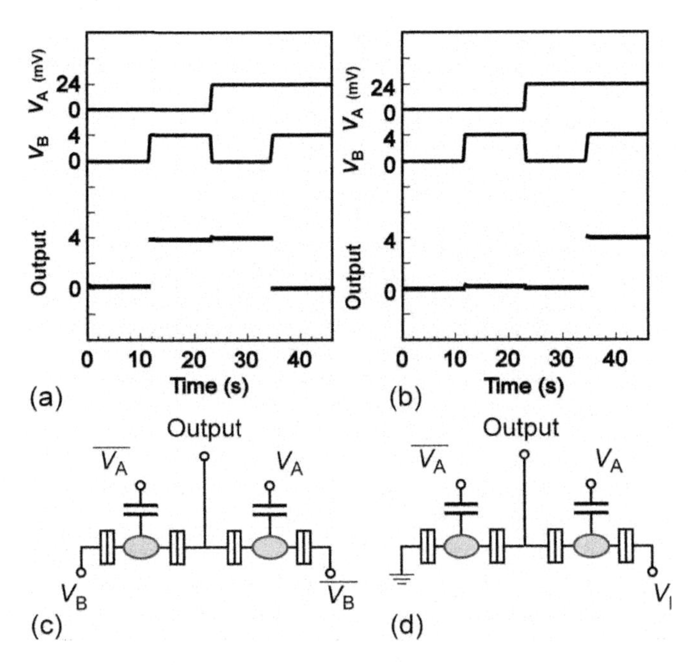

FIGURE 11.17 Experimental half-sum (a) and carry-out operations (b), and equivalent circuits (c) and (d), respectively. Measurement temperature is 25 K. Copyright 2003. Reproduced with permission from Reference 45. The Electrochemical Society.

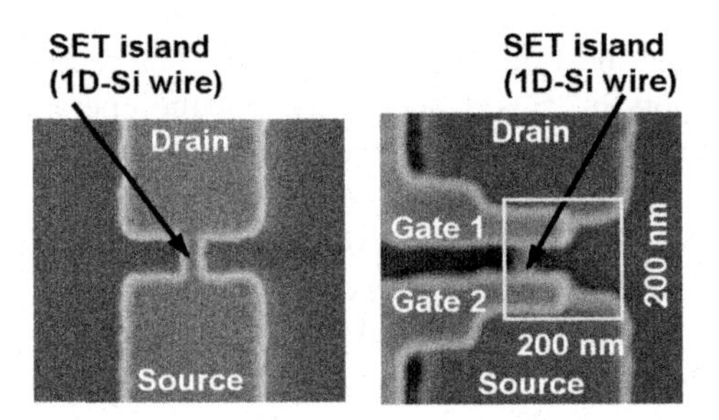

FIGURE 11.18 SEM image of a dual gate SET before (a) and after (b) formation of two ultrafine gate electrodes.

11.4.5 Multigate SET

As described in Section 2, gate-level summation can be carried out using a multigate SET. If all gate capacitances of the SET are the same C_{g0}, and if an input voltage for high-level $e/2C_{g0}$ is used, the function of a multi-input exclusive-OR gate is achieved. We actually fabricated such a device having two equal gates.[46] SEM images of the device are shown in Figures 18(a) and (b). A small one-dimensional Si wire

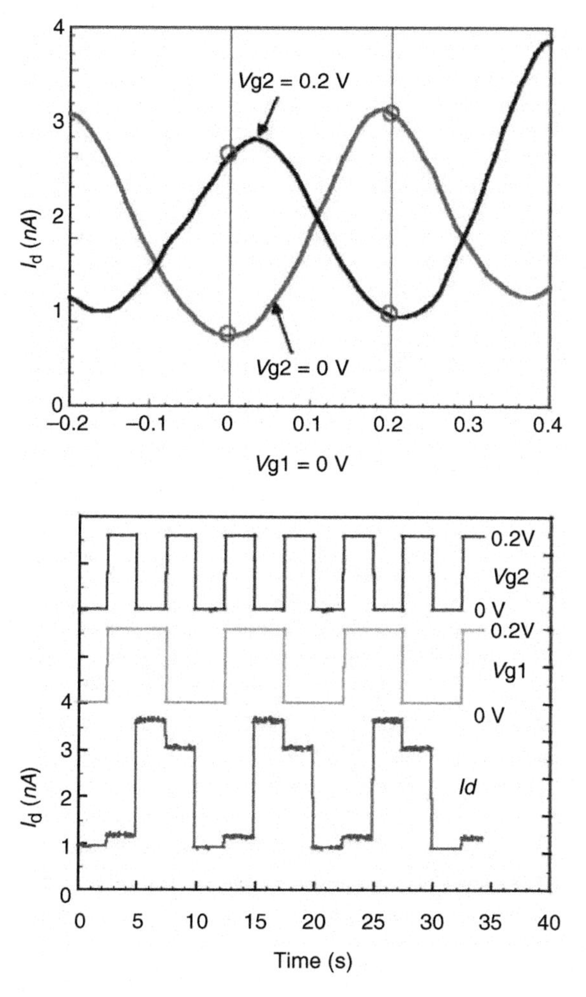

FIGURE 11.19 Drain current I_d vs. gate voltage characteristics of a dual-gate SET (a) and drain current switching characteristics (b). Copyright 2003. Reproduced with permission from Reference 45. The Electrochemical Society.

fabricated on an SOI wafer (Figure 11.18[a]) was converted into a small SET by means of PADOX. Then, using an electron-beam exposure system with a high overlay accuracy, two ultrafine polySi gate electrodes were attached so as to cover part of the island (Figure 11.18[b]). The capacitances of the two gates are almost equal due to the symmetric configuration. The drain current oscillation characteristics as a function of one of the gate voltages are shown in Figure 11.19(a). Since the capacitance of both gates is 0.4 aF, the oscillation phase shifts about π to the negative voltage direction when one of the gate voltages $Vg2$ is changed from 0 to 0.2 V. Figure 11.19(b) shows the drain current switching measured at 40 K in response to the switching of the two input-gate voltages ($Vg1$ and $Vg2$) between 0 and 0.2 V. Low current levels were obtained only when the input voltages were both high or

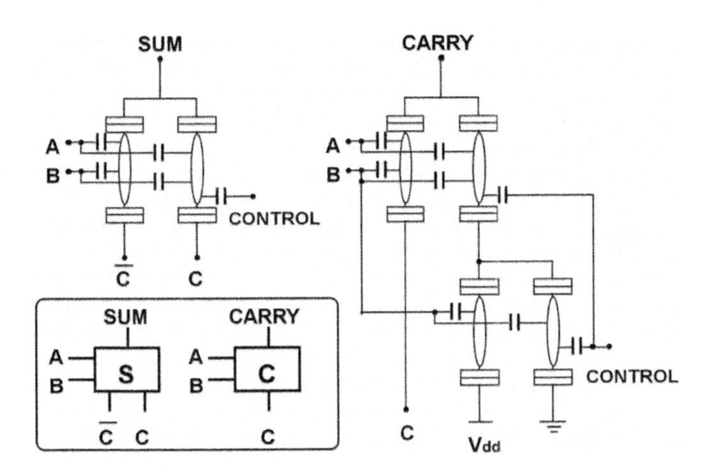

FIGURE 11.20 Equivarent circuits for sum-bit and carry-bit. The inset shows symbols for the sum and carry circuits.

both low. This represents an XOR-gate operation, which can be implemented with just one SET.

This functionality of the multigate SET enables us to make multibit adders with a small number of transistors without any wire crossing.[45] Figure 11.20 shows the circuits for sum and carry calculation. These circuits operate based on pass-transistor logic discussed previously.[44,47] In the figure, A and B represent the addends, and C is the carry-in. We used a pair of dual-gate SETs, one of which has an additional control gate. By applying a voltage of e/C_{con} to the control gate, where C_{con} is the capacitance of the control gate, we can shift the input-gate voltage characteristics of the SET by π. In other words, we can use the two SETs in a complementary way. As a result, we can select and pass one of the two pass inputs according to the input gate voltages.

We can make a multibit adder by using the circuits shown in Figure 11.20. Figure 11.21 shows a four-bit adder. A one-bit calculation is performed using one sum circuit and two carry circuits. The resultant sum of each bit (S0, S1, S2, S3) is boosted before it is transferred to the next stage. In this configuration, we can eliminate crossings of pass-input routes (the fine solid lines) as shown in Figure 11.21. This allows us to connect each circuit without any metal interconnections. The carry is propagated only through small and thin SOI layers that have very small stray capacitance. This is very advantageous not only for reducing device area but also for achieving high-speed operation.

11.4.6 Multiple-Valued Operation

As is widely known, the multiple-valued logic allows us to reduce the number of transistors and the amount of wiring in LSIs. This is another quite effective procedure for reducing the power dissipation in an LSI chip and reducing chip size. The oscillatory conductance characteristics as a function of gate voltage may be applicable to multiple-valued applications,[37,48–50] in which the number of the electrons in the island represents the multiple-valued levels. To realize multiple stability points,

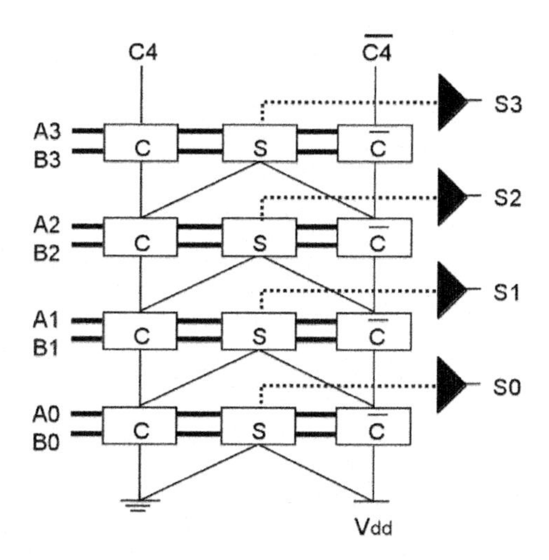

FIGURE 11.21 Structure of the 4-bit adder. The bold solid lines represent the input gates, and the fine solid lines the routes for the pass signals. The sum output nodes are shown by the dotted lines.

we have to attain two-terminal multipeak negative differential resistance (NDR) characteristics, not the gate-controlled resistance oscillation shown in Figure 11.3. For this purpose, we can use an SET-MOSFET circuit like the one discussed in Section 4.2.[37] The equivalent circuit of the device is shown in Figure 11.22(a). The gate of an SET is connected to the drain of a MOSFET, which allows the circuit to have multipeak current oscillation as a function of V, resulting in NDR characteristics.

The measured I-V characteristics of the circuit are shown in Figure 11.22(b). The device is the same as that in Figure 11.10, and the operation condition of the merged SET-MOSFET circuit was also the same. The current I oscillates as a function of applied voltage V, reflecting the I_d-V_{gs} characteristics of the SET. If a current source of 4.5 nA is connected to the circuit, points **a-f** becomes stable. It is advantageous that just a couple of transistors can provide a multipeak NDR device in which the number of peaks is infinite in principle within the breakdown voltage of the MOSFET drain or the SET gate. Conventional multipeak NDR devices, such as resonant tunnelling diodes, require the same number of devices as peaks. These characteristics can also be applied to multiple-valued memories.[37]

We can take full advantage of this NDR device with multiple stability points by applying it to build multiple-valued logic circuits. One of the basic circuits of multiple-valued logic is a quantizer that discriminates an analog input signal into predefined voltage levels. By using the circuit shown in Figure 11.22(a), we have built such a quantizer.[48] Figure 11.23 shows the setup for the quantizer measurement. The transfer gates MOSFET1, and MOSFET2 for probing are connected externally. The input signal (V_{in}) is a triangular wave. The gate of MOSFET1 is driven by short clock (CLK) pulses. Figure 11.24 shows waveforms for V_{in}, CLK, and V_{out}. As shown in the figure, V_{out} is quantized to levels **a-f**, which correspond to the stability points in Figure 11.22(b). Although the operation speed is rather slow, it is not limited by

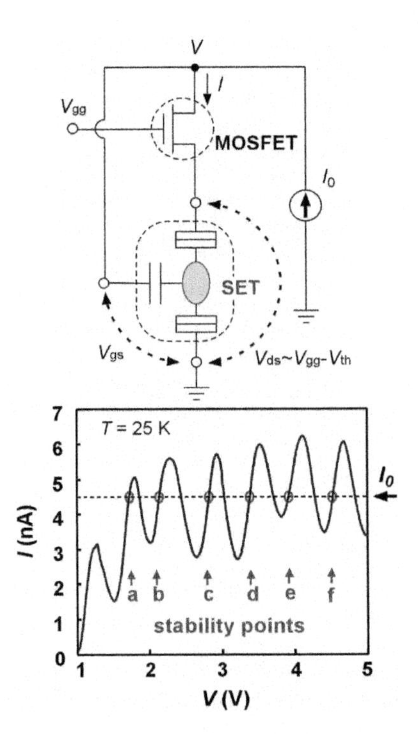

FIGURE 11.22 Equivalent circuit composed of a MOSFET and SET (a), and measured I_d-V_g characteristics (b).[37] Copyright 2001 American Institute of Physics.

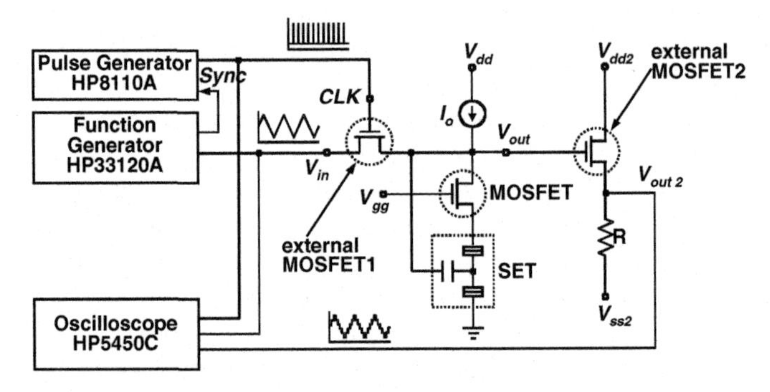

FIGURE 11.23 Measurement setup for the quantizer. The external MOSFET 2 is used as a FET probe to measure the output voltage V_{out}.[50] Copyright 2003 IEEE.

the intrinsic performance of the device, but by the large capacitance of 370 pF existing at V_{out}.

In addition, based on the circuit shown in Figure 11.22(a), we have proposed a full adder for redundant number representation with a very small number of transistors.[48] This is advantageous not only for circuit size reduction, but also for the high-speed operation because we can eliminate carry propagation.

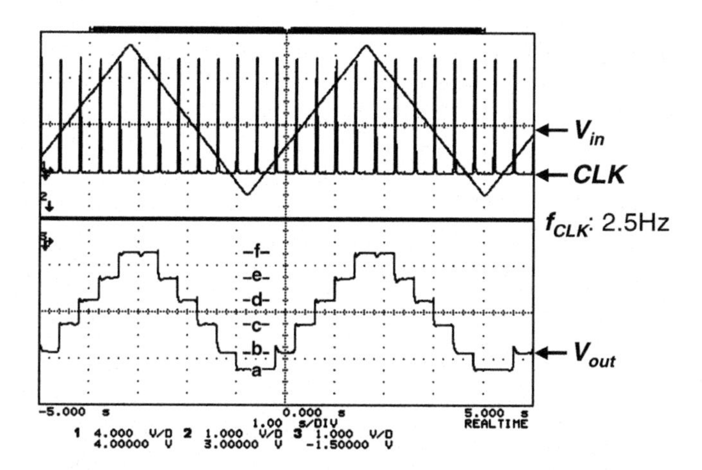

FIGURE 11.24 Quantizer operation characteristics measured using the setup in Figure 11.13, with V_{gg} of 1.08 V and a current load of 4.5 nA. Operation speed is not limited by the intrinsic performance of the device, but by the large capacitance existing at V_{out}.[50] Copyright 2003 IEEE.

11.5 CONCLUSION

The small sizes, low power dissipation, and high functionality of single-electron devices (SEDs) make them very promising. As demonstrated in this chapter, the integration of SEDs has a great potential when we can make SETs from Si on the same chip that conventional CMOS LSIs are made. We have developed a unique fabrication method called pattern-dependent oxidation (PADOX), which is simple and controllable. In addition, the PADOX-fabricated SEDs operate quite stably.

The features of SETs, such as multiple-gate capability and multiple-peak oscillatory characteristics, enable us to achieve special functionalities. In Si SEDs, combining SETs and MOSFETs enhances these functionalities. By exploiting these special features, we can realize complicated functions with a small number of transistors, which will reduce the size and power dissipation of circuits.

REFERENCES

1. Doris, B., Ieong, D., Kanarsky, T., Zhang, Y., Roy, R.A., Dokumaci, O., Ren, Z., Jamin, F.F., Shi, L., Natzle, W., Huang, H.J.X., Mezzapelle, J., Mocuta, A., Womack, S., Gribelyuk, M., Jones, E.C., Miller, R.J., Wong, H.S.P., and Haensch, W., Extreme Scaling With Ultra-Thin Si Channel MOSFETs, *Intl Electron Device Meeting Technical Digest*, 267-270, 2000.
2. Yu, B., Chang, L., Ahmed, S., Wang, H., Bell, S., Yang, C.Y., Tabery, C., Ho, C., Xiang, Q., King,T.J., Bokor, J.. Hu, C., Lin, M.R., and Kyser, D., FinFET Scaling to 10nm Gate Length, *Intl Electron Device Meeting Technical Digest,* 251-254, 2002.
3. de Vegvar, P.G.N., Timp, G., Mankewich, P.M., Behringer, R., and Cunningham, J., Tunnable Aharonov-Bohm effect in an electron interferometer, *Physical Review B,* 40, 3491-3494, 1989.

4. Aihara, K., Yamamoto, M., and Mizutani, T., Three-terminal conductance modulation of a quantum interference device using a quantum wire with a stub structure, *Applied Physics Letters,* 63, 3595-3597, 1993.
5. Likharev, K.K., Single-electron devices and their applications, *Proceedings of IEEE*, 87, 606-632, 1999.
6. Shibata, T. and Ohmi, T., A functional MOS transistor featuring gate-level weighting sum and threshold operations, *IEEE Transaction of Electron Devices,* ED-39, 1444-1455, 1992.
7. Park, H., Park, J., Lim, A.K.L.E., Anderson, H., Alivisatos, A.P., and McEuen, P.L., Nano-mechanical oscillations in a single-C_{60} transistor, *Nature*, 407, 57-60, 2000.
8. Tans, S.J., Devoret, M.H., Dai, H., Thess, A., Smally, R.E., Geerlings, L.J., and Dekker, C., Individual single-wall carbon nanotubes as quantum wires, *Nature*, 386, 474-477, 1997.
9. Fulton, T.A. and Dolan, G.J., Observation of single-electron charging effects in small tunnel junctions, *Physical Review Letters,* 59, 109-112, 1987.
10. Kuzmin, L.S. and Likharev, K.K., Direct experimental observation of discrete correlated single-electron tunnelling. *Pis'ma v Zhurnal Eksperimental'noi i Teoreticheskoi Fiziki,* 45, 389-390, (JETP Letters 1987, 45, 495-497), 1987.
11. Scott-Thomas, J.H.F., Field, S.B., Kastner, M.A., Smith, H.I., and Antoniadis, D.A., Conductance oscillations periodic in the density of a one-dimensional electron gas, *Physical Review Letters*, 62, 583-586, 1989.
12. Meirav, U., Kastner, M.A., and Wind, S.J., Single-electron charging and periodic conductance resonances in GaAs nanostructures, *Physical Review Letters*, 65, 771-774, 1990.
13. Zimmerman, N.M., Huber, W.H., Fujiwara, A., and Takahashi, Y., Excellent charge offset stability in a Si-based single-electron tunneling transistor, *Applied Physics Letters*, 79, 3188-3190, 2001.
14. Takahashi, Y., Nagase, M., Namatsu, H., Kurihara, K., Iwadate, K., Nakajima, Y., Horiguchi, S., Murase, K., and Tabe, M., Fabrication technique for Si single-electron transistor operating at room temperature, *Electronics Letters*, 31, 136-137, 1995.
15. Takahashi, Y.. Namatsu, H., Kurihara, K., Iwadate, K., Nagase, M., and Murase, K., Size dependence of the characteristics of Si single-electron transistors on SIMOX substrates, *IEEE Transaction of Electron Devices,* ED-43, 1213-1217, 1996.
16. Ono, Y., Takahashi, Y., Yamazaki, K., Nagase, M., Namatsu, H., Kurihara, K., and Murase, K., Fabrication method for IC-oriented Si single-electron transistors, *IEEE Transaction of Electron Devices,* ED-47, 147-153, 2000.
17. Fujiwara, A., Horiguchi, S., Nagase, M., and Takahashi, Y., Threshold voltage of Si single-electron transistor, *Japanese Journal of Applied Physics*, 42, 2429-2433, 2003.
18. Nagase, M., Horiguchi, S., Fujiwara, A., and Takahashi, Y., Microscopic observations of single-electron island in Si single-electron transistors, *Japanese Journal of Applied Physics*, 42, 2438-2443, 2003.
19. Takahashi, Y., Ono, Y., Fujiwara, A., and Inokawa, H., Silicon single-electron devices, *Journal of Physics Condensed Matter,* 14, 995-1033, 2002.
20. Field, M., Smith, C.G., Pepper, M., Ritchie, D.A., Frost, J.E.F., Jones, G.A.C., and Hasko, D.C., Measurements of Coulomb blockade with a noninvasive voltage probe, *Physical Review Letters*, 70, 1311-1314, 1993.
21. Paul, D.J., Cleaver, J.R.A., Ahmed, H., and Whall, T.E., Coulomb blockade in silicon based structures at temperatures up to 50 K, *Applied Physics Letters*, 63, 631-632, 1993.

22. Nakazato, K., Blaikie, R.J., Cleaver, J.R.A., and Ahmed, H., Single-electron memory, Electronics Letters, 29, 384-385, 1993.
23. Matsuoka, H., Ichiguchi, T., Yoshimura, T., and Takeda, E., Coulomb blockade in the inversion layer of a Si metal-oxide-semiconductor field-effect transistor with a dual-gate structure, *Applied Physics Letters*, 64, 586-588, 1994.
24. Fukuda, M., Nakagawa, K., Miyazaki, S., and Hirose, M., Resonant tunneling through a self-assembled Si quantum dot, *Applied Physics Letters*, 70, 2291-2293, 1997.
25. Otobe, M., Yajima, H., and Oda, S., Observation of the single electron charging effect in nanocrystalline silicon at room temperature using atomic force microscopy, *Applied Physics Letters*, 72, 1089-1091, 1998.
26. Dutta, A., Lee, S.P., Hayafune, Y., Hatatani, S., and Oda, S., Single-electron tunneling devices based on silicon quantum dots fabricated by plasma process, *Japanese Journal of Applied Physics*, 39, 264-267, 2000.
27. Dutta, A., Oda, S., Fu, Y., and Willander, M., Electron transport in nanocrystalline Si based single electron transistors, *Japanese Journal of Applied Physics*, 39, 4647-4650, 2000.
28. Uchida, K., Koga, J., Ohba, R., Takagi, S., and Toriumi, A., Silicon single-electron tunneling device fabricated in an undulated ultrathin silicon-on-insulator film, *Journal of Applied Physics*, 90, 3551-3557, 2001.
29. Liu, H.I., Biegelsen, D.K., Johnson, N.M., Ponce, F.A., and Pease, R.F.W., Self-limiting oxidation of Si nanowires, *Journal of Vacuum Science Technology*, B 11, 2532-2537, 1993.
30. Shiraishi, K., Nagase, M., Horiguchi, S., Kageshima, H., Uematsu, M., Takahashi, Y., and Murase, K., Designing of silicon effective quantum dots by using the oxidation-induced strain: A theoretical approach, *Physica E*, 7, 337-341, 2000.
31. Horiguchi, S., Nagase, M., Shiraishi, K., Kageshima, H., Takahashi, Y., and Murase, K., Mechanism of potential profile formation of silicon single-electron transistors fabricated using pattern-dependent oxidation, *Japanese Journal of Applied Physics*, 40, L29-L32, 2001.
32. Zimmerli, G., Kautz, R.L., and Martinis, J.M., Voltage gain in the single-electron transistor, *Applied Physics Letters*, 61, 2616-2618, 1992.
33. Visscher, E.H., Verbrugh, S.M., Lindeman, J., Hadley, P., and Mooij, J.E., Fabrication of multilayer single-electron tunneling devices, *Applied Physics Letters*, 66, 305-307, 1995.
34. Satoh, Y., Okada, H., Jinushi, K., Fujikura, H., and Hasegawa, H., Voltage gain in gaas-based lateral single-electron transistors having Schottky wrap gates, *Japanese Journal of Applied Physics*, 38, 410-414, 1999.
35. Smith, R.A. and Ahmed, H., A silicon Coulomb blockade device with voltage gain, Applied Physics Letters, 71, 3838-3840, 1997.
36. Ono, Y., Yamazaki, K., and Takahashi, Y., Si single-electron transistors with high voltage gain, *IEICE Transaction of Electronics*, E84-C, 1061-1065, 2001.
37. Inokawa, H., Fujiwara, A., and Takahashi, Y., A multipeak negative differential resistance device by combining single-electron and metal-oxide-semiconductor transistors, *Applied Physics Letters*, 79, 3618-3620, 2001.
38. Inokawa, H., Fujiwara, A., and Takahashi, Y., A merged single-electron transistor and metal-oxide-semiconductor transistor logic for interface and multiple-valued logic, *Japanese Journal of Applied Physics*, 41, 2566-2568, 2002.
39. Tucker, J.R., Complementary digital logic based on the Coulomb blockade, *Journal of Applied Physics*, 72, 4399-4413, 1992.

40. Ono, Y., Takahashi, Y., Yamazaki, K., Nagase, M., Namatsu, H., Kurihara, K., and Murase, K., Si complementary single-electron inverter, *Applied Physics Letters*, 76, 3121-3123, 2000.
41. Heij, C.P., Hadley, P., and Mooij, J.E., Single-electron inverter, *Applied Physics Letters*, 78, 1140-1142, 2001.
42. Takahashi, N., Ishikuro, H., and Hiramoto, T., Control of Coulomb blockade oscillations in silicon single electron transistors using silicon nanocrystal floating gates, *Applied Physics Letters*, 76, 209-211, 2000.
43. Nishiguchi, K., Inokawa, H., Ono, Y., Fujiwara, A., and Takahashi, Y., Automatic Control of the Oscillation Phase of a Single-Electron Transistor by a Memory Node with a Small MOSFET, *Conference Digest of 61th Device Research Conference*, 135-136, 2003.
44. Ono, Y., Yamazaki, K., Nagase, M., Horiguchi, S., Shiraishi, K., and Takahashi, Y., Single-electron and quantum SOI devices, *Microelectronic Engineering*, 59, 435-442, 2001.
45. Takahashi, Y., Ono, Y., Fujiwara, A., and Inokawa, H., Silicon single-electron transistors and their application to logic circuits, Semiconductor Silicon, 2, Proceedings of the 9th International Symposium on Silicon Materials Science and Technology, Huff, H.R., Fabry, L., and Kishino, S., Eds., 968-978, 2002.
46. Takahashi, Y., Fujiwara, A., Yamazaki, K., Namatsu, H., Kurihara, K., and Murase, K., Multi-gate single-electron transistors and their application to an exclusive-OR gate, *Applied Physics Letters*, 76, 637-639, 2000.
47. Ono, Y., Inokawa, H., and Takahashi, Y., Binary adders of multigate single-electron transistors: Specific design using pass-transistor logic, *IEEE Transaction on Nanotechnology,* 1, 93-99, 2002.
48. Inokawa, H., Fujiwara, A., and Takahashi, Y., A Multiple-Valued Logic with Merged Single-Electron and MOS Transistors, *International Electron Devices Meeting Technical Digest*, 147-150, 2001.
49. Takahashi, Y., Fujiwara, A., Ono, Y., and Murase, K., Silicon Single-Electron Devices and their Applications, Proceedings of 30th-IEEE International Symposium of Multiple Valued Logic, 411-420, 2000.
50. Inokawa, H., Fujiwara, A., and Takahashi, Y., Multiple-valued logic and memory with combined single-electron and metal-oxide-semiconductor transistors, *IEEE Transaction of Electron Devices*, 50, 462-470, 2003.

Index

D

E

H

I

K

L

M

Q

R

S

T

U

V

W

X

Z